Index

After Reading Questions

1. Does reading about influential celebrities change or shape you view of the world and others? Explain
2. How do celebrities influence and inspire change?
3. How did the book change your belief about something?
4. How does Leonardo DiCaprio show good citizenship and contribute to the lives of people around the world?
5. What makes Leo creative, inventive, and notable?

Activity

Grantmakers are faced with difficult choices. They may want to give to many causes, but don't have enough money to go around. Find a partner. Each of you should research a different program that LDF supports. Then, prepare an argument about why your program should receive grant money. Present your arguments to a third person. Ask this person to decide which program should receive the funds and to explain why.

About the Author

Mary Hertz Scarbrough will never forget the time she got to help feed and bathe some elephants at a wildlife refuge. She also saw how many volunteers it took to prepare snacks for the animals. It was a heartwarming reminder that we can all help make the world a better place. She writes, bikes, and volunteers whenever she can from her home in South Dakota.

www.rourkeeducationalmedia.com

PHOTO CREDITS: Cover: ©Chris Pizzello/Invision/AP; page 5, 30: ©Lumeimages.com; page 7: ©luoman; page 7b: ©Sisoje; pages 8-9: ©JoyceMarrero; page 8b: ©SasinParaksa; pages 11, 30: ©Hakim atar / Shutterstock; pages 12-13, 30: ©WLDavies; page 13b: ©NASA; pages 15, 30: ©johandersonphoto; page 17: ©Rafael Ben-Ari; page 16: ©fstop123; pages 20-21, 30: ©ankitsiroya; page 21(b): ©Andrea Raffin / Shutterstocck; pages 22-23: ©Freder; page 25(a): ©a Katz / Shutterstock; pages 26-27: ©SeppFriedhuber; page 27(b): ©Kathy Hutchins / Shutterstock; page 29, 30: ©Pail Piebinga; page 29(b): ©Peter Hermes Furian

Edited by: Kim Thompson
Cover and interior design by: Rhea Magaro-Wallace

Library of Congress PCN Data

Leonardo DiCaprio / Mary Hertz Scarbrough
(Celebrity Influencers)
ISBN 978-1-73161-731-6 (hard cover)
ISBN 978-1-73161-267-0 (soft cover)
ISBN 978-1-73161-743-9 (e-Book)
ISBN 978-1-73161-755-2 (ePub)
Library of Congress Control Number: 2019932359

Rourke Educational Media
Printed in the United States of America,
North Mankato, Minnesota

PROTOTYPING ARCHITECTURE: THE SOLAR ROOFPOD

AN ARCHITECTURAL IDEA IS IN THE END ONLY AS GOOD AS ITS REALIZATION.

The City College
of New York

Christian Volkmann

PROTOTYPING ARCHITECTURE: THE SOLAR ROOFPOD

AN EDUCATIONAL DESIGN-BUILD RESEARCH PROJECT

OSCAR RIERA OJEDA
PUBLISHERS

TABLE OF CONTENTS

KNAACK

PREFACE

by **Barry Bergdoll**

Episodically over the course of the last two hundred years architects have turned to the potentialities of pre-fabrication to project buildings that solve individual programmatic demands free of a specific site and yet also have the capacity to inflect the larger culture of building at a continental scale. Prefabrication always involves both research into the making of things and an exercise in which new systems are designed and existing ones are harnessed. It is, in brief, always built on the understanding that the architectural act links in complex ways to the various waves of industrialization from heavy manufacturing to digital information management but also introduces a building, even without a predetermined site, into the ecology of our occupation of the planet with implications far beyond the drawing board.

Fully aware of the untapped potential of a tradition which encompasses both pragmatic kit houses from Sears and Roebuck in the early years of the last century and advanced building research in the hands of figures like Buckminster Fuller and Konrad Wachsmann in the wake of World War II, Christian Volkmann, with a team of 50 students at the City College of New York, has designed a prototype pre-fabricated dwelling that responds to the unique urgencies of environmental, ecological, and economic challenges of the 21st century. At the same time he has developed his own design research as a potent teaching tool in a new type of training in which students gain hands-on experience of the relationship between designing and making. This resulted in an even more integrated machined fabrication process, along with an experience that serves continually as a concrete point of entry into understanding how design decisions relate to a larger economy and ecology of design. The prototype Solar Roofpod thus stands apart from so many other examples of the recent explosion of prefabrication design in numerous respects, not least in turning to the notion that the site might well be a rooftop in the dense intercity rather than a vacant lot in a new suburban or exurban subdivision.

For Volkmann prefabrication can transform the existing situation, exploiting underutilized potentials and thereby taking an ethical stance by refusing to contribute to the unsustainable addition of new rings of growth around cities with population pressures. Originally realized for a rooftop site in Manhattan – its delivery, assembly, and removal to a new site in Washington are all here carefully documented and analyzed – the Solar Roofpod combines two of the most potent strains of research in urban retrofitting currently underway: the combination of the untapped potential of the flat roofs of our cities to serve as solar energy farms, and the capacity for buildings to sustain additions that are fabricated off site and inserted into urban density with minimal disruption to the complex daily life of cities. The result is a design at once elegant and pragmatic, filled with potentials both for urban living and for architectural education and offering implications for an ethos of responsible building that might pervade both.

The renaissance of research into prefabrication in architecture in the last decade in the United States has largely bi-furcated into two completely independent trends. On the one hand the practices of a number of architects who have reconnected with the mid-century golden age of research into fabrication and delivery of off-site housing units, whether in flat pack or modular logics, have created a niche market in pre-fabricated dwellings, largely produced in small numbers and therefore unable to advance the long elusive goal of realizing great economies through the scale of production.

On the other hand architects fascinated with the potential of harnessing computer design programs to computer-aided manufacturing and BIM systems, have often been more successful in the development of new formal expressions and material experiments without finding the pragmatic framework in which a customizable model can allow for great variety and flexibility and yet remain within the parameters of a sustainable system capable of application in diverse regional settings, and thus accommodating the huge atomization of building codes and local requirements.

Volkmann's Solar Roofpod, developed as an entry in the 2011 Solar Decathlon, the last to be held on the Mall in Washington, D.C., actively works across this divide that has so impeded the development of prefabrication research in the last few years. Like the Philadelphia firm Kieran Timberlake, whose manifesto Refabricating Architecture (2003) underpins this new research, Volkmann changes the traditional units of thinking from components to systems and assemblies, learning much from the techniques of fabrication that have revolutionized the automobile industry, particularly in Japan and Europe, in recent decades.

But where he stakes out his own position is in the pragmatic embrace of the inescapable paradox that pre-fabrication has been, and will continue to be, linked to a degree of hands-on making, adjustment, and finishing. In the architecture of this book the various intersections of the systems of the Solar Roofpod and of the larger economy of the building are not only analyzed in relation to a manifesto on a responsible ecological approach to architects' engagement with the urban environment, they are also continually narrated as a tale of necessary negotiations between fabrication and artisanal making, between designing systems of delivery and honing an individual object for maximum fulfillment of both individual desires and site requirements.

The hybrid embrace of different systems and types of assemblies to produce a building of elegantly simple aesthetic and composite technology and materials is to be sure distinctive in its willingness to amalgamate an enormous range of current technologies, from solar panels to advanced use of new glass products.

But the ultimate contribution of both the research exercise within a college architecture curriculum, and now the careful documentation of the entire process in this engaging volume, is to produce a pragmatic manifesto for an approach to prefabrication where models can evolve their rich potential to educate a new generation of architects and produce a new way of urban living that is at once attractive, responsible, ecological, and, it might even be claimed, ethical.

***Barry Bergdoll** is a Professor of architectural history in the Department of Art History and Archaeology at Columbia University and the Philip Johnson Chief Curator of Architecture and Design at the Museum of Modern Art in New York City.*

EXPLANATION

by **Christian Volkmann**

I have heard this many times: Architects never get their numbers right. They always emphasize the 'big picture', at least while in school. The interdisciplinary project described in this book was designed as an educational project to teach how to connect data and fiction with hands-on knowledge and implementation skills. The goal is to think comprehensively and tangibly. Its ambition is to develop better capabilities to transition the future of New York City. The students involved in this project will become the policy-makers, contractors, architects, developers, engineers, manufacturers, researchers and reporters of tomorrow. The dialogue between interrelated design topics will generate an approach that is far more productive than isolated big-picture-thinking, or its counterpart; specialized niche-thinking. It entails a negotiation process between both worlds, between big and small scales, to achieve what is necessary to think and produce ingeniously. It places weight on the methodology invested in materializing a system of relationships.

The reason for favoring this approach is that change and invention – improving efficiency, increasing productivity, avoiding error – is based on 'thinking it through'. This approach has to happen in many scales and genres: sometimes simultaneously, sometimes analytically separated from one another, but always with the goal of fitting into a holistic framework, and always with the purpose of realizing something tangible and practical.

The following description of the Solar Roofpod's design process is not linear: It zooms in and out, starting from an urban and environmental scale, rooted in statistics and overall observations, then moving closer and closer to a haptic material and assembly scale. Many viewpoints are needed to develop an idea, with input from different realms. This input is often processed by intentionally changing focus and subsequently reacting. The incorporation of all viable criteria by negotiating their meaning, both artistic and pragmatic, led to this design-built result, an overall prototype with its flaws and strengths.

For now, the analysis of results is secondary: Numbers are constantly changing within each context. For a 'Solar Roofpod's' performance, for New York City as a metropolis, and for the world beyond. With the delays involved in analyzing and publicizing new findings, and given their subjective interpretation, data can never be totally accurate. However, trends can be comprehended, and coherencies can lead to a better understanding of facts and subsequent action. Collaborative skills, and the ability to look beyond individual contributions towards an interdisciplinary approach is needed. Therefore, in this book, initial statistics are only used for the purpose of developing a comprehensive 'vision'.

Action is required to transition New York to a city of the 21st century. Though New York has always been a center for global innovation and creativity, certain infrastructural deficiencies ring nostalgic, belonging to the past rather than to the future. These deficiencies include aging and outdated infrastructure, unnecessary pollution and resource and energy wastefulness, to name a few. The current period of addressing these detrimental issues and finding the right solutions for transitioning into a cleaner, more sustainable urban environment, will define which cities evolve as preferred places of 'urban dwelling' versus the cities that will be left behind. Our design process was scrupulously conscientious in its attention to real world specifications. The ideas presented can be applied to metropolises worldwide.

Metropolises are the subject of major architectural activity and innovation: The way we build and rebuild our cities has changed dramatically in the past century. Industrialization has affected how we build, and the profession of the architect has changed along with it: How can we proceed in the best way possible? How, for instance, can we take advantage of prefabricated systems without compromising creativity or the city's multifaceted and positively curious appearance? How can we use architecture as an educational and communicative tool, by giving the engaged citizen food for thought? How can we strengthen our ability to produce this 'comprehensive vision' in a materialized approach? Comprehensive thinking opens the door for more sophisticated implementation.

In the past decade, under Mayor Michael Bloomberg's political administration, the city has done a great deal to initiate the discussion of how to approach the challenges of sustainability in our urban environment. It is clear that these efforts will be measured against their implementation as real projects.

The project described in this book responds to the demand for implementation: It is a concrete, realistic, and provocative hands-on manifestation of a prototype for transitioning urban life and infrastructure.

In this respect, the materialized big-picture prototype brings everything together. By constantly interrogating many of its features, we will now be able to verify how much this prototype can contribute to a greener and greater New York City. The project exists and is grounded in reality: Its performance can be measured, its ambience can be experienced, and it thus contributes to and can be held up against the idea of how we should approach the complexities of future urban living. It fits into a comprehensive framework, which considers all the pragmatic conditions and realistic means to transition New York to a protagonist of contemporary City Living.

The knowledge gained during the process, especially in the detail and hands-on fabrication, can now be converted in order to implement design towards this greater goal.

URBAN ENVIRONMENT: AN INFRASTRUCTURAL CHALLENGE

1.1

1.2

1

POWER GENERATION

75% of our global population live in urban or at least suburban environments, and this trend is rising[1]. However, most of the cities in the western world were set up for modern life about 100 to 120 years ago, in the dawn of the industrial revolution.

The origins of these no longer 'modern life' conditions can be immediately experienced in our New York subway system (1.1). In this case, much of the logistical infrastructure used to transport people around the city was built at the beginning of the 20th century. A constant reminder of this is the prevalence of riveted steel girders and columns. Fortunately, we kept up with the challenges of transportation, adjusting to flow rates from several thousand to the present figure of upwards of three million people per day. With an extended and more effectively managed system, this improved network is able to serve to capacity.

Getting from one point to another still works pretty much the same as it did some time ago. (Unfortunately, even the noise level is close to the same as it was a century ago).

Another less visible but equally as important urban amenity is the availability and distribution of electrical energy, which has also not kept pace with the demand. Due to the ever increasing dependence on electrical power per capita, the distribution grid has issues with satisfying our hungry need for lights, TVs, appliances, air-conditioners, computers, mobility etc., to power our modern lifestyle. When our 'grid' was set up a long time ago, nobody could have envisioned how much electricity we consume nowadays[2].

The infrastructure serving this task has systemic deficiencies. Its distribution system, centralized amidst a long-distance transmission grid, does not seem the right answer to a problem that keeps growing. In August 2003, a significant summer power outage on the North American East Coast demonstrated this shortfall: More than 55 million people in the US and in Canada ended up without power[3]. The problem exposed is that the electrical power produced by the power companies cannot easily be stored and is generally consumed less than a second after being produced[4]. Therefore, the load on any network must be precisely matched by the supply to it. Any overload of a power line, or underload/overload of a generator, can cause hard to repair and costly damage, such that the affected device is immediately disconnected from the network if an imbalance is detected. In an emergency more power must be obtained from other generators, or they will 'shed load to protect their equipment, effectively cutting power to some areas. This is usually done by 'system operators'. It is their job to bring the system into a perfect balance with the help of computer programs.

One would think that the bigger and more centralized the system, the easier it is to manage its supply. In the case of power, the advantages of having a bigger conglomerate of suppliers are not as advantageous as expected, since the power produced doesn't have a long 'shelf life'.

In the case of the 2003 black-out, more than 250 'overloaded' power plants shut down in a cascading effect within an hour. The strategy of balancing the system did not work on this particular day, when an initial power plant in Ohio encountered significant technical problems. The balancing of power by means of mega-networking a hierarchically organized system clearly failed. If one does the math right, each power plant was responsible for an average of 200,000 people without power.

The set of questions that remain include: Whether dependency on this 'Big Brother' system of energy is the most efficient and most elegant way to power our cities? Why is power produced far away? Under what circumstances is it produced? How does this affect us?

1.3

First of all, there are substantial transmission losses in any kind of energy distribution: the farther away the source, the more energy gets lost in transmission. It is estimated that approximately 20% of all energy produced in the United States is lost in transportation[5].

The artist Richard Box brought attention to this issue when installing a 'field' of more than a thousand flourescent light bulbs underneath a low-overhead power pylon. The installation was entirely powered by this 'waste energy' (1.4).

Hypothetically speaking, if we all had a 'mini-generator' in our apartment windows, and these devices only produced power as necessary, the savings would be huge. Furthermore, we would have an immediate relationship to our consumption of fossil fuels.

In conclusion, one fact is clear: the current, centrally planned and controlled distribution system doesn't work as efficiently as intended; it produces a lot of wasteful 'left-overs'. Secondly, the means with which we produce energy is environmentally unconscious and not very sophisticated.

In the US, most of the energy used to produce electricity is made of non-renewable materials: It is effectively fuel we burn using the resulting heat to generate electricity. These natural resources are gone for good after they are consumed. Instead, they could be used much more productively for chemical or pharmaceutical applications, for instance.

It is worth noting that mankind has consumed more fossil fuels in the past 50 years than in its entire prior history, and it is estimated that we will run out of oil in ca. 50 years, out of natural gas in ca. 100 years, and out of coal in ca. 200 years.

Even though there is potential to extend the supply of fossil fuels, for instance by means of fracking to extract shale gas, or by means of processing oil sands, these additional sources of fossil fuel production are questionable: They cause massive irreversible environmental damage. Certainly, these resources, which usually have taken at least hundreds of years to form naturally, are consumed at a rate that is clearly heading towards exhaustion.

1.1 NYC, subway system, riveted columns.
1.2 NYC, 1st electrical streetlight, 1882.
1.3 NYC, power consumption: On peak days, power up to 11,000 MW is common, the same rate as entire countries such as Chile or Portugal.

1.4

1.5

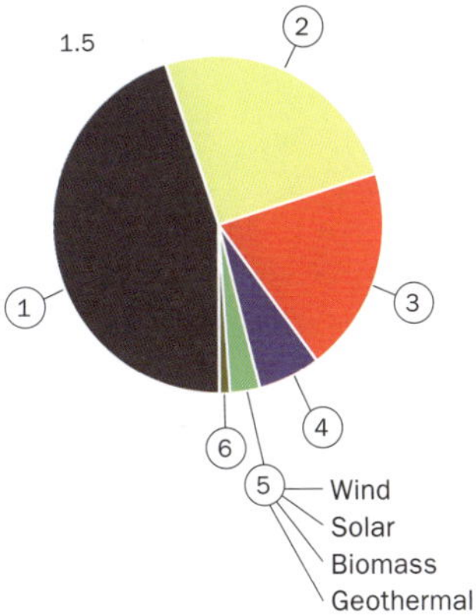

1. COAL 42%
2. GAS 27%
3. NUCLEAR 19%
4. HYDRO 6%
5. RENEWABLES 5%
6. OIL 1%

1.6

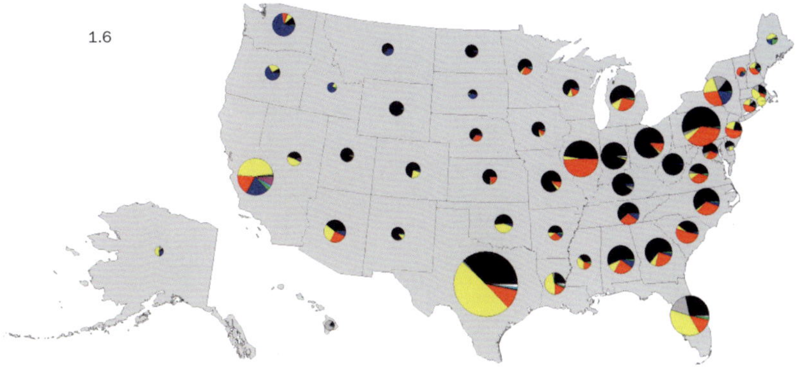

eGRID2007 year 2005 Generation (MWh)

- COAL
- OIL
- GAS
- NUCLEAR
- HYDRO
- BIOMASS
- WIND
- GEOTHERMAL

1.4 "Field", Richard Box, Skenfrith/Wales, 2004.
1.5 USA, Energy generation mix by fuel type.
1.6 USA, Energy generation mix by fuel type, differentiated state by state, 2007.

About ¾ of the US's energy supply is based on a mix of fossil fuel 'products': currently, a bit less than 50% of this mix is coal. Natural gas is the new favorite, approaching a 25% margin, and becoming increasingly popular. This is due to the fact that gas plants operate more efficiently, have an easier permit process and produce fewer CO_2 emissions than coal plants[6].

About 90% of power plants constructed in the last ten years are either gas plants or 'dual-fuel' plants, which commonly use a mixture of two fossil-fuels, one of them being gas. What this has led to is the seven-fold increase of natural gas prices in the last ten years, before fracking technologies caused substantial price decreases. This is evidence of the fact that diversification is just as important in the energy industry as it is in any other business.

In New York State, the energy generation mix is not as dependent on one single fuel as the countrywide average (1.5, 1.6). No one fuel is used to generate more than a third of the state's power; the leading sources in the NY mix are natural gas (30%) and nuclear energy (30%). Thanks to a high amount of nuclear energy, only 50% are sourced from fossil-fuels[7]. The relatively high percentage of nuclear energy comes with its own risks, however. It is known that the 'Indian Point' nuclear plant, only 40 miles from New York City, is built to outdated seismic specifications. Furthermore, geologists have located a previously unknown active seismic zone that intersects another fault line perilously close to the plant. Consequently, one of the two reactors has the highest risk of earthquake damage in the country according to the Nuclear Regulatory Commission (NRC)[8]. However, since both reactors provide about a quarter of the power needed in New York City the mayor is a strong proponent of this energy source, declaring that "Indian Point is critical to the city's economic viability".

Zooming in from New York state data to New York City's generating capacity, the dependency on fossil fuels is cause for consternation. As it stands, natural gas and 'dual-fuel' plants make up 95% of the city's generating power[9], and are preferred for reasons of security, practicality, flexibility and the relative ease of the permit process. However, these plants emit several million tons of CO_2 into the environment annually and are considerably more destructive when methane and other dangerous greenhouse-gases are not fully combusted in the generation process. The effect of these gases on ozone depletion in the atmosphere is estimated to be 20 to 50 fold more than the CO_2 emissions often publicized. Therefore, their operation is not nearly

1.7

1.8

as clean as often publicized. Thus, the dilemma appears: We can either power our city with relatively clean nuclear energy, along with the related transmission losses, risks of failure and contamination that come with this choice, or we draw on locally generated fossil-fuel based energy, and the resulting damage from greenhouse gas emissions.

Power generation accounts for about one-quarter of total emissions of CO_2 worldwide: the main culprit in global warming. The US and China are collectively the largest emitters of CO_2. Hosting over 8,000 of the 50,000+ power plants worldwide, our power-related emissions account for about 25 percent, or 2.8 billion tons, of CO_2[10]. There is an overwhelming amount of evidence that these greenhouse gas emissions are the main contributor to our changing climate patterns, with 'global warming' as its main result. Although the forecasted damage to our civilization is horrendous – the expected sea level rise alone will cause infrastructural upgrades and repairs costing trillions of dollars. Despite annual United Nations (UNO) climate change conferences attempting to push legislation on this issue, progress has been very slow, particularly in the USA. Simply put, governments are not capable of deciding on appropriate measures fast enough.

The flooding of New York City due to sea level rise was predicted in computer simulations years ago (1.7). However, the consequences for New Yorkers were experienced at first-hand when superstorm Sandy struck the city in Fall 2012: flooding-related power losses made it clear that the city's infrastructure is extremely vulnerable to such occurrences. The level of comfort and quality of life, often based on fossil-fuel combustion, was actually cancelled out by the consequences of generating energy in that exact fashion. Especially the loss of power, lasting for days for major parts of the city, and the damage suffered to the electricity grid and the subway system due to saltwater flooding (1.8) generated widespread awareness of the need for resiliency.

The goal of an emission-free power generation mix for New York City would logically lead to the activation of renewable energy sources, known as 'renewables'. These are sources of energy that are not depleted by use, such as water or wind-driven, or solar-activated devices generating electricity.

The diversification of emission-free urban power has been the starting point for an initiative by the Department of Energy called 'Solar America Cities'. Through this program, certain East Coast cities (New York, Boston and Philadelphia) have been working to accelerate the adoption of solar energy technologies for a cleaner and more secure urban energy future. The City University of New York (CUNY) is acting as one of the main research facilities for this program. It has focused on creating a 'solar map' of New York City, which evaluates the solar potential of each of the city's one-million-plus

1.9

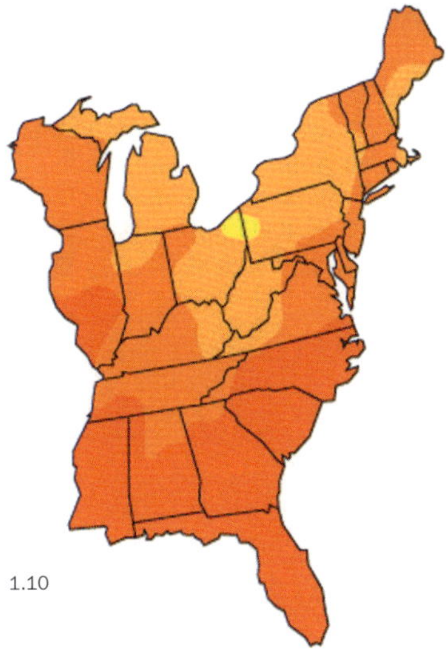

1.10

buildings as an outreach tool: If policymakers, real-estate owners and entrepreneurs were to focus on developing the roofs of these building, solar energy could have an impact on New York's power-related economic and infrastructural trajectory.

The 'solar map', launched in 2011 (www.nycsolarmap.com), concentrates on two viewpoints: 1) the real time account of photovoltaic energy generated in NYC, and 2) the potential for future development, focusing primarily on individual property owners with an interest in their individual property yields[11] (1.9).

While the current production of solar electrical energy in the city is listed at the 10,000 kW level, its potential is actually much higher. Related studies assume that the suitable roof areas provide enough space for solar panels to power half the city during hours of peak electricity use, which occurs in summer months (as seen in the 2003 black-out). The resulting cooling loads and our appetite for air-conditioning during these months greatly increase demands on our grid. Thus, instead of burdening the air quality of Manhattan with additional CO_2 emissions due to utilizing power of the dirtiest, most outdated power plants, emission-free solar power would be the perfect choice for this season, since solar irradiation in our geographic latitude is plentiful and predictable (1.10).

The energy clearing price (the cost of a kWh) is most often defined by peak power plants. Therefore, New Yorkers could potentially "cash in", saving millions in electricity costs and creating thousands of new jobs in this industry, if we relied more exclusively on solar power.

During the last century, the country's electricity system has been centrally planned and controlled. As a result, utility company monopolies have delivered electricity in return for guaranteed profits from state regulators. Americans accepted the environmental consequences because economical and technological alternatives were limited in availability when the electricity grid emerged.

However, the rapidly and recent falling cost of renewable energy offers a dramatically different electricity future today. Citizens could become power generators instead of just consumers, sharing in the economic benefits of clean energy and processing the relationship between production and consumption. This can become possible only if policymakers shift control of the electric grid back to the consumer.

1.7 NYC: Flooded City, simulation, 2006.
1.8 NYC: Flooded subway station after superstorm Sandy, 2012.
1.9 NYC: Solar Map, data visualization and analysis, 2011 (CUNY).
1.10 USA: East coast, solar radiation; 1000 kWh/kw-yr are sufficient for PV power generation. All of the Eastern US exceeds this requirement.

2

HEAT ISLAND

1.11

When driving from New York City to the countryside in the midst of the summer, one immediately realizes how much hotter the city is relative to those more remote places that we crave at that time of the year. We city dwellers are getting hit with heat in ways our counterparts in suburban and rural areas are not. On average, our ambient temperatures are about 5-10°F warmer than they are in the countryside, especially at night.

What makes matters worse is that we prefer air-conditioning as a solution, which actually exacerbates the problem. Air-conditioning is heat exchange between interior and exterior, such that we are merely shuffling the heat to the outside of our houses and increase the ambient temperature in doing so (1.14).

There are several reasons why the phenomenon of an 'urban heat island effect' exists, making us feel miserable, especially at night:

Thermal inertia: Cities absorb more solar energy during the day and slowly release it at night. The reason for this is an increased amount of 'thermal mass'. Simply put, our cities are built of brick, concrete and asphalt, all mineral-based materials that have a high heat capacity and therefore absorb heat exceedingly well. They effectively work as heat storage, releasing the absorbed heat at night. Thus although there is no solar load during the night, the heat stays on like a radiator that you cannot turn off.

Evaporation: While the countryside is predominantly natural land surfaces, soil and vegetation, the city is composed of predominantly sealed surfaces. Unfortunately, these surfaces can neither absorb nor release moisture in a space where it is helpful to do so. Absorbent matter delays the peak of drainage, i.e. during a rainstorm, and hence prevents flooding. This moisture absorption is also the antidote against the heat island effect. Eventually, absorbed water is released through evaporation, which takes place when the air is capable of absorbing moisture. As water evaporates, energy is lost from the air, which reduces its temperature. Because of this effect, also called enthalpy of vaporization, air is cooled by any moisture absorbed from soil and plants.

Wind: The buildings in an urban environment effectively create a screen against winds. During lower wind speeds, a hot air dome is perceivable, which captures and accumulates hot air within a zone extending about two to three times the height of neighboring buildings. Above this height, the difference between urban temperature and rural temperature is actually about the same.

Wind also enhances evaporation, as it transports the moisture content away more quickly. However, it is not just the temperature of the air, but also its quality that causes problems when air circulation is prevented. For instance, we inhale more particulate matter, tiny dirt particles that can cause diseases like asthma, respiratory allergies or pneumonia. Many of these fine particles are the consequence of burning fossil-fuels in vehicles, power plants and industrial processes.

Anthropogenic heat: Many activities present in urban environments contribute to heating up its ambient temperature: traffic and transportation, year-round climatization of most office and commercial buildings, as well as industry and manufacturing processes contribute to heating up its ambient temperature. Traffic and transportation, year-round climatization of most office and commercial buildings, as well as industry and manufacturing processes all contribute to that effect. The release of heat

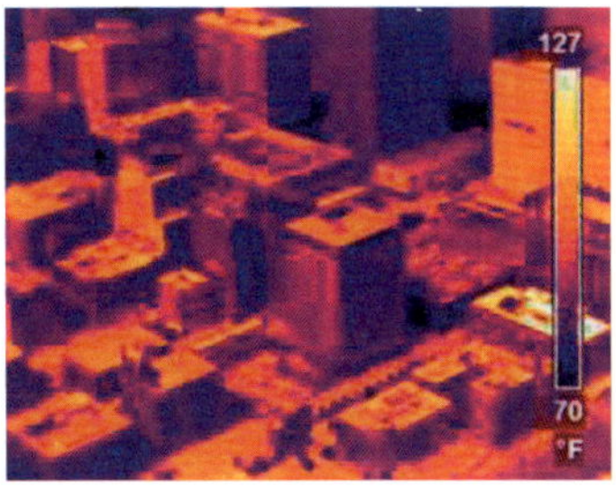

1.12

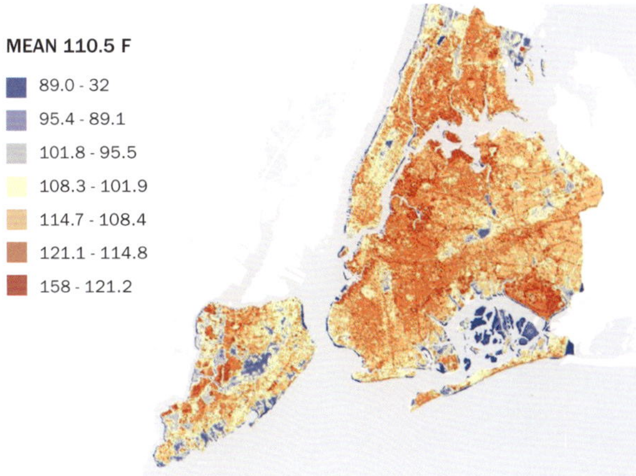

1.13

1.11 Cadiz (E): Seasonal urban shading, reducing heat gain.
1.12 Infrared image of typical urban environment: highest surface temperature; the roofs.
1.13 NYC. Surface temperature: August 2002; (Landsat satellite image) Many microclimates are detectable; note Central Park in Manhattan.
1.14 Paris (F): Heat Island, effect on air-conditioning; highest peak in densest neighborhoods.

related to these activities is called anthropogenic or waste heat. In this respect, performing an exergy analysis in any related field could improve the way we power these processes, with less waste heat released into our environment as a result.

One can identify two categories that influence the heat island effect. On the one hand, there is human intervention, a behavior that we are directly responsible for. On the other hand, there are numerous meteorological influences: climate zones, related seasons, wind speeds and directions, cloudiness, precipitation patterns, etc. While we can only adapt to some of these known meteorological conditions, we are obligated to make active and prudent choices in the field of intelligent architecture or passive design. Some examples of proactive choices in favor of passive design include influencing material specifications, on many different scales and approaches. This includes choosing material colors according to their emissivity, which determines their absorption rate.

We can also reintroduce vegetation and capitalize on the resulting evaporation as a built-in cooling device. Another passive design technique is to shade 'hard scape' regions, which significantly decreases the amount of heat that they would usually absorb under exposed conditions (1.11). Finally, we can adapt to wind patterns for passive cooling (in NY, the dominant wind direction in summer is south-west, in winter north-west), and respond by manipulating density, shading and accurate orientation. Passive design is indispensable as it relates to intelligent architecture. Education on the behavior necessary to accurately operate the designed environment is closely related to it.

A Heat Island is a man-made effect. It is up to us how we balance it. However, to burn fossil fuels to produce cool air within our apartments, and simultaneously worsen the issue by doing so, is not an intelligent response. We have to create an environment that establishes favorable long-term conditions.

New York is still a growing and densifying city, thus the heat island will become more significant in the future. It will further impact air quality, public health and the demand for energy. We also run an increased risk of blackouts as the heat island effect elevates demand for electricity. If we want to stay well, we must come up with viable solutions.

1.14

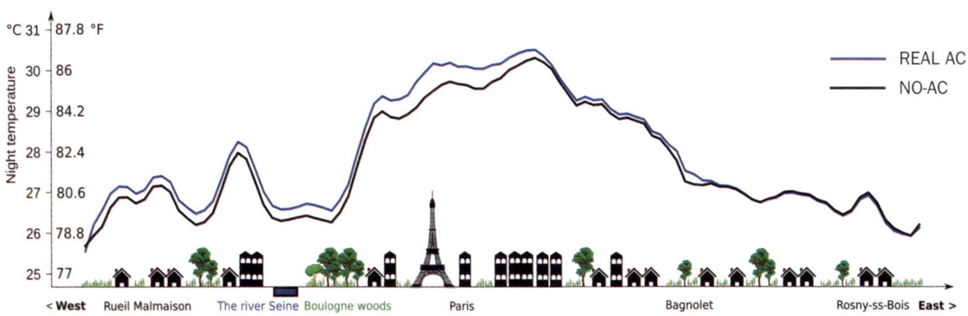

3

BUILDING ENVELOPES

1.15

Many of our buildings' interior spaces heat up in the summer, due to both urban environmental factors and our building's inadequacies. As for our buildings, the efficiency of the envelope is of utmost significance in preventing infiltration of heat or cold.

During the winter, these building envelopes separate our interior climate fully from the exterior and must withstand leaking heat to the outside. During the summer these requirements are quite different: Considerate ventilation can benefit efforts to balance an interior climate, if humidity and peak ambient temperatures are avoided. Night ventilation can be utilized to avoid the latter. Generally, for natural ventilation purposes the building has to be evaluated as a whole, focusing on thermal convection paths through the building, considering the façades as one of the essential in- and out-takes to discharge heat. If buildings are built too deep, with floor areas too distant from operable windows, it is difficult to achieve these conditions. For example, cross ventilation usually works for floor plate depths of less than forty feet, which are, due to the need to maximize rentable square footage, commonly exceeded in New York's contemporary building stock. The consequence is that buildings need to be mechanically ventilated; the building is literally on a 'breathing machine', like a patient in a hospital, dependent on operational energy to provide comfortable conditions. To stay in this metaphor, the building's vital organ, the skin, cannot function anymore. Operable windows have become obsolete. With rising energy prices, the concept of only using mechanical ventilation and air-conditioning will become increasingly costly and burdensome; moreover, the greenhouse gas emissions related to this unnecessary energy expense are detrimental.

In the winter, these envelopes would be predominantly sealed. Therefore, thermal conduction is now governing the loss of energy, and possibly the loss of comfort as well. The quality of the building's winter coat is thus essential. And equally in summer, this performance bears conductive losses, now in the form of conditioned cold. The evaluation of these conductive losses is essential to improve our buildings' skins. The question thus remains to be asked, how efficient are the conduction-related thermal properties of our 'envelopes' in NYC's building stock?

The fact that about 75% of primary energy consumed in New York City is used for buildings indicates that their envelopes are not particularly efficient and inevitably contribute to unnecessary operating expenses[12].

There are many reasons for the inefficiency, but even in our recent past, this loss of energy has not been recognized as the dire issue it is. Energy has been relatively inexpensive, and the environmental damage caused by unnecessary fossil fuel combustion required to heat buildings has not been the center of attention. Since much of the real estate market is investor driven, it is more profitable to also save on the initial expenses of envelope insulation during construction. The developer's strategy is often to let the future building owner manage additional expenses in the operating budget, which are then reimbursed through the property tenant's rent. These tenants have no idea that they ultimately

1.16

1.15 NYC: Mainly curtain wall. Even sophisticated buildings such as the LEED 'platinum' One Bryant Park tower have an inferior thermal resistance compared to opaque envelopes.
1.16 NYC: View from Brooklyn. Eyes cannot perceive what an infrared camera can; the skyline showing substantial heat loss.

1.17

1.18

take on the burden of extensively heating and cooling their energy-inefficient buildings. They are fundamentally the weakest link of this economic chain.

City, state and federal agencies have clearly demonstrated their interest in creating a more energy-efficient environment. In the early 1990s the Department of Energy launched its Building Energy Codes Program (BECD), in order to increase energy efficiency in America's residential and commercial buildings. It coordinated with other government agencies, state and local jurisdictions, national code organizations, and industry to promote stronger building energy codes and help states adopt, implement, and enforce those codes. However, this enforcement is riddled with problems. Despite the existence of quite stringent energy regulations, it is still possible to build with systems that circumvent the set requirements.

For example, New York City's most notable skyscrapers are typically clad with light-weighted curtain wall technology, the predominant material of which is glass. The insulated glass units that clad these building have negligible thickness, typically the glass units being less than an inch thick. While this is profitable as it maximizes the rentable floor footprint, the substance essential to providing adequate thermal insulation is missing. However, corporations often favor this kind of approach to building materials, since it promises an image of transparency, and thus communicates accountability and trust (1.15). It is also known that the installation of these façade systems is both expedient and economic, with an early completion date as a favorable result. Although daylighting is noted as an advantage of glass facades, it is unlikely that this can offset the costs of climatizing the building due to heat losses and gains encountered throughout the year. As a rule of thumb, the upper third of a floor's façade provides sufficient daylight to the interior. However, many glass-building facades are covered with a ceramic frit pattern to essentially reduce the amount of light and heat captured by the glass skin. It is important to note that the façade area below desk level, along the façade's spandrel panels, contributes nothing to daylighting qualities despite the fact that it is often left undifferentiated from the rest of the building's skin. Differentiation along the building skin is, however, critical when considering the multiple requirements of daylighting, glare, heat gain and loss, fresh air, views and physical connection to the outside. However, this is often clearly overlooked, especially in urban environments.

One of the main reasons to apply a curtain wall system is that it is a readily available; a practical and perfected assembly solution within the building industry. Design issues are often outsourced to the curtain wall supplier so that the architect is only marginally involved in the technical feasibility and performance requirements of this system. Yet, what is critically lacking in this lightweight exterior wall system is its thermal efficiency due in part to its development during the era of cheap oil in the mid-twentieth century. If our eyes could see the thermal inefficiencies of our urban environment, as made visible with infrared technology (see 1.16), the issue would certainly come to the fore.

Another very popular, but more old-fashioned envelope type in New York is the masonry façade, often used for residential buildings due to their low maintenance needs, durable advantages, and contextual qualities (1.17). These systems, although

1.19

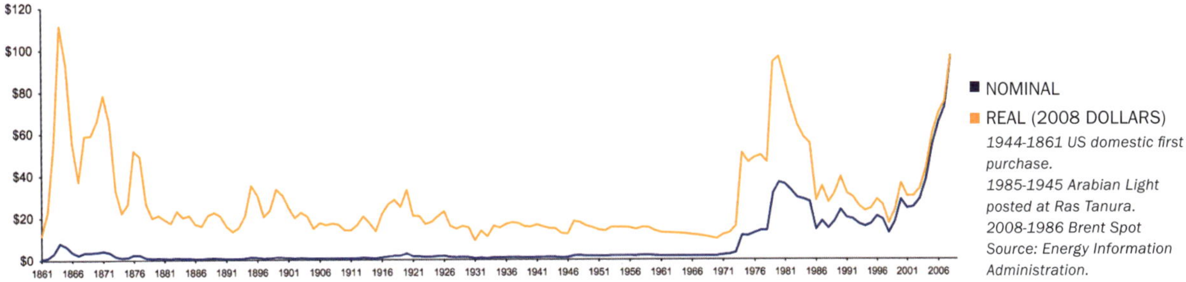

slightly more thermally resistant, have their problems, too. Most of them are built as cavity wall systems with different layers, effectively keeping the inside dry, while separating the loadbearing structure from the facing brick. However, when these façades are being repaired, in accordance to NYC's mandated Local Law 11, one particular layer often remains missing: modern rigid foam insulation. This material, which was readily available in the early 50's, did not become popular nor deemed necessary until the oil crisis of the 70s. However, most of the aforementioned brick buildings were built well before these materials were accessible, and retrofitting to modern energy efficient standards is highly problematic.

Today's foam insulation has a much higher thermal resistance than brick: about 35 times the resistance, to be precise. Consequently, in order to achieve the equivalent insulation value of 2" foam board, a brick wall would have to be 6 feet thick; unfeasible for countless economical and technical reasons. Still, one might question the use of an oil-based material, considering its embodied energy and the production processes necessary to manufacture it. However, despite this observation and the rising cost of oil (1.19), the break-even point for this investment would be achieved in a few years, as compared to the energy that would otherwise be spent heating the building continuously through combustion, without any lasting benefit.

While it would make sense to retrofit the existing brick façades with efficient insulation, this is much easier said than done. In order to access the cavity to install this insulation, the entire facing brick layer would have to be removed, weighing in at a considerable 40 pounds per square feet.

After completing the new insulation layer, the heavy facing brick would then have to be reinstalled, and resupported. A less complicated and more economic option to replace the facing brick with a lightweight mounted system in the course of the renovation would be advantageous. These lightweight brick façade solutions might have to be similar to curtain wall technology to be effective, and any research in this direction would thus be crucial to the progress of envelope efficiency.

Both envelope applications, curtain wall and old-fashioned masonry, require improvements to outperform their older counterparts. Most envelope systems currently feature very low thermal resistance values, called 'r-values'. It is reasonable to estimate an r-value of 2 for curtain walls, an r-value of 5 for un-insulated masonry walls, and an r-value of 15 for insulated masonry walls.

For the following envelope performance investigation, we will extrapolate the r-values of insulated and un-insulated masonry walls to R-10, and assume that these walls include windows as part of the assembly. The roofs of these buildings are a part of the envelope, and are often better insulated than the façades, as it is easier to retrofit these surfaces with insulation. For the roofs, we will assume an average of R-20. 'Thermal transmittance', which is the reciprocal of the thermal resistance, is measured in BTU/s.f,˚F,hr. These values correspond as follows: U-0.5 for our curtain walls, U-0.1 for masonry walls, and U-0.05 for the roofs. These U-values determine how much energy (BTU) is transmitted through the exterior walls, relative to square footage, amount of time and temperature difference.

1.17 NYC: Mainly brick. 'Prewar' as well as 'Postwar' buildings feature often uninsulated brick façades.

1.18 Brick repair, exposing thermal deficiency. No insulation exists; retrofits are difficult.

1.19 Graph of oil prices from 1861-2007, showing a sharp increase during the 1970's energy crisis (orange line adjusted for inflation).

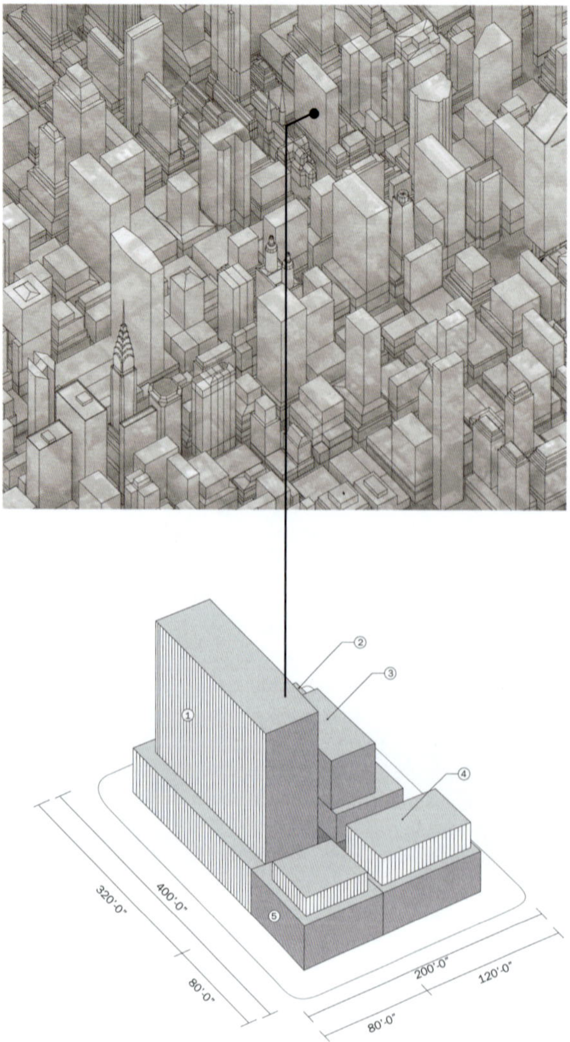

				C	M	R
Bldg. 1	(base)		60 x 320	19,200		
			60 x 80	4,800		
			60 x 260	15,600		
	(roof)	2x	15 x 320			9,600
			20 x 50			1,000
	(top)	2x	250 x 300	150,000		
			250 x 50	12,500		
			250 x 50		12,500	
	(roof)		50 x 300			15,000
Bldg. 2	(base)	2x	60 x 120		7,200	
			60 x 40		2,400	
			60 x 60		3,600	
	(roof)		60 x 120			7,200
Bldg. 3	(base)	2x	60 x 220		26,400	
			60 x 80		4,800	
	(roof)		30 x 220			6,600
		2x	15 x 50			1,500
	(top)	2x	60 x 190		22,800	
		2x	60 x 50		6,000	
	(roof)		190 x 50			9,500
Bldg. 4	(base)	2x	60 x 120		14,400	
			60 x 80		4,800	
		2x	15 x 120			3,600
		2x	15 x 50			1,500
	(top)	2x	90 x 25	4,500		
		2x	50 x 25	2,500		
	(roof)		90 x 50			4,500
Bldg. 5	(base)	2x	80 x 60		9,600	
	(roof)	2x	80 x 15			2,400
		2x	50 x 15			1,500
	(top)	4x	50 x 20	4,000		
	(roof)		50 x 50			2,500
TOTAL				213,100	114,500	66,400

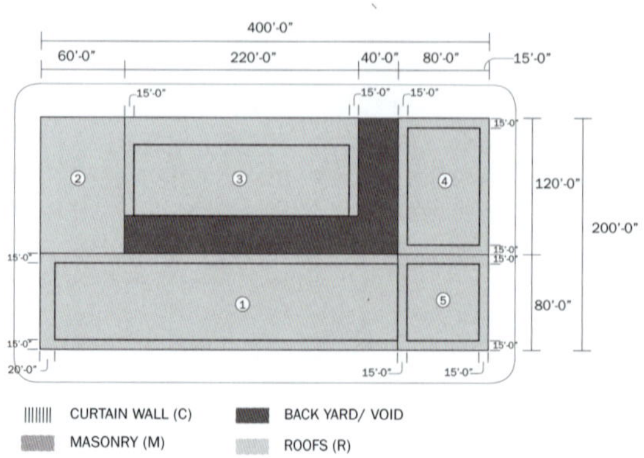

	C	M	R
U-value (BTU/s.f.,°F, hr)	x 0.5	x 0.1	x 0.5
Transmittance (BTU/°F,hr)	106,550	11,450	3,320
Per day (BTU/°F) [x 24 hr]	2,557,200	274,800	79,680
TOTAL (BTU/°F)			2,911,680
Heating Degree Days, NYC = ~5,000 °F, day/year			
TOTAL Heating required (BTU/year)			**14,558,400,000**
TOTAL with Weekend/night savings (x 0.7)			**10,190,880,000**
OR, in kWh [x 0.0002928]			2,983,890
OR, in tons of OIL equivalent (TOE) [x 8.6 x 10^{-5}]			257 tons

1.20

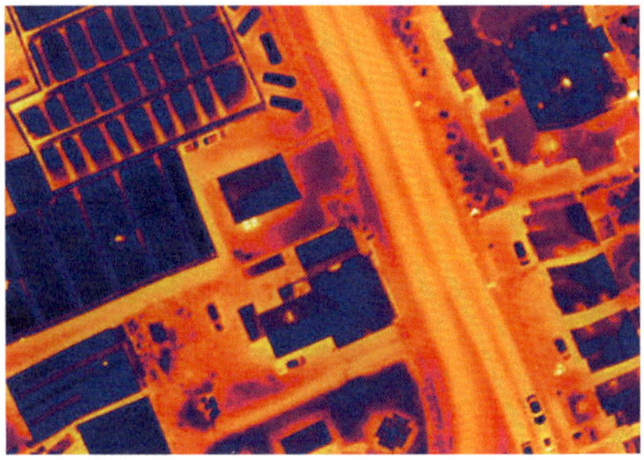

1.21

1.20 NYC: How much heating energy is consumed by our building stock? Sample block calculation, rough estimate. The TOE (tons of oil equivalent) does not take into account any efficiency loss during combustion.
1.21 Thermal imagery surveys by air are increasingly popular in improving development policies and incentives.

By taking one generic street block of the more than three thousand in Manhattan (1.20), and calculating the amount of energy wasted in a heating season based on the two common inefficient envelope systems previously discussed, the effect of these inefficiencies on our environment would be far better understood. In this simplified calculation, the estimated energy necessary to reconcile the thermal losses for this block alone in a full heating season would amount to an impressive 10 trillion BTUs. This is the equivalent of more than 250 tons of oil. It is clear, especially when looking at all of New York simultaneously – or even globally – that much of this excess could be saved if the façades were renovated and thermo-proofed. Taking into account that curtain walls make up 88% of the losses in this sample, a simple upgrade with today's technology would be advantageous and feasible, since the dry-mount system is easy to retrofit. Some solutions would be to insulate spandrel panels efficiently, utilize glass-integrated vacuum insulation and shading devices, and reduce the amount of glazing to the third that is really necessary for sufficient daylighting. More research should be generated to facilitate this approach.

Unfortunately, thermal resistance surveys of entire urban building environments are not yet available, thus the impetus behind the former simplified calculations. Presently, laser systems technology produce aerial thermal imagery that can effectively monitor roof performance (1.21). This kind of technology could potentially expand to include systematic surveys of vertical façade performance. This method can ultimately identify areas that demonstrate high heat loss for the purpose of targeting grant assistance and policy strategies. While legislative agencies intend to push this route, energy code requirements unfortunately have higher standards than what has actually been achieved. This discrepancy might be related to loopholes in the system, especially those related to glazing. But even more so, they are likely associated with the tendency to embrace change. As it stands, if incentive voluntary programs are the preferred means of policy, then the target will most certainly move slowly. In Europe, mandatory code requirements have had a much greater impact. For instance, the current discussion to bind an owner's property tax with the actual efficiency of the building envelope, manifested as an obligatory 'Energy-Pass' is currently under review for implementation. It is doubtful that this effort is conceivable for New York, since this culture tends to looks down on economic meddling by the government.

With such high-energy loss transmitted from building envelopes into the environment, along with subsequent concerns of carbon pollution and global warming, these building skins are closely connected to our overall energy infrastructure and in dire need of renewal. The efficiency of envelopes yields huge potential to lower our carbon footprint. All parties in the building sector, one of the most important economic subsectors in the US, have to commit to eagerly join forces to secure progress towards such efforts, and move the target forward continuously.

The architect is said to be a protagonist, delivering proficient research while collaborating with policy makers and the industry as necessary to provide superior and economically viable solutions. Politics must approach their perspective with the promise of long-term sustainable growth, not simply the most economic solution. Legislature needs to remove loopholes from energy codes, and monitor efficiency, to create a reference tool for progress. Finally, investors must commit to developing transferable efficient solutions with higher short-term investment but better operational value. Many of our architectural jobs in the near future will involve thermo-proofing our buildings, and the subsequent comprehensive research necessary for effective performance renewal. Our envelopes in NYC's building stock need an upgrade on all these fronts.

4

URBAN VEGETATION

1.22

New York City has lost 9,000 acres of vegetative cover in the past 25 years[13] (1.24). This is more than 10 times the surface area of Central Park, or about 100,000 tons of carbon sequestration per year. Carbon sequestration relates to the absorption and long-term storage of atmospheric carbon dioxide (CO_2) in trees and other kinds of vegetation. Unfortunately, as the city grows, the green lots continue to disappear. In response to the city administration's awareness of the benefits of a greener urban environment, it launched the 'million trees' initiative in 2007 to upgrade New York's urban forest by about 20% (www.nycgovparks.org).

Central Park (1.23), designed more than 150 years ago, remains one of the greatest achievements of urban planning, and stands as New York City's most prominent vegetated real-estate. This park serves not only as the green lung of the city – nowadays breathing a little harder – it also provides far more complex benefits. As discussed before, soft scapes are able to reduce ambient temperatures due to the evaporative effect of their moisture absorbing surfaces. Furthermore, shading contributes to the same evaporative cooling effect by preventing absorption of the sun's radiation. Temperatures in Central Park are therefore considerably lower than elsewhere in the city.

Here we find another centralized system: Central Park is the heart of the city, effectively one instance serving all. A more even distribution would not only be more efficient, but would also provide benefits that everyone could enjoy. These benefits are as follows:

Trees produce oxygen. One tree produces the amount of oxygen needed for ten people to breath in one year. We discharge CO_2 when breathing (...and release even more CO_2 during all our fossil-fuel driven processes). Since trees absorb CO_2, at a rate of about 40 – 50 lbs./year per tree, they essentially act as carbon sinks[14]. They do this by removing carbon from the air and storing it as cellulose in their trunks and branches meanwhile releasing oxygen back into the air, and providing circumstantially a beautiful building material. In fact, trees are one of the cheapest, most effective means of drawing excess CO_2 from the atmosphere. They, like all other vegetation, are instrumental in diminishing the 'greenhouse effect' that we are exposed to. The principle among all plants – not just trees – is the same: carbohydrates are formed through photosynthesis, and with it return oxygen back into the air as a by-product.

Plants remove air pollutants. Among the many major pollutants in urban environments are Sulfur Dioxide, Ozone and Nitrogen Oxide. All three are the result of burning coal, oil and gas for electricity, as well as home heating, refining oil, and emissions of automobiles and industries, to name a few. The many fine particles in the air that harm our health are called 'particulates'. These include heavy metals such as cadmium, chromium and lead, which are part of typical emissions. Plants can remove these air pollutants mostly through their leaves. Therefore, they improve the air quality and our health.

Plants provide shade for buildings. This may not sound very effective, since buildings are typically tall in urban environments. However, trees can also grow considerably tall in these environments, even up to a height of 40 – 50 feet. Shading is most efficient if placed on the east or west side of the building, since the angle of the sun is lower at these times of the day. When the sun strikes the tree, it reflects the radiation before it hits and is absorbed by the building's façade (1.22). Therefore, it diminishes the cooling loads of the building.

1.23

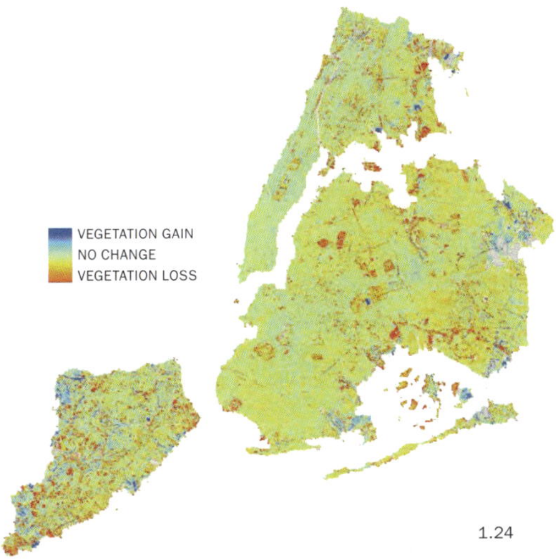

1.24

1.22 NYC: Street tree, planted as part of the 'million trees' initiative; reducing cooling loads.
1.23 NYC: Central Park, New York's 'green lung'.
1.24 NYC: Vegetation loss, over the last 25 years; 9000 acres of vegetative cover.

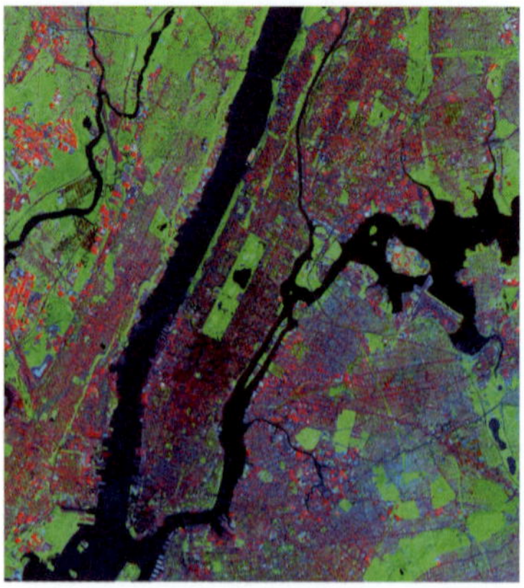

1.25

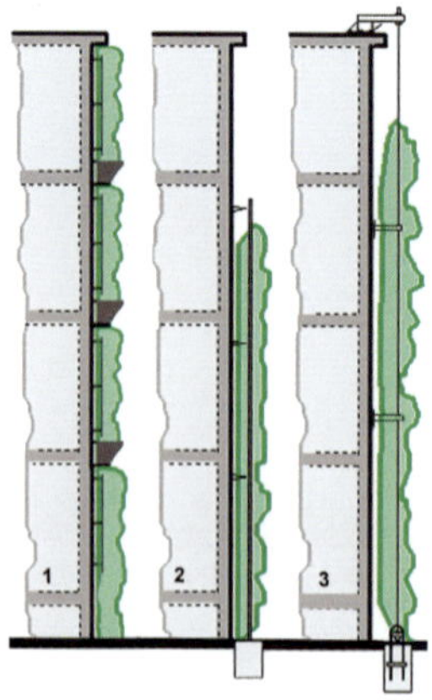

1.26

1.27

Trees come with their own structural system, comprised of a trunk, branches and roots, which defines their shape via genetic coding. Other plants, not structural but equally as efficient in CO_2 and pollutant removal, can also be used for effective shading configurations. In recent years, façade trellis systems that utilize vines have become relatively popular. If placed in front of the façade (1.26, 1.27) facing the sun, this strategy maximizes the benefits of shading in summer, while adjusting its filter in the winter months by losing or thinning its leaves to take advantage of solar passive gains.

These greening systems are sometimes vulnerable since they need maintenance and irrigation support every now and then. But, in a city where every square foot of real estate is precious, the vertical application of plants could be an effective instrument to revegetate our environment.

The strategy can, however, be utilized in more than just the façades of a building. If the trellis is applied to the roof, its shading function becomes more effective, since the roof is exposed to higher environmental loads (1.28). The cumulative reduction in cooling load is then even higher. The roof also constitutes a living space that we could enjoy, like a mini-park for every building, providing space for relaxation in our hectic urban lives. Instead of the suburban front lawn, we would have the urban roof garden to enjoy.

If the roof space is configured as a green roof, its benefits would also include a longer life span for the membrane. Studies have shown that temperature differential exposures of up to 120˚F for an 'ungreened' roof greatly impacts the life span of the roof membrane. The most important processes responsible for this impact are the constant expansion and contraction caused by these extremes of temperature and freeze-thaw cycles, as well as ultraviolet light degradation[15].

In the case of a green roof, plants protect the membrane from exposure, resulting in a much smaller temperature differential of about 40˚F, and absolutely no exposure to UV-light. It is worth noting that roof membranes produce a considerable amount of waste after their time, with a life cycle of only about 15 years for old-fashioned asphalt roofs, constituting 5% of registered landfills, or 50 billion pounds of waste annually[16]. Extending the membrane's life expectancy by covering and protecting it would significantly avoid unnecessary waste.

Plants protect our water. In rural environments, the root system of plants prevent soil erosion caused by rain. In urban environments however, vegetation is of much higher significance in its capacity to influence stormwater management by absorbing water in soft scape areas. For example, trees retain a high amount of rainwater on their surface during light rainfall. Better yet, the soil they grow in is able to absorb the water and slowly release it over time via evaporation, or by any supporting drainage system.

New York is mostly hard scaped surfaces, made up of concrete sidewalks and drained rooftops, all sealed areas. In certain Manhattan neighborhoods, such as Midtown, the 'sealed area' percentage is close to 100%[17]. When it rains, all of the water collected on these surfaces is immediately drained into the public sewer system, which

1.28

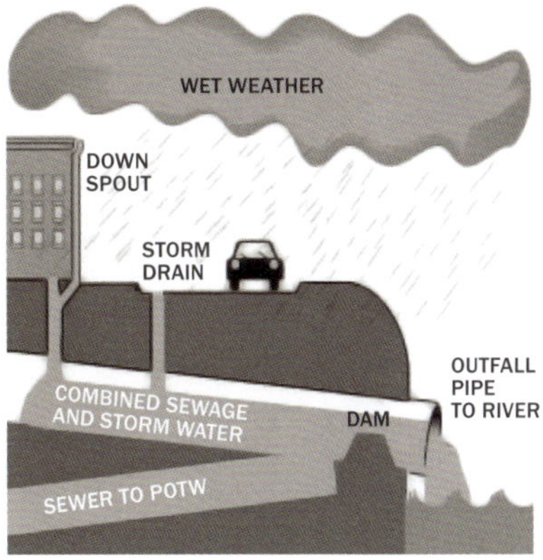

1.29

1.25 NYC: Vegetation intensity. Sparse in Manhattan (except Central Park), denser in the outer boroughs.
1.26 Trellis support systems, alternatives.
1.27 Santiago de Chile: Consorcio Building with 'vegetal skin', E.Browne.
1.28 Green roof garden in urban setting: roof vegetation is most effective for shading.
1.29 Combined sewer system, discharging into rivers; schematic.
1.30 Comparison of run-off: Ungreened and green roof.

was designed more than a century ago. This system combines both sewage and urban runoff, and was sized to carry three to five times the average dry weather flows during downpours. The system has been significantly improved, but if these loads cannot be handled, water is either released into a stormwater overflow system, or if necessary even into the rivers (1.29). Anecdotal evidence of this occurrence is that the Hudson River Swimmers call off their swimming sessions immediately following a significant downpour. Of course, solving this dilemma by increasing the capacity of the sewer system is an expensive undertaking. Instead of enlarging the infrastructure, one could simply delay these loads, and lower the peak of accruing rainwater loads (1.30). This is where vegetation cover, like green roofs, kicks in as a feasible solution: for every 5% of vegetation cover added, stormwater runoff is reduced by approximately 2%[18].

Vegetation also acts as a natural filter for pollution. It removes polluted particulate matter from the flow as it reaches the storm sewers. Reducing the flow of stormwater reduces the amount of pollution that is washed into a drainage area. Trees use nutrients like nitrogen, phosphorus, and potassium, all byproducts of urban living, that would otherwise pollute and overfertilize the rivers.

Vegetation reintroduces biodiversity. The question remains whether cities are intended to be an urban environment comprised exclusively of human beings, dogs, cats and other pet animals, or if it should be considered an ecosystem, additionally hosting a variety of interdependent and interconnected life forms. This query probably goes back to

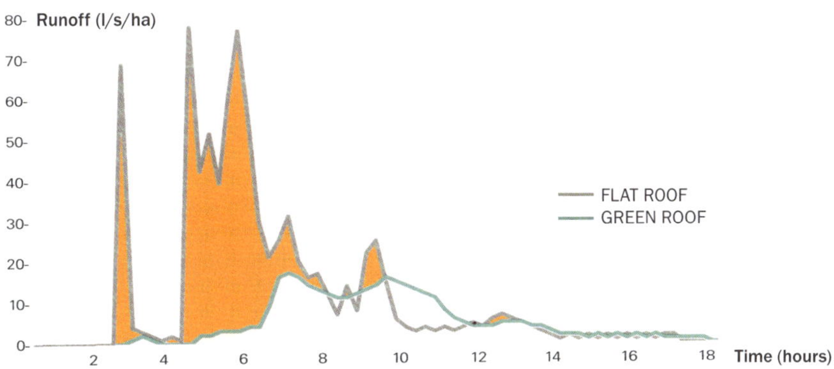

1.30

1.31

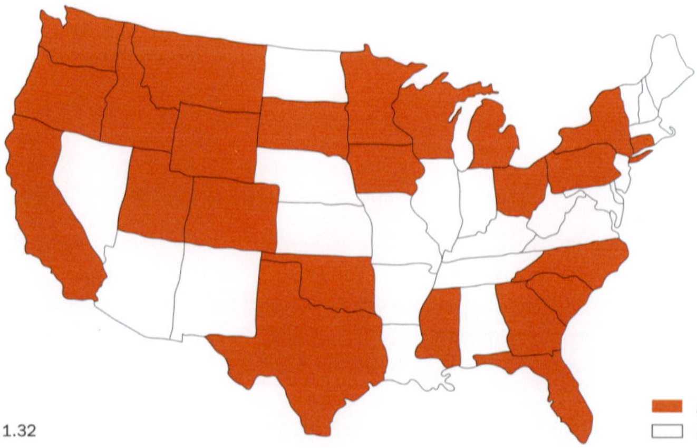

1.32

the 17th century philosophy of Cartesian dualism, which describes the dichotomous separation of humans from nature. In this case, the countryside features nature, the urban setting features us and our capacities to manufacture and manage goods and services. However, it is clear that parks and green spaces not only create relaxation, (bird watching as is the fastest growing leisure activity in New York's larger parks) but also contribute indirectly to our health and well-being in other ways (1.31). For instance, it has been scientifically proven that as species disappear, infectious diseases escalate in humans. An example of this phenomenon, called the 'dilution effect', is the West Nile Virus, which was rediscovered in New York during an outbreak in 1999. The virus corresponded to a decrease in bird population density, which acted as a 'buffer species' between the mosquitos carrying the disease, and us humans who suddenly ran the risk of being infected with encephalitis. Since most bird species have developed resistance to the virus in order to survive, their feeding on and eliminating mosquitos indirectly protects humans from infection[19].

Birds have been distinctively identified as one of the main indicators of a healthy ecological balance. Rachel Carson's 1962 groundbreaking book, *Silent Spring*, presented the idea that if birds disappear the entire ecosystem, including humans, would deteriorate.

Unfortunately, the reflective glass towers that make up the skyline of Manhattan, are not exactly the perfect habitat for our bird population. As New York City hosts partially migratory birds on an Atlantic flyway, an estimated 90,000 birds are killed each year flying into buildings[20]. In the country, the higher estimate of annual deaths is one billion. This manmade issue is one of the largest threats to the bird population.

Birds, bats and bees (the three 'B's) are identified as protagonist species of biodiversity. They are also all potential inhabitants of an urban ecosystem. Their significance originates from the fact that they are all involved with the biological control of both insect and pest populations and more importantly, they are the chief 'pollinators'. Pollination is the transfer of pollen from one flower to another, a critical

1.31 NYC: Central Park birdwatching. One of the fastest growing leisure activities.
1.32 USA: Bee colony collapse disorder (CCD), affected states, 2007
1.33 NYC: Bat population in Central Park: Up to a thousand.

1.33

component of fruit and seed production. Bees, butterflies, birds, and bats pollinate 75 per cent of the world's staple crops and 90 per cent of all flowering plants. According to the U.S. Department of Agriculture (USDA), we are facing an "impending pollination crisis," in which both wild and cultivated pollinators are disappearing at alarming rates due to habitat loss, pesticide poisoning, diseases and pests. In fact, we have the fewest number of cultivated bee colonies in history, declining from around 4 million in the 80s, to 2.5 million in the 90s, a 20% decrease per year[21]. These losses became even more dramatic at the end of 2006, during the outbreak of Colony Collapse Disorder (CCD), which resulted in an even more mysterious and steeper decline for unknown reasons. Reports of colony losses varied widely, ranging from 30% to 90% and affecting at least half of the States in the country (1.32). This trend continued in following years, and with a rate of decline of approximately 30% per year, pollination is becoming a critical issue.

In certain European cities, for instance Paris, London and Berlin, urban beekeeping is increasing in popularity. It is considered an alternative ecological niche for bees: because the use of pesticides is forbidden within city limits, the honey production rate is higher and both the animals and the product are healthier.

The idea of 'Greening the city' should lead to an agenda that responds to current acute issues. The decrease of urban biodiversity, emission-related downfalls as well as the infrastructural short-comings need to be addressed accurately. Instead of reverting to schemes that favor the implementation of centralized instances, wherein one intervention is a placeholder for political good will, the approach must extensively saturate the entire urban ecosystem in a way that satisfies all of the above factors. Urban vegetation is related to quality of life and public health.

1.34

5

URBAN DENSITY

It is undisputed that Manhattan is one of the densest cities in the western world. Millions of tourists visit the city, all with their own preconceptions about density; for example that certain street canyons never see direct daylight. A certain enchantment of this environment is generated by these clichés. However, it might be that we are used to this higher building density and therefore accept it more easily than a non New-Yorker would. The way we deal with density is also related to a centralized behavioral pattern. During weekends, Central Park is the place to be when searching for openness (1.34). On a busy weekend, this park, which makes up 6 percent of Manhattan's total acreage, is frequented by more than a million people. It attracts 38 million visitors annually, an estimate drawn exclusively from the 1.5 million inhabitants that actually live on Manhattan Island. This number of people in a park may seem overcrowded, but in an urban context openness and solitude are not necessarily the same phenomenon. Solitude is not easy to find here. Openness, in the case of Central Park, is.

In its morphologic appearance, the growth of New York seems to be less restricted than old world cities, such as Paris, Berlin or London. Here in New York, the height of the buildings, which define the height of the street spaces, are less rigorously mandated by building departments. However, there is still a full array of regulations in effect that determine both the 'type of use' relative to zoning districts, and the 'intensity of use' relative to lot sizes. The concept of zoning was introduced in 1916, after complaints about new skyscrapers casting acre-long shadows and depriving neighboring properties of light and, of course, value. One prime example was the 42-story Equitable Building built in 1915[22].

The immediate reaction to these complaints was the announcement of zoning district types: Residential, Commercial, Manufacturing, and Parks (as left-overs), and their related allowable building sizes. Lower Manhattan and Midtown were commercial areas and became the densest quarters (1.35), a direct result of the geological bedrock formations underneath, and essentially generated the extreme heights.

The predominant residential quarters grew around Central Park, which started at its south edge, along 59th Street, dividing the Upper Eastside from the Upper Westside. In these neighborhoods some quarters are quite homogeneous with similar cornice heights, as a consequence of these early zoning regulations. Other areas, however, feature more and more high-rise apartment towers, especially where real estate developers see market potential.

Ultimately, a building's intensity of use is regulated not only through its zoning district location, but also by anomalies such as 'incentive programs' that allow an increase of 'bulk', and therefore extra

1.35

floor space. Many of these amendments were generated and overhauled in the Zoning Resolution of the 1960s[23]. As it stands, we are now faced with a complex regulatory system that has many twists and turns. As a planning instrument, regulations have to be flexible to adjust to persistently changing conditions. Thus, there is a continuous stream of amendments that respond to building technologies, shifts in land use, population migrations and other socio-cultural trends.

Floor Area Ratio (policy-driven regulations): The main planning instrument determining the maximum size, or 'bulk', of a building on a lot is determined by the Floor Area Ratio, or the FAR. This ratio is the total floor area permitted on a zoning lot, divided by the complete lot area of this parcel. For example, if a building contains 20,000 s.f. of floor area on a zoning lot of 5,000 s.f., then it has a floor area ratio of 4.0.

The typical Floor Area Ratio for moderate to high density residential use (called 'R6 - R10') is around 4.0 and 7.0 in New York, resulting in a building with roughly 6 to 10 stories (due to partial lot coverage and set-backs). In commercial districts, the 'FARs' are much higher.

However, the Floor Area Ratio is not related to exact science. Our well-being is not that sensitive. We neither physically benefit nor suffer from an additional floor of building height. Therefore, the city introduced a revolutionary zoning idea in the 1960s to benefit urban dwellers in a different way, still in use today. These are 'incentives' that permit additional 'bulk' for certain developments and improvements that are in the best interest of the city and general public, and that are not easily attainable by state entities.

Such projects include infrastructure, social equity, cultural stimulation and public health. In order to get support from the private sectors, the city promises FAR bonuses in exchange for the provision of public amenities. These amenities are comprised of many different visions: privately owned public spaces, performing arts spaces, subway improvements, preservation efforts, fresh food stores, and affordable housing through the Inclusionary Housing Program, to name a few. The latter is an interesting commitment. Affordable housing fosters diversity, since it encourages developers to set aside a portion of the building at below-market rates as affordable units in exchange for an increased maximum floor area. This is a collectivist idea originating from the 60s that financially supports and integrates underprivileged citizens by giving them access to affordable urban housing, and therefore to city life.

If these 'amenities' are proposed, the FAR increases by about 20-30%, resulting in a considerable amount of additional floor space, and, therefore, real estate income. This makes developers sit up and take notice.

1.34 NYC: Preferred weekend activity. Experiencing Central Park's openness.

1.35 NYC: Aerial photograph of Midtown Manhattan. One of the densest environments, generated in part by geological conditions, but even more by zoning regulations. Also note: No significant vegetation is detectable.

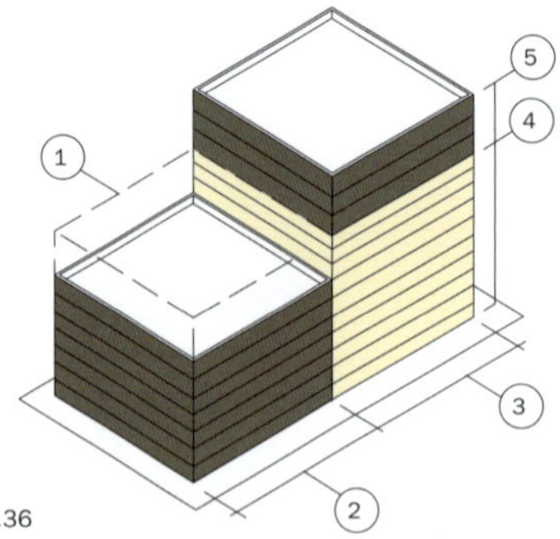

1.36

1. UNUSED DEVELOPMENT RIGHTS
2. ZONING LOT A
3. ZONING LOT B
4. PERMITTED FLOOR AREA ON LOT B
5. ADDITIONAL FLOOR AREA FROM LOT A

The idea behind the incentives is to respond to certain needs of a city or neighborhood, by rewarding developers who move in an intended direction. These incentive programs are paramount to the support of New York's political vision for the future. The same instrument could be used to create cleaner energy for our environment. One has to be aware that the most powerful tools for change in a democracy are often related to politics and its associated policymaking.

Air-right trading (economy-driven regulative): There are other planning mechanisms in place, called air-rights, that are controlled more by financial power than political instances. What they do have in common with the incentive programs, however, is their influence on the typically fragmented urban spaces. New York's infamous air-right trading regulations are a reflection of a socio-economic culture. In a city that claims to be the world center of trade, it is no surprise that even the air above buildings will be a good in demand if they are not built to their allowable maximum height. Currently, air rights are worth between roughly $100 and $300 per s.f. depending on the building's and the city's commercial activity. Air-right trading is part of the unused development rights that apply to possible bulk and unrealized maximum FAR on a property. The City allows for many more stories on a development site, whose lots have merged. By purchasing an adjacent lot that accommodates a building with unexploited maximum floor area, one can transfer the unused bulk to the developable property (1.36). One is still constrained by set-backs and building height maximums, but there is a lot more floor area to be conquered.

This rule typically applies to adjacent lots. But in limited circumstances, the city allows for the transfer of air-rights from one lot to another, when the lots are not side-by-side; rather they are across the street or, for a corner lot condition, on another corner lot at the same intersection. This is called a TDR, short for Transfer of Development Right. This rule applies to sites with historic buildings in particular. As an example, Christ Church at Park Avenue took advantage of a TDR, enabling them to sell their air-rights for $30 million at the peak of the real estate market in 2005 (1.37)[24].

Another mechanism to increase FAR, called a District Improvement Fund Bonus (DIFB) can be launched by the City to strengthen the transformation of a district that is crucial for the city's growth strategy. This bonus allows the city to encourage higher density in order to boost the district's speedy development. However, any additional FAR comes at a high cost to the developer. For instance, an additional square foot at the 'Hudson Yards', between 30th and 42nd Street, and 7th Avenue and Hudson River in Manhattan, currently costs close to $120. The city adjusts these prices according to demand and economic development, much like a stock trader.

Since the City is sometimes struggling to meet its complex budget, this is a welcome and substantial source of capital. However, the results are a permanent determinant for the city's morphology. The question remains whether this is indeed the best long-term strategy for a functioning urban environment. While the original idea of zoning laws was designed to ensure more equitable access for people to light, air, and sun, it is debatable that they do so. In fact, the aforementioned economy-driven incentive programs do the exact opposite. Instead of generating homogeneity in an effort to create healthy or more socially equitable conditions, air-right and bonus programs create conditions that exacerbate existing distorted conditions.

1.37

1.38

The city's efforts at controlling density have to be considered alongside the cause and effect of a dynamic overlying development strategy, which often relates to specific popular trends. For example, the 1916 zoning laws that responded to certain control deficiencies involving how access to air, sun and light depreciated property values. Moving forward, the 1961 incentive program reacted to a deficiency in social equity and a need to balance the social environment; the development of air-right trading has been a reaction to boost the building economy. Finally, the 'Bonus Fund' program directly relates to the city's lack of funding for important infrastructural projects.

It can be concluded that responses related to controlling density are a reaction towards the particularities of a time period. It can also be presumed that the imminent era will involve the renewal of infrastructure and the quest for efficient renewable energy sources to power our cities. If solar energy is the favored choice, then all buildings must have access to these solar loads. Thus, it is implied that possible shadow patterns must be considered and possibly regulated.

Higher density would only equate to greater efficiency if evenly applied. It must be recognized that New York's extremely fragmented urban terrain, hosting vastly different scales of buildings within close proximity of one another, is thus quite unsophisticated. The shadows cast by taller buildings significantly reduce the total solar loads of neighboring buildings, and deny other buildings the fundamental right to equal solar access. Thus, a more homogenous environment would be favorable despite the fact that it contradicts many provisions and incentives to air right regulations, in particular, those that involve economic mechanisms. The overall city's zoning plan is currently being revised, and will have the opportunity to consider and incorporate solar access rules more seriously.

1.36 NYC: Air right trading rules. Department of City Planning.
1.37 NYC: TDR trading. Christ Church, Park Avenue.
1.38 NYC: Air rights. Highrise at 3rd Avenue and 23rd Street.
1.39 NYC: Virtual Zoning plan diagram of Manhattan, 2011, depicting maximum FAR. Color code: Residential; cyan; Commercial; yellow; Manufacturing; purple (Proxy Design Studio/a+u).

1.39

6

TRAVEL

1.40

A metropolis is always judged on how easy it is to get around. New York, especially Manhattan, has an astonishingly efficient transportation system. The street grid system, developed in 1811, dominates the entire environment, and consequently also the traffic. Its main arteries – the avenues – run in the longitudinal direction, while the cross streets are responsible for local distribution and bridge and tunnel connections to the outer boroughs. This grid system has revolutionized urban orientation. However, as the volume of cars has been consistently increasing, these traffic arteries have become congested. This is a common problem elsewhere in the world. London, for example, has launched road tolls in the form of congestion charges that demand heavy fees for drivers in certain traffic zones during work hours.

New York has reacted differently, making through traffic on certain arteries (especially Broadway) so difficult that people are discouraged to drive. The city's efforts to close lanes and replace them with bike lanes or pedestrian zones establishes the idea that vehicular traffic is often not the best way to get around. It comes as no surprise that bike and pedestrian traffic has increased. Comparative travel times among the different transportation methods in Midtown have been published, and the car is often the clear loser. The people of New York are adjusting, as the city currently constitutes the only locality in the country where more than half of all households do not own a car. In Manhattan, the figure is over 75%. This is a significant change with respect to carbon dioxide emissions. New Yorkers emit 20 million tons less CO_2 per year than they would if they drove as much as the average U.S. urban resident.

Currently, two-thirds of New York commuters travel to work by transit or by walking. This is the highest number of any large city in the country. If compared to Atlanta, the difference is stunning (1.43). Cities that have based their transportation system on the car will have to reconsider. The era of inexpensive petroleum-based transportation will only fade along with the exhaustion of its resources, as will the morphology of cities whose infrastructure is based on travel along permanent highways and ending in parking lot expansions. This kind of travel has been detrimental to urbanism, and deemed unattractive by many of its inhabitants. Not to mention the fact that asthma rates are higher in cities heavily dependent on vehicular traffic.

The greener way to travel is clearly public transportation, and New York City has made its preference clear. After almost two decades of only maintenance work on the existing subway grid, the construction of the 2nd Avenue subway line was finally launched in 2007, an addition that has been in the planning stage since 1951. The commitment to long-term environmentally friendly transportation will be a financial challenge for our present times. While this kind of underground construction is expensive, in the long run it will add much efficiency. By promoting the use of subway transportation instead of car travel, greenhouse gas emissions can be significantly reduced. A car ride that entails driving the 13 mile length of Manhattan will produce 5 times as much CO_2 than covering the same distance with the subway (1.44). And you can bet that the same trip will take less time if one takes the subway. But, what if traveling was not an option at all?

Even fewer emissions, energy and stress would be produced as most transportation relates to the individual's typical commute between work place and home. However, due to the increasing possibilities of virtual connectivity, also called telecommuting, physical commuting is becoming more and more obsolete. A rough estimate of the number of home offices in the US is currently around 30 million households[25], and this trend is rising.

For the most part, we have accepted the necessity of commuting to and from work everyday.

1.41

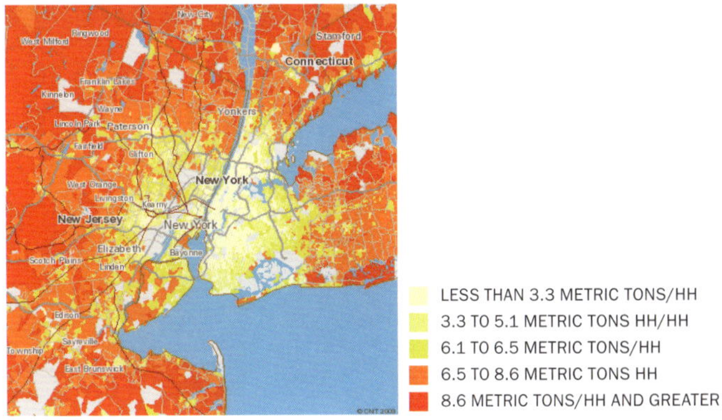

1.42

1.40 NYC: Typical traffic congestion at a street crossing.
1.41 USA, Atlanta: Traffic at night.
1.42 NYC: CO_2 emissions from car use per household.
1.43 Commuting comparison: New York and Atlanta; charts show a different commuter life style.
1.44 CO_2 emissions per traveler for fewer different means of travel.

The idea that work areas and residential areas should be separate from each other originated in early modernism, when these conditions were deemed healthier and more practical. However, types of work were often manufacturing-based at the time, and have since shifted drastically towards the service industry. The original reason for separating work from home was to protect against emission plumes and noises. In many instances, this separation is no longer necessary. It is questionable why we invest so much time getting to and from work (80 minutes on average in NYC), and why separation between the two environments is still an instrument of urban planning.

Mixed use developments and live work environments are increasingly favored over bedroom communities and office monocultures in an effort to keep neighborhoods vibrant. In urban environments however, it is more difficult to add to such an intended mix, since they are already so densely built. The options that seem most plausible are either retrofits of unused buildings, or identifying existing underutilized spaces that can adapt to the building stock.

The more complex the mixed-use neighborhood, the more pedestrian traffic they will attract, which equates to local and short distance movement. Thus, choices of goods and services are both more accessible and plentiful. Encouraging people to walk more will decrease fossil fuel consumption with direct benefits to public health. In this respect, mixed-use is even better than simple sense construction, as it embraces complexity of use, while contributing to a healthier urban environment. This relationship understandably functions at a larger scale too, and directly relates to CO_2 emissions (1.42). City dwellers produce far fewer greenhouse gas emissions than their counterparts in the sprawling suburban environments.

1.43

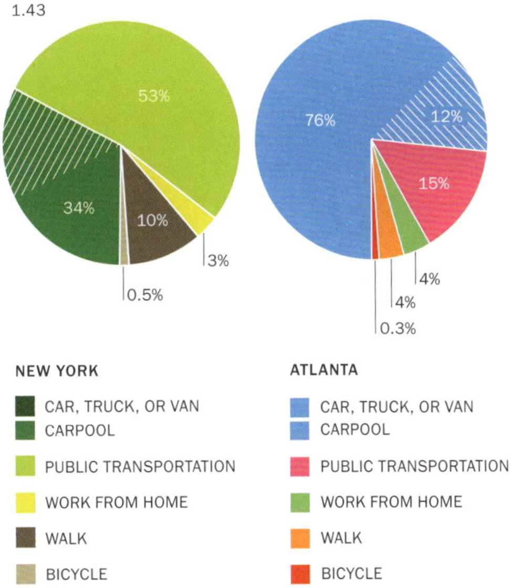

1.44

FOR COMPARISON IF YOU TRAVELLED 13 MILES BY:

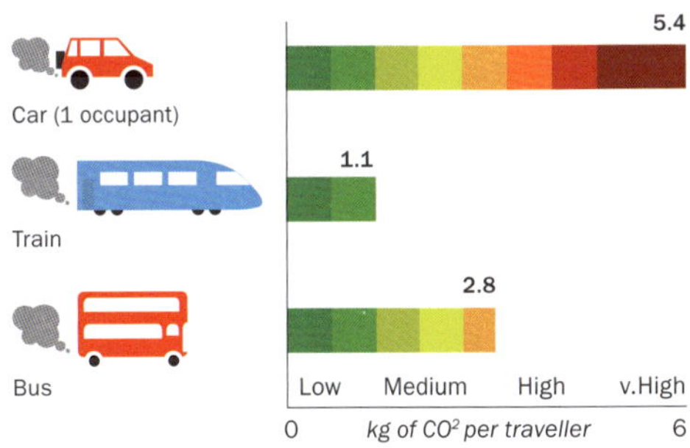

7

WASTE/GARBAGE

1.45

New Yorkers produce an average of nine pounds of combined commercial and private garbage per capita every day, which results in ca. 40,000 tons of trash[26]. This amounts to roughly 400,000 waste cubic yards, which translates to a fleet of many thousands of large garbage trucks (1.45), typically seen on the streets. This is well above the nationwide average, not astonishing for a city that has ample commercial and tourist activity.

The question remains: what do we do with all this garbage in an environment of 27,000 people per square mile – the most dense population in the USA – and no space to dump it? Until 2001, the city was highly dependent on a landfill on Staten Island named 'Fresh Kills'. It was opened in 1947 as a temporary measure for three years, but became the largest landfill of the United States in the proceeding 50 years. New Yorkers were relieved that there was a waste disposal infrastructure in place, but the Staten Islanders were not happy with the entire city's garbage being dumped in their borough. After years of negotiations, the political pressure and influence the islanders applied to both the State and City governments led to the landfill's closure in 1996. In 2002 'Fresh Kills' was sealed and closed for good, and since then there has been no localized garbage disposal for NYC. This has become a big problem, as the City's long-term garbage disposal strategy is based on a complex export system to other states, without sufficient acknowledgement and integration of local solutions. This lack of strategy is bewildering. For example, the success of New York's water supply infrastructure, built over a century ago, is grounded in its original predictions for future infrastructural needs and demands, and subsequently 'building it big'.

This has been New York's strength in becoming the metropolis that it is today. However, since garbage is not the preeminent topic to make friends with, politicians tend to avoid the issue, which involves ample intolerance and concern from voters, and culminates in the droll abbreviation 'NIMBY': 'Not In My Back Yard'.

But, what to do with all of this waste if nobody wants it? We seem to have adopted the same attitude and response as our politicians: Ignore it, out of sight; out of mind! We are busy people and casually toss away our garbage. We sometimes neglect to sort it either, preferring not to think about garbage or where it will end up. However, without the motivation to comprehend and manage our own waste, we will never figure out how to make less of it, and how to best process the rest of it.

'MUNICIPAL SOLID WASTE' (MSW)

Of all the waste that is produced daily, generated by residents, public agencies and non-profit corporations, the City takes care of roughly 35%, or 14,000 tons[27]. The rest is handled by carting companies. Since New York no longer has its own landfills, (the closest landfill in New York State is more than a 100 miles away) most of the non-recyclable waste must be moved elsewhere. More than half of the waste goes to Pennsylvania (up to 5 hours away), and another quarter heads to Virginia (about 7 hours away). These numbers fluctuate depending on typical economical characteristics of supply and demand. There are other states involved in our trash disposal as far away as South Carolina and Ohio.

About 80% of New York's municipal waste, is first collected by the city's compactor trucks and before being sent to its final destination by a hauling fleet of trailers that consume diesel at a rate of as little as 3 miles/gallon[28]. It is estimated that these trucks travel a collective 135,000 miles and use about 35,000 gallons of diesel per day, which represents an impressive environmental and financial impact[29]. Consequently, disposal costs per ton of waste have risen considerably. When 'Fresh Kills' was still in operation, one ton of garbage cost about $40. Since the closure of Fresh Kills over the last decade, this cost has increased to over $100 per ton[30].

1.45 NYC: Compactor truck picking up garbage in Harlem.
1.46 NYC: Composition of Municipal Solid Waste (MSW).

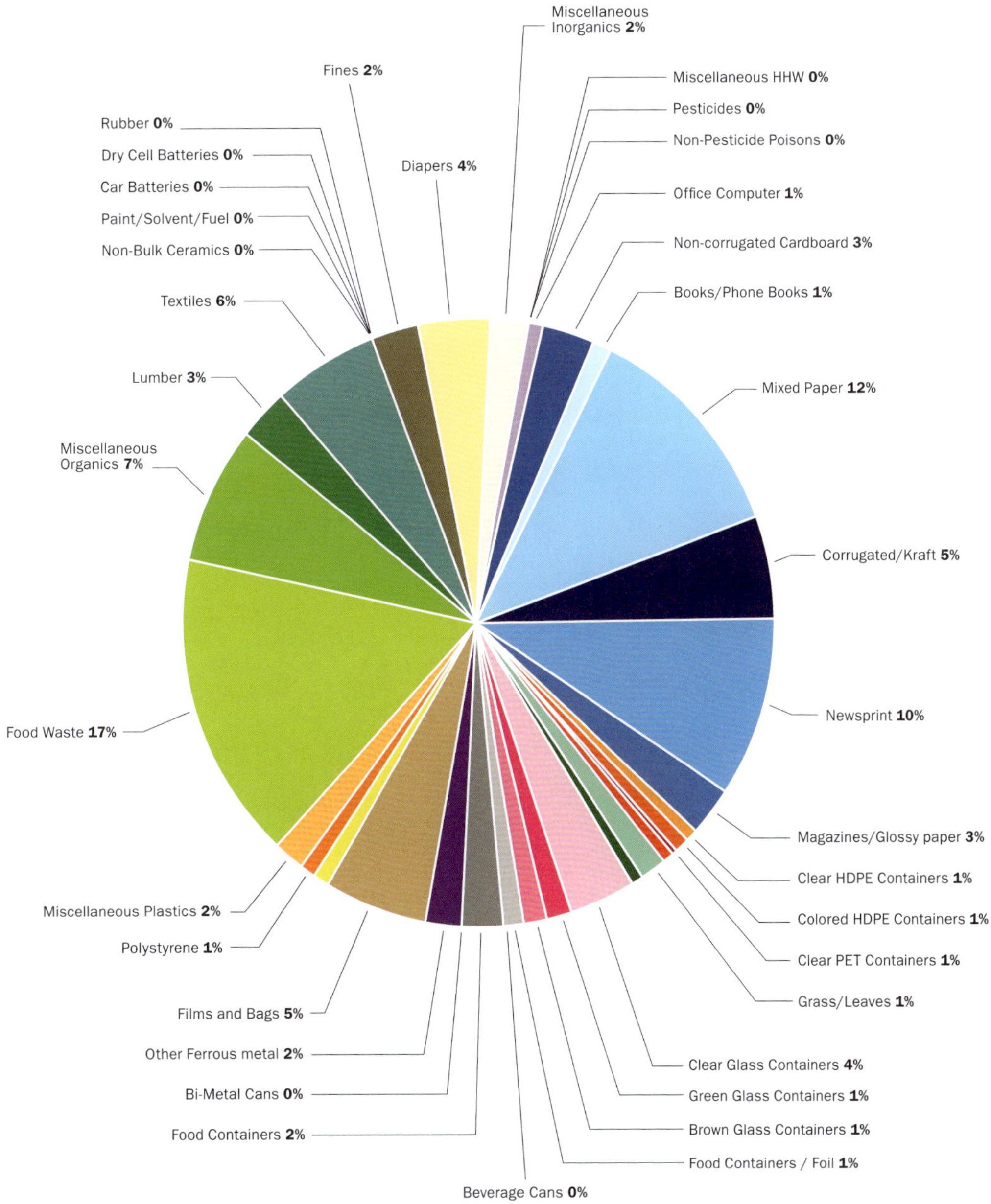

1.46

TIME IT TAKES FOR GARBAGE TO DECOMPOSE:

Glass Bottle	*1 million years*
Monofilament Fishing LIne	*600 years*
Plastic Beverages Bottles	*450 years*
Disposable Diapers	*450 years*
Aluminum Can	*80-200 years*
Foamed Plastic Buoy	*80 years*
Foamed Plastic Cups	*50 years*
Rubber-Boot Sole	*50-80 years*
Tin Cans	*50 years*
Leather	*50 years*
Nylon Fabric	*30-40 years*
Plastic Film Container	*20-30 years*
Plastic Bag	*10-20 years*
Cigarette Butt	*1-5 years*
Wool Sock	*1-5 years*
Plywood	*1-3 years*
Waxed Milk Carton	*3 months*
Apple Core	*2 months*
Newspaper	*6 weeks*
Orange or Banana Peel	*2-5 weeks*
Paper Towel	*2-4 weeks*

1.47

1.48

1.49

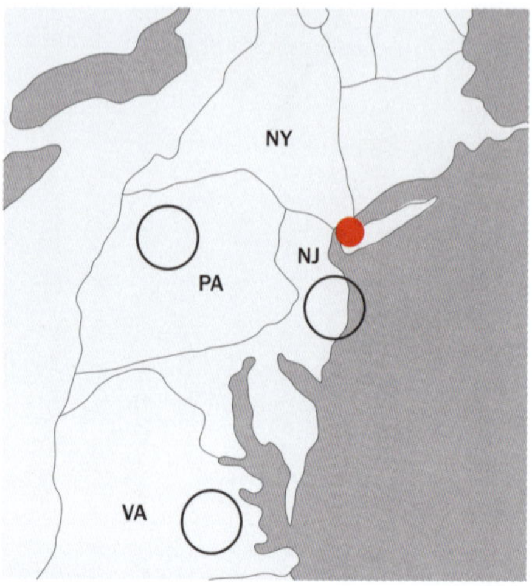

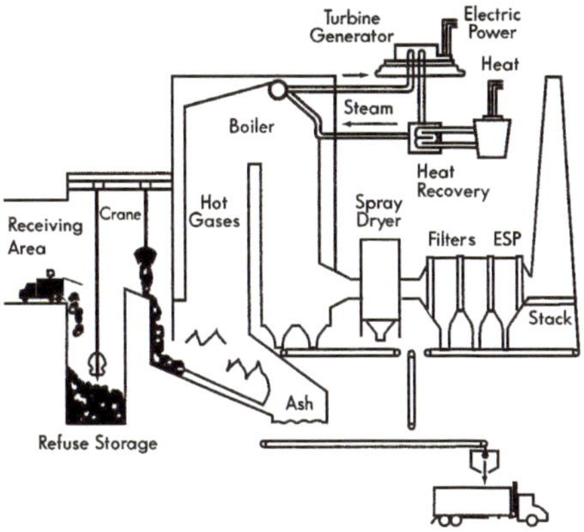

1.50

1.51

1.47 Typical decomposition times for landfilled garbage.
1.48 Pennsylvania landfill: NYC garbage imports.
1.49 Pennsylvania, New Jersey, Virginia: Preferred landfills for NYC.
1.50 Schematic of incinerator plant.
1.51 Incinerator plant, refuse storage.

Yet worse for New York, there are several petitions in Congress (Interstate Waste Legislation) that are attempting to restrict the export of garbage over state borders. If this bill passes, New York might be in a precarious situation.

Beyond financial costs and space issues, landfills are quite a dirty solution to disposing of our garbage in terms of long-term sustainability. Garbage does not just go away, especially in in our era of oil-based materialization. Rather, it can take several decades or more for it to decompose (1.47).

While traditional waste, including food scraps and similar organic matter, decomposes relatively quickly, our modern plastic-based packaging and chemically treated composites can take hundreds of years to 'disappear'. Moreover, the release of methane gas and the pollution of soil and groundwater leaking into the environment are detrimental to both our health and environment. One of the most pressing issues is the presence of methane gas (CH_4) in landfills, which is released in the decomposition, and has a heat-trapping potential 20 - 50 times more powerful than CO_2. Landfills are therefore a substantial component to the global warming effect. A few landfills are able to collect their methane and generate energy, but as this is a flammable gas and landfills are vast open spaces, this is not easily achieved.

Another option for handling waste are incinerator plants or Waste-to-Energy (WTE) plants, which some countries in Europe and South-East Asia have already transitioned to. Since the late 90s common waste policies (i.e. Landfill Directive 1999) have changed over concern about shortages of landfill capacity. Often the preference is incineration because of its potential as an energy source for power and heating. Modern technology behind waste incineration, releases, if accurately filtered, far fewer emissions and is therefore less harmful to our environment than the uncontrolled exhaust that is typical of landfills. In short, burning trash is cleaner than storing it in our environment for hundreds of years. Developed countries like Japan, Sweden or Denmark, have relied heavily on local incinerator plants that, in turn, provide a substantial amount of heat and power for their buildings. Suddenly, waste becomes a resource for energy production. The principle being (1.50) that after the recyclable waste is sorted, the unrecyclable materials are burned in a furnace at 800 - 1000˚F. The garbage is then converted into ash, flue gases and heat. After passing through several controlled steps, the flue gases produce steam via a boiler that powers a turbine and generates electricity. The waste heat from this process can also be recovered and used for district heating schemes. The ash, a waste product of this process, can be used for building material supplements, if it is not too toxic (for example, as an ingredient in concrete to replace lime). If it is too toxic, it must be landfilled. But any disposed ash still only takes up a tenth of the space as the unprocessed material. In addition, waste used as a fuel source ultimately replaces the very fossil fuels that constitute a high percentage of imported goods.

What makes these plants feasible, is their filter technology that remove toxins with chemical and electrical reactions integrated into their emissions control. At present, the cleanest incinerator plant, which is in Rahway, NJ, emits only a fifth of the maximum emissions allowed by state legislation. To put this into greater perspective, backyard burning of just ten pounds of trash can produce more dioxins, furans, and PCBs than 100,000 pounds of trash in an incinerator plant. However, all this cleanliness comes at a price. Strict pollution controls, and of course manufacturing and maintaining the equipment itself, cost millions of dollars, while moving trash hundreds of miles to a landfill is most often cheaper. As yet, the factor of 'cleanliness' has not found the right price tag. The effects on

1.52

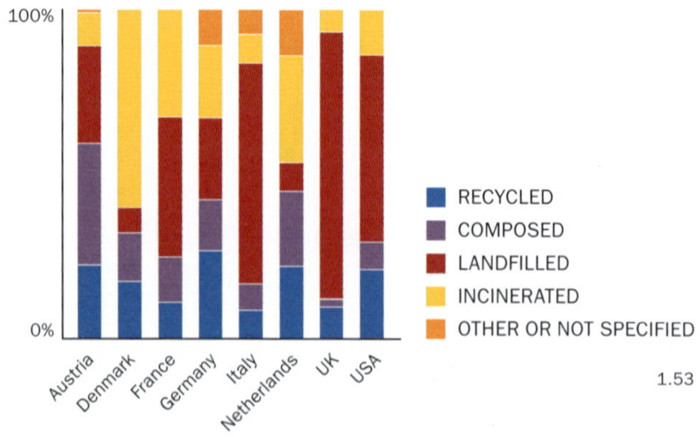

1.53

our environment are mainly ethical considerations, while decisions are most often based on financial constraints. Laws that restrict the export of waste materials across state lines can take a step toward being environmentally conscious, and would make incinerator plants more economically competitive with landfills. This step would provide the volume of waste which is necessary to run these plants more economically. Currently, only a fraction of trash collected in Manhattan, approximately 3% of the city's total waste stream, goes to incinerator plants in New Jersey[31]. The rest is transported out of state, involving much longer transportation distances. Therefore, the city has proposed a Solid Waste Management Plan (SWMP) to enable more efficient long-distance waste disposal in the future. Unfortunately, it intends to maintain its current strategy of long-distance hauling. But it is planning to transfer the waste by barge or rail to container stations, where it will be transported to out-of-state landfills. This modest but beneficial change is an effort to reduce the amount of truck traffic in local neighborhoods and Interstate highways.

Exporting garbage by rail and barge will ultimately be more economical than by truck, and it is intended to distribute garbage to less expensive landfills than previously possible. However, rail and barge destinations are more limited than truck accessible destinations, with the risk of economic dependencies. Even though the fuel used for transportation of garbage will be reduced, not much else has changed conceptually. Instead of investing in more technically advanced and locally applied solutions, which New York often favored at the height of its development, it continues to endorse the far away landfills.

RECYCLING, 'ZERO WASTE'

Both landfills and incinerator plants are centralized systems that need elaborate transportation to collect the garbage at different places and funnel it towards transfer stations. The greater the number of households, and the higher the amount of garbage in general, the more trucks are necessary to collect the garbage. This makes the entire endeavor 'dirtier', with rising traffic-related emissions and the expansion in infrastructure necessary to process this waste. Thus, efficient waste management should focus first on the source (the producer of garbage). Even more specifically, the consumer of household waste should be the focus, as it is linked more to consumerism than necessity, which is the case with disposing of attractive, but typically obsolete, packaging and the general behavior surrounding throwing away our goods.

The closer the solution is to the source, the greater the possibility of reducing the stream of waste, and the less effort involved at the larger scale. Therefore, it is important to identify areas where reduction at the level of the consumer is possible. What we call garbage could actually shift from focusing on end-of-the-pipe disposal to understanding waste as an essential part of our materials resource management.

1.54

The first recycling programs in New York were started in the mid '80s. Since then, recycling has been branded effective for paper, plastics and metal, since they are collected separately from non-recyclable garbage, again there is a focus on and an increase in the transportation effort. In 2002, the city-run recycling program was briefly suspended, due to the fact that the additional infrastructure necessary to operate the program could not support itself economically despite the value inherent in the recovered materials. If the economic venture is more important than how we culturally deal with the problem, we are effectively giving in in to wasting our finite material resources and with it the basis of our existence. Politics must encourage recycling via policies that present recycled materials with an economic edge over virgin materials. One example of a favorable policy would be to tax recycled materials differently. As is evident in the most mundane recycling programs, even the coordination of bottle deposit rules cannot be politically coordinated, as it varies from state to state. Clear administrative signals are necessary to make recycling a mainstream mission on the federal level.

New York has not exactly been a role model for recycling. While the national recycling rate is around 30%, NYC is trailing slightly behind[32]. Larger cities typically have the advantage due to density and the relative means of collection as is evident in the high recycling rates from other metropolises throughout the country: Philadelphia at 35%, Los Angeles and Chicago at 45%. However, few states or local governments define recycling in the same way, which is again evidence for more necessary federal involvement. Often, different kinds of materials are included in the percentages, related to the strengths of the particular programs. Generally, although comparison with other cities might be inaccurate, it is clear that New York, and we as its consumers, could do more to reduce waste in our way of life.

One successful example of policy encouraging the reduction of consumers' waste stream emerged in Germany in the 90s. In this paradigm, grocery and department stores were required by law to recycle their packaging and provide the opportunity for customers to leave their packaging behind and thus take only their acquired merchandise home with them. This opportunity was habitually used by customers, and stores assumed the cost of recycling the extra disposal. By shifting the responsibility to the merchant, pressure was applied to the industry to come up with more efficient packaging methods. In addition, the end-consumer learned how to save effort, while bringing visual attention to the excess of unnecessary packaging in our culture of consumerism. This had an effect on the industry's strategy, and immediately reduced the amount of one-way packaging.

1.52 NYC: Waste transport by barge, preferred by the city's new solid waste management plan.
1.53 Comparison of different waste processing strategies: European countries and the USA.
1.54 Dirty White Trash (with Gulls), 6 months' worth of artists' trash, 2 taxidermy seagulls, light projector; Tim Noble & Sue Webster, 1998.

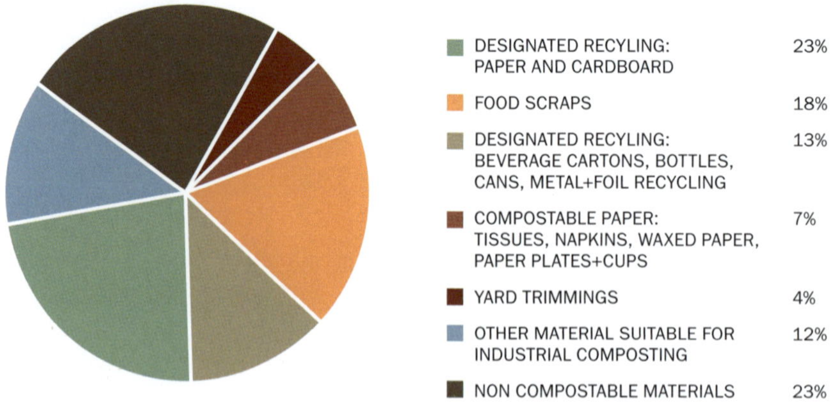

1.55

COMPOSTING, 'ZERO WASTE'

Different materials have different properties and thus processes for recycling. As soon as they enter the industrial processing stream, converting them to another marketable product requires further additional effort from the infrastructure. The emissions from collection trucks and other related efforts make the entire undertaking less sustainable. However, this elicits the need to categorize waste by the time required for its decomposition. Organic matter, in the form of food scraps or yard and brush trimmings, has the shortest decomposition time. Biodegradable paper products and packaging are also suitable for composting. If these two types of waste materials remained at their household source to be composted in place – meaning: if the decomposition time were in an acceptable range for keeping them in residential areas – effectively removing them from the municipal waste stream would have a great impact on processing efforts. Food scraps alone constitute 17% of our municipal waste, which is one of the largest portions. If combined with the other miscellaneous organics (1.55), more than a quarter of the municipal waste stream could be diverted.

Another element of research to keep garbage within the consumers' premises is the development of plasma torch incinerators; very different from those common in the 80s. They operate at much higher temperatures and are capable of gasifying virtually all waste with minimal by-products while offering the benefits of power and heat production. It is anticipated that they might become compact enough for local installations, perhaps finding their way back into larger building basements for local waste processing.

Composting 'organic garbage' into rich soil could easily be used in back yards, community gardens and public parks, or marketed as potting soil. In fact, certain community programs and gardens have already started this 'composting revolution' by accepting food scrap donations in their local designated areas, or for instance at farmers' markets. The decomposition 'turn-over' of food scraps is faster then it would be in landfills, since the organic matter is exposed to air, light and moisture. Furthermore, methane doesn't accumulate under these conditions.

The city of New York has started to look into municipal urban composting programs, but so far only two compost facilities are being operated, for backyard waste. However, other cities and regions have developed this enterprise much further. In Europe, Austria leads with about 30% of its suitable waste being composted (1.53), focusing in particular on industrial-scale composting, which is an emerging technology. Our neighbors in Canada are also a forerunner, with the world's largest MSW co-composter in Edmonton, Alberta. This plant converts more than 200,000 tons of residential solid waste per year into 80,000 tons of compost[33].

However, even though such industrial scale waste processing is admirable and appropriate, the missing contact with the consumer is less productive. As a long-term strategy we must aim to develop a broader understanding of waste cycles and the efforts to manage them. To accomplish this, communities should be involved in tangible educational

1.56

efforts. Especially in composting, where the product of a simple process becomes the starting point for new crop production and gardening, this tangibility is obvious.

CONSTRUCTION WASTE

Household waste is only a fraction of the waste produced, as there are other categories processed exclusively by private processors rather than a municipality. Construction waste in New York, also called 'C+D' for Construction and Demolition, constitutes over half of the total waste stream by weight. In other American cities, this fraction is typically between 25% and 40%[34]. There are many valid explanations as to why New York has a higher amount of construction related waste. One consideration is that older cities typically have older building stock, and thus invest more in construction. Furthermore, wealthier cities have a higher degree of renovation activity.

However, some of these renovations are questionable, as I have seen newly built, never used kitchens, or store furnishings, ripped out and demolished just to personalize an acquired space and supposedly generate resale value. Additionally, commercial show rooms often do renovations just to entertain their customers, to become more attractive and produce eye-candy.

At a larger scale, one could also argue that density is typically accompanied by higher complexity and effort, which in turn creates more waste: for every new building being built, an older building has likely has to come down. This is particularly the case for New York.

It is estimated that approximately 50% of construction waste is recycled. However, these estimates vary since private sector waste is not as easily controlled as its municipally carted counterpart. There have been several models of governmentally supervised C+D waste disposal. Most notable are local Californian programs, and also the LEED program under the supervision of the United States Green Building Council (USGBC) that focuses on controlling the constructions sector's waste stream. It is clear that C+D recycling can be more effective if all the parties involved collaborate: contractors, subcontractors, architects, suppliers and disposers (1.57). Since the C+D industry operates in an economy where return prices and competing landfill disposal costs constantly change, accurate information about recycling rates and costs are often not available. Therefore, committed and creative action is required to face this challenge.

Of all construction waste, clean fill materials typically constitute the bulk of what gets recycled, accounting for 30% - 40% of the total. These are mineral materials, most often rubble and dirt from excavation jobs, which can be reused for many different applications, from road construction to landscaping. This segment of C+D recycling is relatively easy, since complicated processing is not necessary. Other, more difficult materials to process include asphalt, ferrous and non-ferrous metals, wood, drywall, insulation, packing materials (cardboard) and many miscellaneous items that are dependent on local economic interests. However, it is likely that one would profit from these recyclable resources, and more likely that one could economically break even

1.55 NYC: Compostable fractions of MSW waste.
1.56 NYC: Typical street dumpster; C+D waste, a major contributor to landfills.

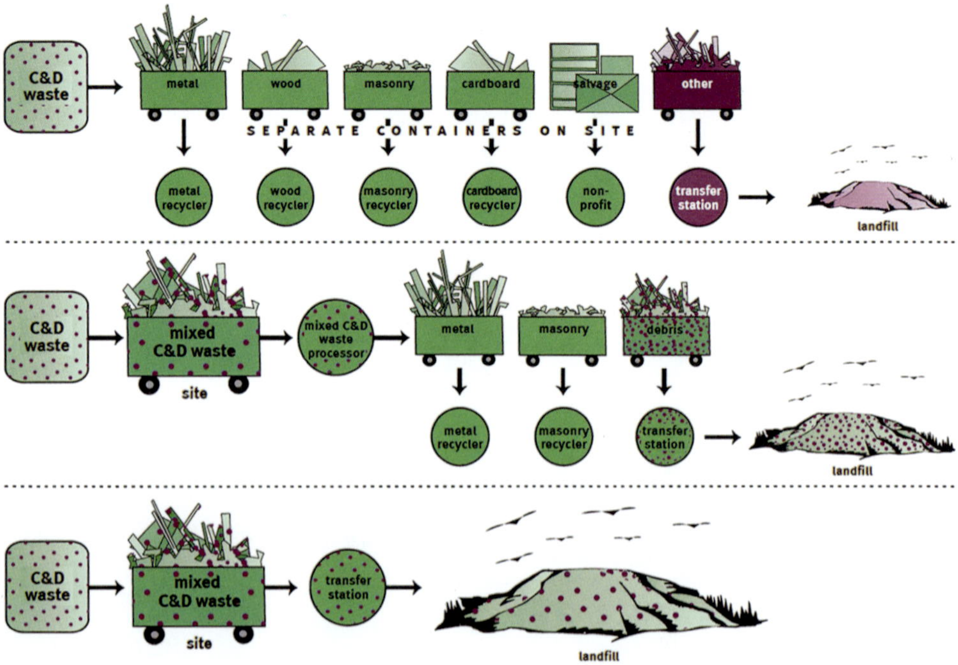

1.57

by simply carting them away from the jobsite, if the municipality would support these efforts. Early source separation is generally the best strategy, as it yields the highest recycling rates and also the highest prices for materials. However, since many modern building materials are fundamentally composites, such as electrical wire or certain glazing or panel systems, it is not possible to separate them. Material homogeneity or at least the ability to separate the composites would be preferable for the act of recycling. Another problem occurs when additional bins for separating multiple categories of recycling is required. In dense environments, where space is a luxury, staging areas are often limited in size and thus early separation cannot always be pursued.

In this respect, the architectural profession has to better strategize with respect to life cycle projections of building products. It is not just the newly finished building's appearance that is important, its anticipated aging over its lifetime is also an important factor. Product specifications should anticipate the disposal, or better, reuse of these material resources. Maintenance and repair issues also need to be thoroughly addressed during the planning phase, so that building components do not get discarded because their replacement was too difficult, but only if they were indeed at the end of their possible use. As in all other fields of waste processing, the aim must be to reduce the waste stream through knowledgeable decisions and thus preserve our resources.

LONGEVITY

All of the waste disposal alternatives discussed above do not address the issues at hand in their entirety. There seems to be no win-win situation for recycling as consumption and processing inherently involves energy, emissions and pollution. In whichever location one decides to acquire something new, it is likely that one discards something old, especially in space-limited New York City. Therefore, the ultimate way to avoid garbage is to take care of one's possessions as long as possible. In this case, the consumer's decision must place an emphasis on quality, the manufacturer must produce and profit from quality, and the designers of goods, for example architects as the designers of buildings,

1.58

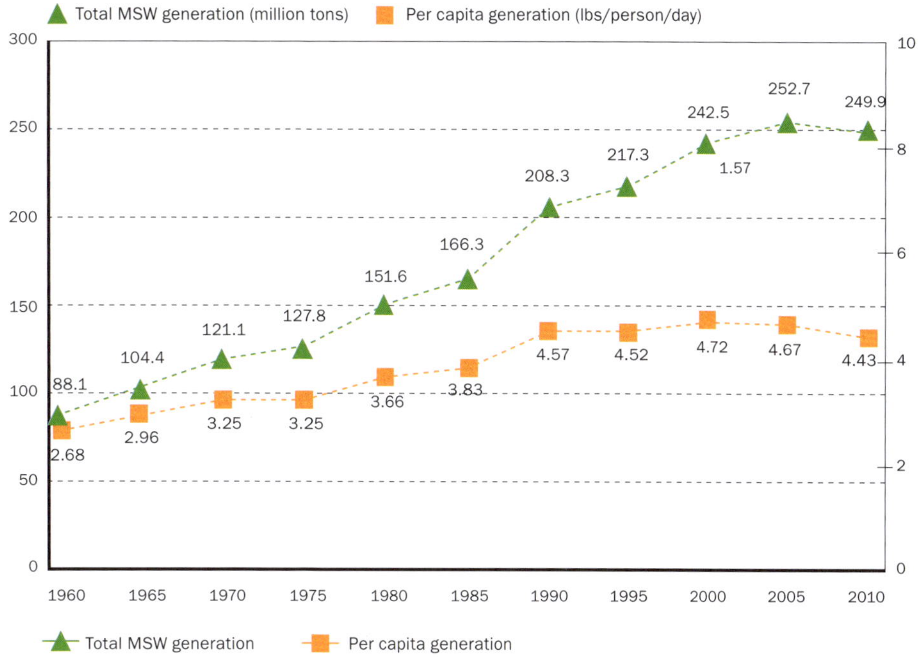

1.59

1.57 NYC: Different C+D waste pathways and their estimated recycling success rate; Construction and Demolition Waste manual, Dept. of Design and Construction.
1.58 Material management strategy for waste, NYS Dept. for Environmental Conservation.
1.59 USA: MSW generation rates; total and per capita.

must specify quality. As with the design of a building, an architect or builder begins this specification by selecting the material for optimal durability, with the least impact on our environment.

A significant reduction of waste could be achieved if common procedures were available to quantify a material's impact over its lifespan, with facts that are easily comprehensible and implementable. Designers and architects, but also the professions involved in the buildings' upkeep and maintenance, have to be able to assess the aging process of building systems in order to preserve their operability and beauty as long as possible.

Similar assessment tools currently available in the building industry are called 'Environmental Product Declarations' (EPDs) and can be likened to 'nutritional facts' for construction materials. These make it possible to conduct Life Cycle Assessments (LCA), most often by means of computer software.

However, is this approach really getting to the point? Isn't living with and enjoying the newest gadgets embedded in our culture? Isn't trendiness a critical factor in boosting our economy? The long-term effects of this behavior might in fact exploit the resources we have, and undermine the basis for our existence. For generations we have been over-consumers of our finite resources. The values related to consumption have to change.

Currently, the USA expends nearly double of what it produces. We source from other countries to produce the goods for our own consumption.

We consume more clothing, metals, plastics, electronics, and chemicals than any other country. This lack of consumer discipline must be addressed politically and culturally. This can be achieved through proactive policies, incentive programs, and taxation, to name a few, in order to effectively adapt our behavior.

Primarily, we have to acknowledge that reducing our waste stream will have the greatest impact, and that we as the consumer are ultimately responsible for this impact, thus sustaining advantageous conditions for the future. Education builds this intellect and the related comprehensive knowledge, which is the optimal method of developing the mindset necessary to effectively tackle over-consumption.

URBAN ENVIRONMENT SUMMARY

1	ENERGY	Inefficiency due to centralized power generation Fossil Fuel dominance for U.S. power generation Natural gas + 'dual-fuel' plants provide 95% of NYC's generated power CO_2 / GHG emissions worldwide: Related to power generation = 25%
2	HEAT ISLAND	Increased heat drives higher power consumption for air-conditioning Vegetation loss: Reduced evaporative effect boosts ground temperature Anthropogenic Heat: Traffic, air-conditioning, etc. cause heat output Exposure to sun of our roofs and streets increases heat absorption
3	ENVELOPE	Inferior façade performance regarding heat losses Heating emissions from heating oil linked to respiratory illnesses Curtain wall technology (all windows) with unsatisfactory r-values 75% of primary energy consumption related to buildings in NYC
4	VEGETATION	Urban vegetation loss: 9000 acres in past 25 years in NYC Necessary increase of carbon sequestration Benefits of green roof applications Air quality related to vegetation and public health records
5	DENSITY	Unevenness of urban building massing Missing FAR incentive programs Fragmentation of zoning district uses Lower emissions per person
6	TRAVEL	Traffic congestions due to over-use of vehicles 23 Mil. tons less CO_2 compared to average US driving emissions Fossil fuel based travel will fade with exhaustion of resources Majority of travelled miles related to live-work commuting
7	WASTE	Up to 10 pounds of garbage daily per NYC citizen More than a quarter of municipal waste is organic Development of clean Waste-to-Energy incineration plants NYC waste disposal dependent on other States' landfill capacities

- Decentralized power generation; Urban photovoltaics (PV)
- Activation of renewables in urban applications
- City driven policies/incentives
- Emission reduction goals; Predict/analyze effect of installations

- Encourage passive design strategies
- Utilize urban shading strategies
- Reduce anthropogenic heat sources
- Utilize vegetation cover to lower ambient urban temperature

- Encourage passive design strategies
- Embodied energy/materials research
- Energy Code reconciliation
- Energy-efficient building strategies

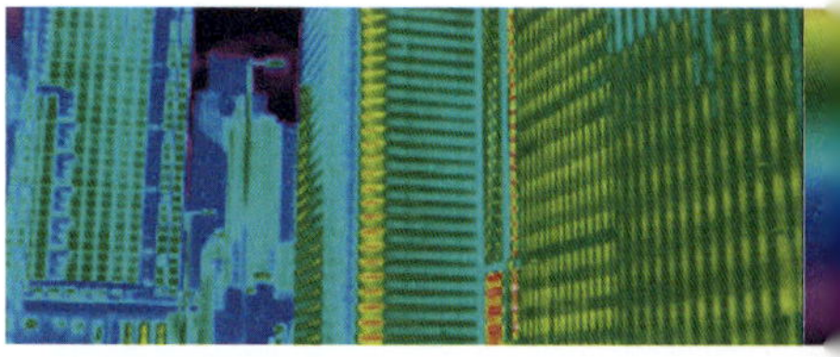

- Re-greening' the city in even distribution patterns
- Roofs as possible greening sites; 'Urban farming'
- Carbon sequestration, storm water and pollution filter
- Biodiversity, and life cycle increase of roofing materials

- PV efficiency based on shading patterns/air-right regulations
- Amending FAR incentives to encourage PV/solar installations
- Mixed Use integrations to socially upgrade neighborhoods
- Densification results in smaller 'footprint' per capita

- Encourage public transportation / walkable neighborhoods
- Upgrade public transportation infrastructure
- Conservation policies
- Encourage 'tele-commuting' and live-work environments

- Reduce waste, encourage recycling
- Composting as materials management
- Power generation by means of waste processing
- Avoid elaborate transportation: local processing and jobs

DESIGN CONSIDERATIONS: 'SNAPSHOT IN TIME'

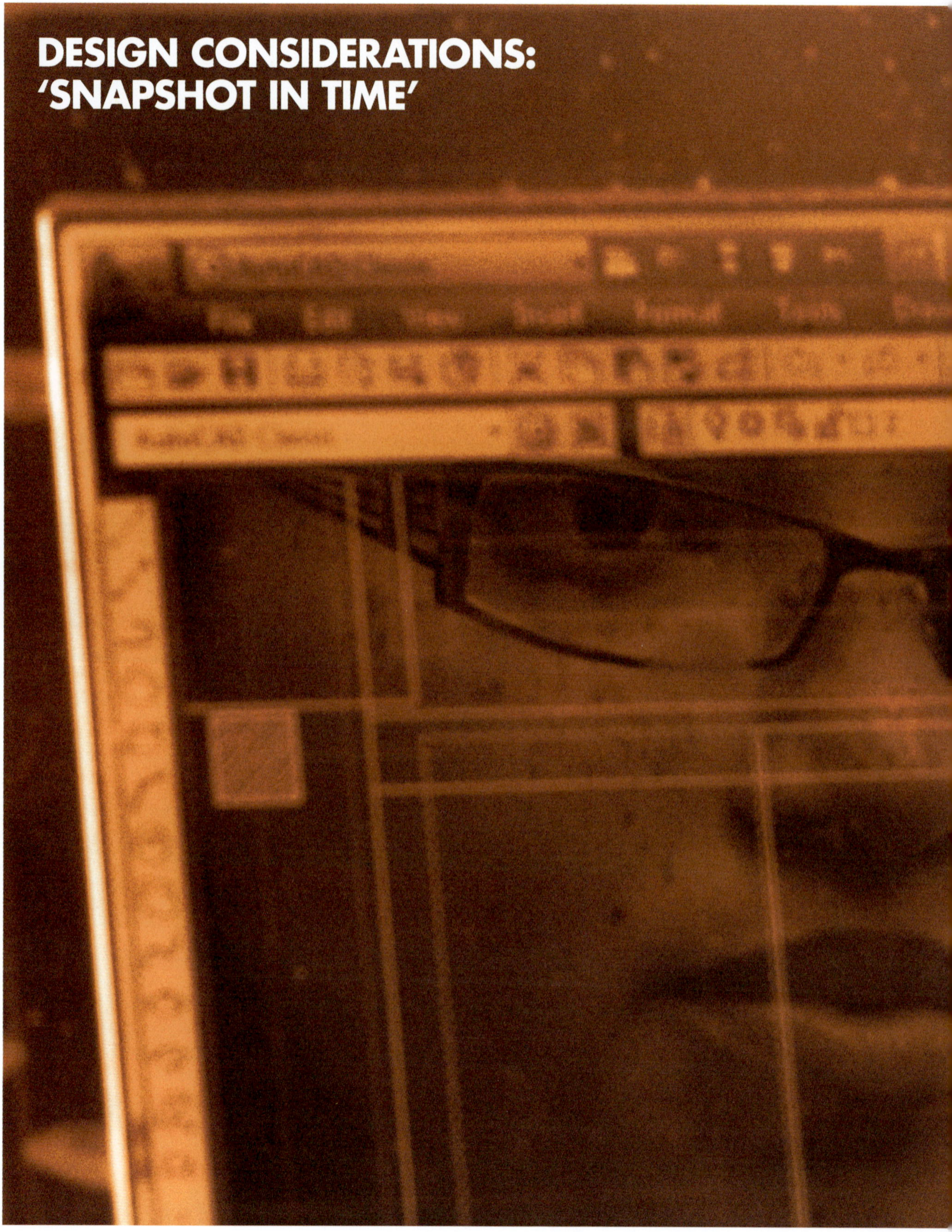

HP TrueVision

0

FIRST APPROACH: "WHY?"–'VISION'

The coming decades will – or must – witness infrastructural changes, if New York wants to be capable of transitioning to a more sustainable future. The issues discussed in the last chapter will only become graver if New York's population grows by close to 10% as is estimated in the next 20 years[35]. The magnitude of the issues at stake is expected to increase proportionally.

WHY?

When designing, one has to relate to the 'ectoscopic' framework of one's environment, to the cultural context that surrounds us: The way we power our cities, the way we build, how we travel, how we cope with recreation and work, and how we deal with extensive consumption and its consequences, are some of the resounding questions.

Design decisions have a imperative to relate to the future. Some of the questions to be asked therefore consider why certain solutions are culturally advantageous. They are an articulation of intention. Architecture is an adapted art, and has to solve problems. Therefore, any building design that wants to be truly contemporary must, in some way, address the infrastructural challenges already enumerated.

Many different criteria, including economic, ecological, socio-cultural, legal, functional, technological and aesthetical factors, link the building to its culture and its future.

However, since the design approach is broad, the relative scale at which one can look at the issues is small, limited to an overall context. It is an approach that relies on zooming-out to understand and affect its holistic cultural environment.

The approach taken is critical, couched in terms of its period, but also within the set of popular questions that are starting to evolve. Any vision is subjective, relying on acceptance and recognition by many people in order to be influential. The 'Solar Roofpod' project was conceived as an educational project, intended as a responsive conceptual proposal to provoke discussion.

The small-scale approach has to grow in scale. It has to address how the design intention can be realized: the transition between the overall observations and a material strategy is thus crucial. The following are the criteria, derived from observing contemporary urban life, which were instrumental in shaping this architectural project.

2.1

2.1 NYC: "The Cosmopolis of the Future": Prediction by Moses King, 1911
It is taken from a guide to how the city would develop, imagining layers of new structures and traffic ways built atop the existing city; mobility was equated with value creation, as a response to current cultural development.
(Illustration: R. Rummell)

2.2

1

SITE: ROOFTOPS

It is undisputed that choosing the right site for a project is one of the most important decisions. In terms of context and orientation, this choice is seminal for the performance and attractiveness of any future building. A smart site response will amortize itself over its entire life cycle, not only experientially, but even more so through lower operational and maintenance costs. In terms of monetary considerations, this decision promises the most payback for the least amount of money.

INFRASTRUCTURE

The trick is in the optimal utilization of existing infrastructure. Seen in these terms, any environmental impact on a site is part of a positive infrastructural impact, since it would help to 'power' a building. But, beyond the interplay with natural forces, an urban environmental impact now includes the infrastructure that is integrated into the built environment. Accrued over years, this infrastructure becomes an asset, while any addition tapping into it is also challenged to improve its environment.

Any urban environment constitutes therefore dense infrastructure, which has to be taken advantage of. Utilities, traffic systems, structural overcapacities, in short: any developed and serviced plot of land has high value.

Often, the least expensive solution is also the least sustainable solution. However, when taking advantage of beneficial site conditions, these two factors can in fact develop synergy. The ability to influence sustainability is at its highest early in the design process, as cost impacts are the lowest (2.3). When combining the two, one thus must consider both the existing infrastructure of the built environment and the available environmental forces. What advantageous sites are available in New York?

ROOFTOPS

For now, any of the largely unoccupied roofs of residential mid-rise buildings could serve as our site. These roofs, even though sometimes depicted in individualized contexts, are probably the most overlooked location in our dense environments.

Although many discontiguous 'mini-projects' can be found, where individuals take over some interstitial spaces (2.2), they are most often residual leftovers dotted with air-conditioning compressors, vent stacks, smoke hatches, skylights and other utilitarian infrastructure; most of these interventions are improvised. A systematic approach to occupying this underutilized 'real estate' is missing. The spaces available constitute an undiscovered world in a city that famously defines its growth vertically, with breathtaking views at its upper level. There are tens of thousands of such roofs available for a typological development. A design that can be multiplied will yield much greater effect. The effect of a singular prototype can easily be extrapolated to understand the greater consequences of the intervention.

While these urban roofs are attractive, the question of why they remain unoccupied needs to be investigated. Who in New York would not like to live penthouse-style? But, there is more to this location. Roofs are exposed to the highest environmental loads. A roof membrane can be subject to solar

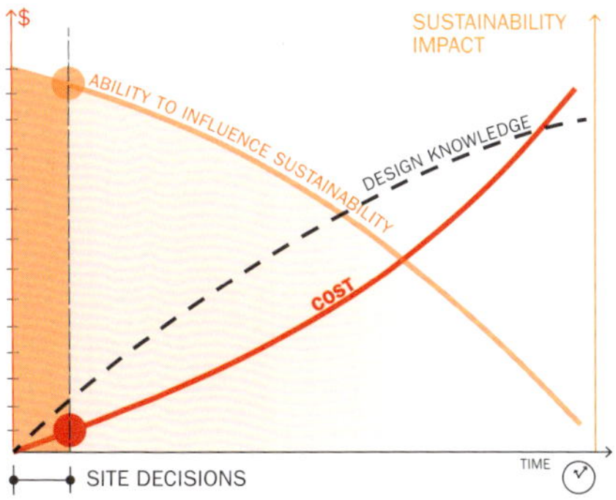

2.3

2.2 NYC: Common interstitial roof oasis, Brooklyn NY.
2.3 Sustainability impact on design decisions: smart site decisions yield higher economic return.
2.4 NYC: Roofpod utopia: schematic rendering of an urban Roofpod environment.

2.4

insolation in summer that results in membrane surface temperatures exceeding 150˚F[36]. Material expansion and contraction, not to mention ultraviolet radiation, negatively impact the durability of the roof. We may yet learn to take advantage of these loads, but at present they are wasted or even deleterious to our roofs and buildings.

POTENTIAL

The sun's overall energy potential is about 6000 times greater than global annual energy consumption (2.4)[37]. Although we lack perfect technical means to convert all of this potential energy, we actually don't need to: the amount of energy raining down from the sky is so plentiful that a fraction of it would cover even our considerable needs. Wind is also present at greater heights atop buildings, and can also provide additional renewable energy.

Therefore, the rooftops, in addition to being our most underutilized urban space, have the best access to our greenest and most abundant energy resource, the one responsible for the creation and sustenance of all life on Earth: the Sun!

2 POWER DECENTRALIZATION

Most of NYC's electrical power is produced either by nuclear power or by burning fossil fuels, predominantly natural gas. The former involves high risks in operation and disposal, the latter involves the emissions of greenhouse gases. Both methods rely upon centralized systems, which traverse considerable distances between the place of production and the place of consumption. This means transmission losses.

DECENTRALIZATION

We chose to create a decentralized energy infrastructure in the rooftop environment. The idea is similar to the system of rooftop water towers, developed in the late 1800s to modulate water pressure at a local scale as it entered each building from the central municipal system. These water towers are not only an effective solution to this particular problem; they have also become an icon of the typical New York city-scape.

RENEWABLES

The solar arrays envisioned for this project have this same potential to feed buildings from the 'top down'. Renewables have been developed by now to the extent that they can contribute efficiently to energy collection, although solar thermal systems are about four times as efficient as photovoltaic systems[38]. In other words, the use of heat as an energy source rather than photovoltaic electrical power offers a much higher yield. Thus, we chose to drive our air-conditioning system using heat, directly harvested as needed from the sun. Heat can also be used in adsorptive and absorptive cooling processes to generate cold. And since there are always electrical energy loads from appliances, lighting fixtures, computers and other electronics, a combination of both solar thermal and photovoltaic panels optimizes energy supply.

The least green element of many photovoltaic systems is conventionally the battery array used to store energy for use offset from the moment of harvesting. An alternate to energy storage in batteries is to feed locally-collected energy into the larger system and retrieve your energy deposit when needed, if permitted by the central grid provider.

In Germany, power providers are legally required to prioritize absorbing the power generated by individual photovoltaic arrays over their own energy production, even if it means a greater cost per kWh than they can charge for. This is part of the German government's subsidy program to push for the widespread application of renewables. The stated goal is to run 80% of all electrical production off renewables by 2050[39]. Given the numerous power producers in a recently deregulated market, the main challenge is logistical. Understandably, research worldwide currently concentrates on 'smart grid' solutions to optimize distribution of power harvested by all means.

MINI-GRID

The idea of applying renewables, such as solar-thermal and photovoltaics, to the roofs of our cities has a clear advantage relative to the distribution logistics. Since every 'Solar Roofpod' would sit atop a host building, any surplus energy produced would go straight into the host building's distribution system, and be consumed locally. By providing power from the top down, not from the bottom up, the central grid is relieved, and transmission losses avoided. The host building replaces the battery since it absorbs the surplus not immediately consumed by the Roofpod. As a result, the smart grid is disentangled and more manageable!

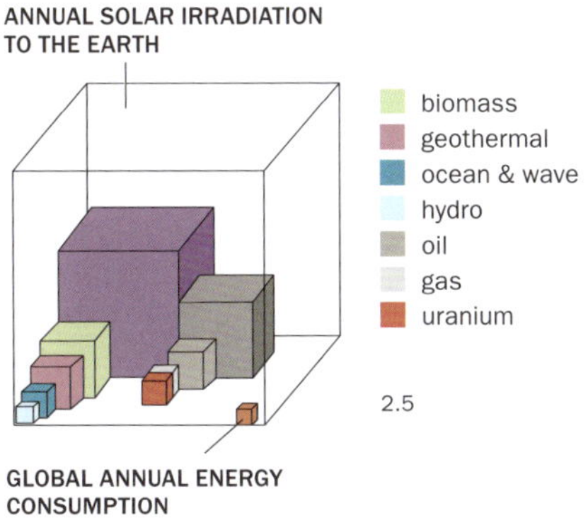

2.5

2.5 Solar Irradiation vs. established global energy resources; plenty of potential for solar power.
2.6 Passive energy façade for a smart envelope system, early design proposal of 'building blocks'.

PASSIVE DESIGN – SMART FAÇADE

In our early observations of New York's building stock, we also noticed that a common approach to providing comfort in buildings is to input energy unnecessarily. For instance, building floor plates are often too deep so that they need to be constantly acclimatized to provide sufficient air exchange; air-conditioning is run when spaces are unoccupied and heating can often not be regulated except by opening windows individually; and, electric lights are on constantly, even at times when daylighting would have been possible.

We therefore decided that the entire envelope would need to integrate smart environmental responses, potentially contributing to power generation, but more importantly, managing environmental loads positively to avoid energy consumption. This kind of passive energy façade strategy could adjust to diurnal and seasonal changes as well as to personal preferences. Many diagrams (2.6) were executed to address the operation. We have to relearn how to design our buildings in a way that manages their environmental loads not by applying energy supplied by the electric grid, but by smarter means of design.

2.6

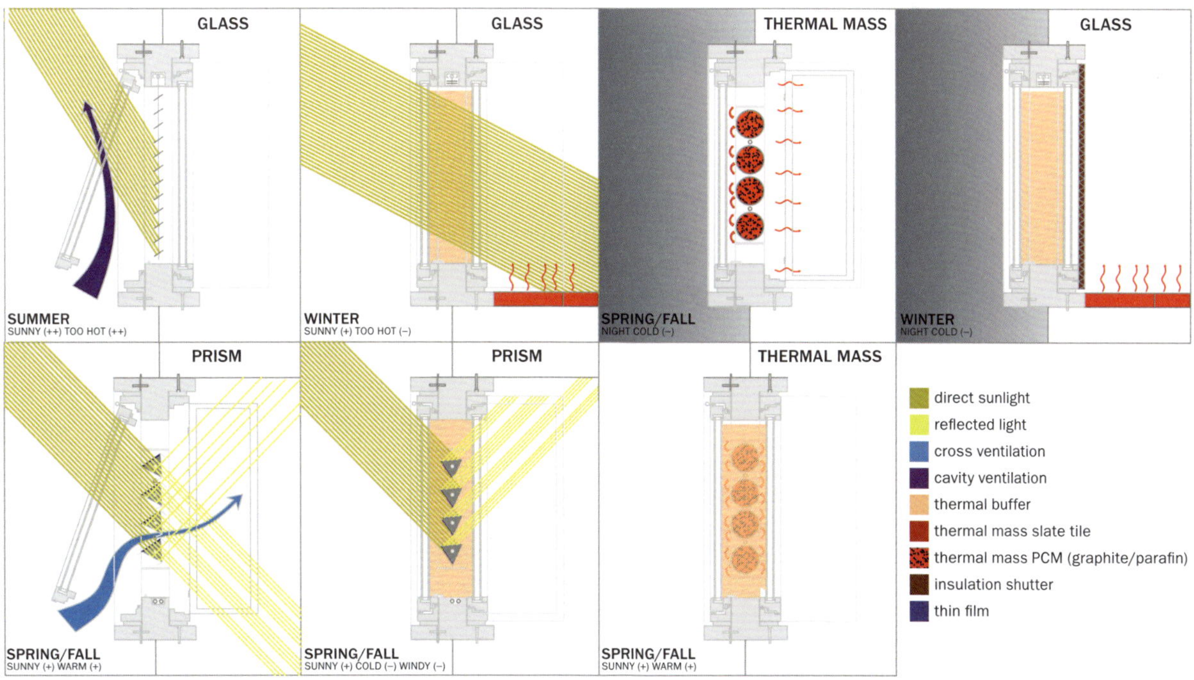

3

DENSIFICATION

The city is one of civilization's most successful inventions; unifying dwellings and making them more efficient through interrelation proves the sophistication of our organizational skills. An existing city depends on its infrastructure. Setting up such systems is not a one-time undertaking but a repeated process of growth and renewal, by nature piecemeal and multi-layered.

The purpose of infrastructure is to increase the city's productivity, but it also improves living standards. In fact, the more people take advantage of the same infrastructure, the less their impact on the environment. The per-capita carbon footprint of an urban dweller is thus a fragment of that of a suburbanite or a country dweller (see 2.7).

Technical progress and socio-cultural change demand adjustment and additions. Infrastructure as a general term designates everything that serves our ease of living: energy, water, gas, heat, food supply, public transportation, disposal of sewage and garbage, facilities for health, for education, for events, for recreation and so on. Tapping into or complementing the existing infrastructural systems is most often more effective and more economic than initiating new systems. It is also more sustainable and less wasteful.

We New Yorkers live in the densest urban environment of our country, with 27,000 people per square mile on average[40]. But urban population densities in Europe, Asia or South America are often even higher.

How much density is advantageous for a more effective infrastructure? Is there a limit to density, after which conditions become unbearable? Invention and renewal of infrastructure must balance the effects of densification with the necessary respect to quality of life. But, if London, Tokyo or Buenos Aires can tolerate population densities twice ours, then add-ons to our city certainly can be managed.

GROWTH CONDITIONS

It is paramount to provide conditions that allow optimal infrastructural growth during densification. That depends to a great degree on effective policies, such as zoning regulations. This means that the heterogeneity of New York's urban massing is problematic, because it is generative. The uneven and fragmented cityscape comes with disadvantages for the conditions one finds on urban roofs. In particular, shadows cast by other buildings constitute an issue in an urban setting. While commercial neighborhoods are shaped by this uneveness, many residential quarters are not yet affected. It is important to preserve these conditions with smart zoning regulations.

In New York, particularly heterogeneous neighborhoods, many of which are commercial and built up with skyscrapers, such as Midtown, are disadvantageous for photovoltaic applications since airright trading and other building height incentives produce such uneven massing that certain roofs do not get much solar exposure at all. Since the access to environmental loads will gain importance for power generation as photovoltaics and solar thermal devices become more effective, future policies and regulations should favor more even massing: buildings' exposure to environmental loads is a vital precondition to clean energy, and consequently as democratic a right as the right to eminent domain.

INCENTIVE PROPOSAL

We decided to hypothesize about the regulatory directives we felt would positively influence the Roofpod installation in general, but particularly the introduction of photovoltaic installations in the city.

We presumed that many buildings still have enough available FAR to extend their square footage, or, that they would be awarded extra FAR for

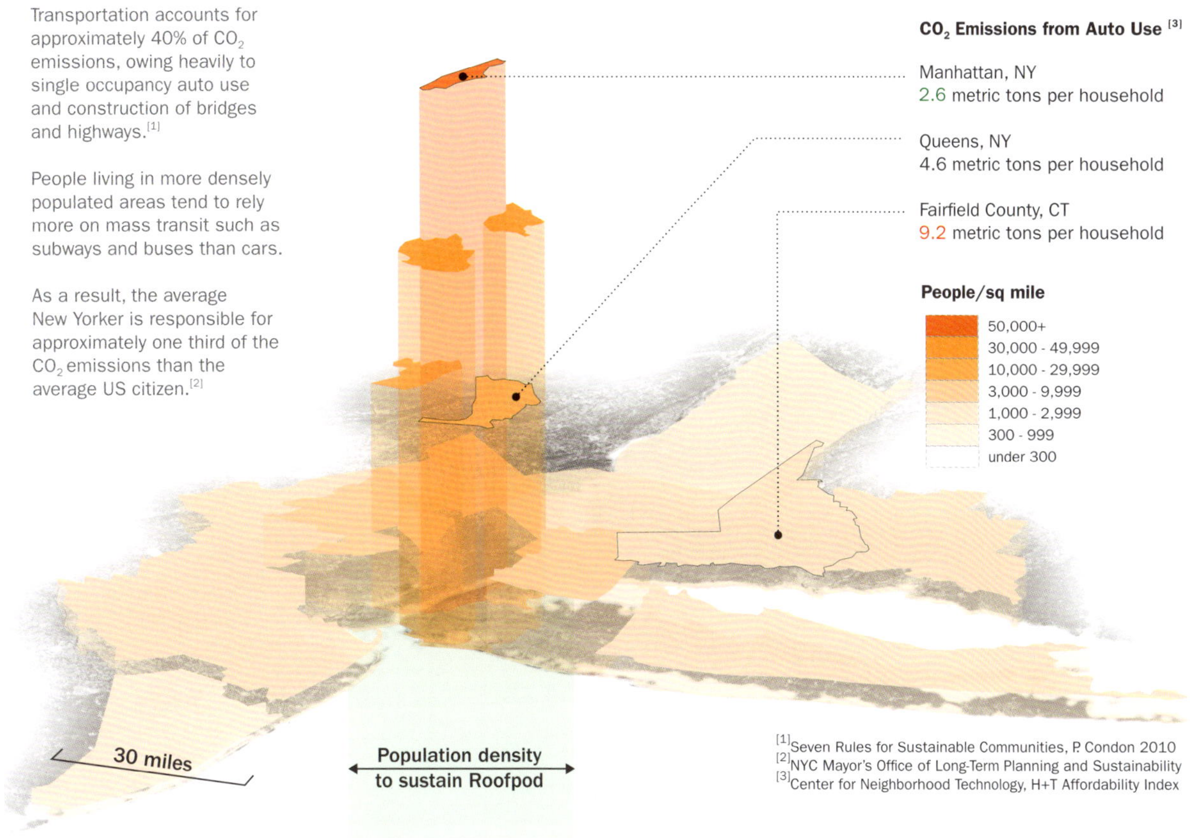

2.7

photovoltaics, so that the city would incentivize the related one-storey rooftop extensions. We feel that these extensions would be justified for several reasons:

1) Cleaner energy production is to the city's advantage;

2) A socio-cultural learning opportunity for the city's citizens goes hand in hand with the idea of using our roofs to locally balance energy deficits, and harness environmental loads for our comfort;

3) Densification is a likely development in the coming years, as population increases. If handled prudently, adding density in height would be the most sustainable way to grow the city and avoid sprawl, because it would require less infrastructure to integrate more people;

4) With less sprawl, greenhouse gas emissions would be reduced to help mitigate the effects of global warming.

5) Photovoltaic installations would make the city's infrastructure less vulnerable against climate change related natural disasters. During superstorm Sandy in 2012, significant parts of the city lost power due to flooding, with economic losses in the billions of dollars.

6) New York's rooftops and cityscape are fascinating spaces. They have great experiential potential to attract people, who in turn would invest in the new recreational and environmental projects, benefitting the local building industry and thus the city as a whole.

A roof extension incentive would thus improve many aspects of urban quality of life.

2.7 Urban living: Population density contributes to sustainability. Per-capita CO_2 emissions in comparison; student drawing, communications design.

4

ELEMENTARIZATION —SYSTEM-BUILT CUSTOMIZATION

In order to create a design that can be applied universally, not only on New York's roofs but also on comparable building-top situations elsewhere, the building system must be reproducible and standardized for machine production. This comes with the following advantages: 1) Costs will be reduced through economies of scale in fabrication, tools and jigs; 2) Site installation times will be reduced and the impact to the residents of the host building during construction minimized; 3) Waste is reduced by coordinating material dimensions during production to optimize modules; 4) Future refurbishing/ retrofit of 'Roofpods' is facilitated; 5) Elements can be transported to the roof with greater ease, using small hoisting equipment, or even elevators, creating cost savings; 6) Standard connections can be developed for simplified assembly; 7) System-built prefabrication can generate local businesses.

INDIVIDUALITY

However, this is not just a technological or logistical problem. Its success is driven by user acceptance. Although pursued by most illustrious architects and inventors, from Thomas Edison in the early 20th century, to Richard Buckminster Fuller or Frank Lloyd Wright in the 30s and 40s, prefabricated buildings in the United States were often not as sought after as expected. Wright himself proclaimed 'Individuality a national ideal,' symbolized by the American House. What he understood is that people do not want to live in the same house as their neighbors, but instead want to make their houses their own. This observation is even more pronounced in building design than in car design or consumer marketing.

In developing his system-built house type, Wright came to the conclusion that although the same details can be reused, the overall appearance must develop individuality. The 'Usonian House' – of which some sixty were built – bears witness to this strategy. Although many building elements, such as interior walls or built-in furniture, are based on the same detail sheets, the way in which these elements were combined to form the whole varies widely. It is not solely a question of style, but is responsive to different programs and contextual or site requirements. In addition, certain features are often used in a unique way, which contributes to the buildings' individual richness.

FABRICATION STRATEGY

We decided to conceive a building system drawn on both prefabrication and individualization. These two poles mark at one end the generic, which is valuable for prefabrication, and at the other, the specific, which strengthens each project's individual expression and its responsiveness to performance criteria.

We decided to develop a modular system with several subsystems that provide a mix-and-match strategy to satisfy individual preferences as well as possible contextual requirements. Many schematic alternatives of a design approach were generated to understand and improve the aesthetic potential

2.8

2.8 Modular design alternatives, schematic, generated by the kit-of-parts strategy. The main design elements are represented: solar roof structure, modular envelope and the surrounding garden.

Top: Family-of-four; Middle: Bachelor pad; Bottom: Community space.

(2.8), and identify the necessary components. In this approach, we tried to restrict ourselves. We did not want to be in the position of constantly inventing more parts for the kit, but instead, to keep the investigation focused on a limited number of essential design items.

How to fabricate an energy-efficient solar house prototype? We started to look at the one prefabricated item that was defined already: the solar panel. Most of these panels are proportioned the same way, in a one-to-two width-to-length ratio. This is to make them easily combinable. The particular size is related to the ease of handling, too.

We decided that the solar panels had to be featured in an emphasized fashion, since they were responsible for generating the renewable energy needed to power the house. The geometric system that the panel installation was based on, aroused our interest: Would it be possible to base the entire geometric order of the project on the module strategy articulated in the panel assembly?

Our focus shifted to the envelope of the house. Its purpose was to conserve and manage the energy captured as efficiently as possible. While the solar panels, an already manufactured item, had to be integrated in the roof design, the envelope system had to be invented and built from scratch to develop a unique and superior approach. The analysis of its design parameters were crucial.

The surrounding roof garden related to both aspects of fabrication: integration and invention. The planters and the transfer beams were already prefabricated, while the framework to tie the assembly aesthetically together had to be invented.

Our task was to coordinate between the multiple design aspects of these main elements.

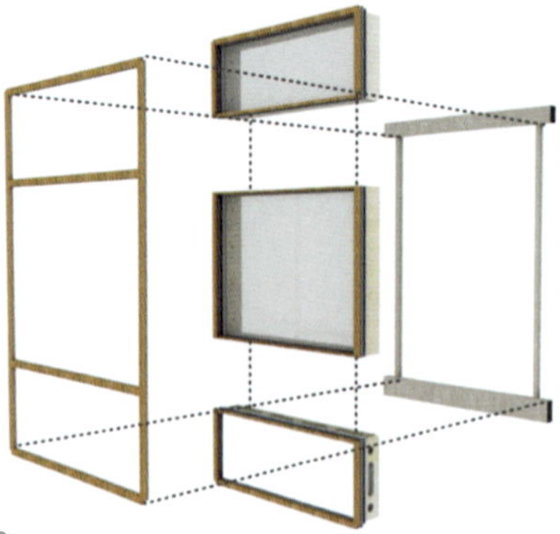

2.9

2.10

'BUILDING BLOCK' FAÇADE

Our particular interest was the development of an exterior envelope system that could use typical curtain wall technology, but deploy it for greater energy efficiency. The envelope system had to provide superior thermal resistance to conserve energy; it also needed operable features to provide the desired passive design regulation, for contributing to interior comfort. And it was equally important that it express the idea of modularization aesthetically. It had to allow for playfulness and individuality (2.9).

The solution was the development of different modules, defined in a catalog that make a whole. There were blocks for ventilation, blocks for light redirection, blocks for framing views and blocks for providing privacy. These features were related to the passive design strategies. However, they also expressed the desires we have when experiencing architecture. They were filters to awaken the senses.

These blocks can be used in countless combinations according to the user's desires. This exterior wall system would be dimensionally suitable to allow for an array of different sized buildings and alternative plans, as required by different functions.

The system should be simple enough so that future mass-production would be feasible. The necessary requirements related to this objective had to be analyzed during the design process. What requirements would make a good product?

'LANDSCAPE' MODULE

As a second subsystem, a roof landscape system was proposed (2.10), that primarily had to be compatible with the overall building system and had to serve as the transitional structural device between 'Roofpod' and the host building.

By doing so, it created a crawl space beneath the Roofpod through which to feed electrical, plumbing or mechanical media into the host building. The landscape modules were therefore covering an installation layer. The modules should thus be removable.

Architecturally, this constitutes the immediate context for the building's interior-exterior relationship; it is the foreground against which the distant skyline is viewed. This subsystem carries the roof vegetation to provide areas for gathering, garden work, etc. Because the depth of the transitional structural system, called dunnage, is determined by the loads it receives from the Roofpod, its deeper dimension allows for more intensive greening than typical green roofs. This determined the plants that could be planted within this landscape. While usually only small plants are possible for green roof applications, on our roof we would be able to plant even small trees. Thus, the modularized roof landscape can transform the roof into the building's secret garden, with walkable and planted surfaces.

Again, the way this system must be produced was related to prefabrication conditions. It had to be simple enough that repetitive fabrication would make sense.

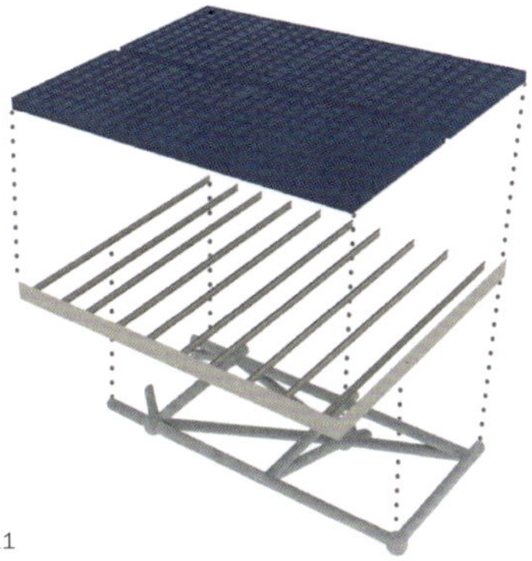

2.11

2.9 'Buidling block' façade module, schematic design.
2.10 'Landscape' module for intensive green roof, schematic design.
2.11 'Solar trellis' module for energy-harnessing and shading, schematic design.

'SOLAR TRELLIS'

While the subsystems related to the building and the garden were conceived to benefit the user, this subsystem, the solar trellis (2.11), responds primarily to the building's energy-harnessing system by carrying the photovoltaic and solar-thermal collecting devices and simultaneously shading the house.

Since solar panels are still undergoing technical changes, we want to ensure that outdated or defective devices could be easily replaced so that the 'Roofpod' would not lose value, or even become outdated. Thus, the *trellis system* works as a mounting scaffold for any solar equipment, directing it towards the sun while the shadows cast shelter everything below it. This could be either the Roofpod's roof membrane, which would be protected from direct radiation, infrared material degradation and heat gain; or, it could be an outdoor space beneath a free-standing trellis – a pergola offering a shady place from which to enjoy the garden.

The system needs to be stable enough to withstand strong winds. Usually, the wind speeds at the top of buildings are higher than on the ground. Therefore, all connections have to be considered.

Wind could generate even more energy for the Roofpod, if mini-turbines were installed. However, the competition rules of the Solar Decathlon didn't allow for such income. Any research on the topic had to wait.

OTHER BUILDING ELEMENTS

We were aware that there were other components of the assembly that could not be easily modularized. For example, the programmatic requirements of bathrooms, kitchens and closet spaces are so diverse, even based upon personal preference, that they remain somewhat unpredictable. Life cannot only rely on modules. We chose conceptually to organize these serving functions in a customized interior core. A more systematized core element, which we would complement with other features, would be a task for future refinement, after sufficient customer feed-back. It also would relate to different occupancy and use.

The calibration of a system to occupy New York's roofs is a tricky task. On the one hand, the approach must be generalizable: installation-friendly, cost-effective and beautiful. On the other hand, standardization cannot be depended upon to resolve everything, since the interventions are site-specific. Not only do roofs come in different sizes, with different conditions such as views and orientations, but the subjective preferences of future roof-dwellers are also diverse. Any kit-system must thus be able to adapt to specific requirements. Roof properties with their spectacular skyline backdrops are very desirable; a certain standard of living is required to satisfy market expectations. It is questionable whether a 'one-size-fits-all' prefabricated system can meet these standards. Smart standardization therefore has to absorb specificity and variation within its system.

5 URBAN VEGETATION —MULTI-PURPOSE ORGANISM

Modern green roofs have been around for quite some time now, and there are more and more of them in urban environments. Cities such as Chicago instated policies that make green roofs mandatory for certain types of public buildings.

The many benefits yielded by this extra layer of roof cover are well known. Benefits to the building owner include: 1) A longer roof membrane life span; 2) Heating and cooling cost reductions; 3) Noise prevention; and, if water catchment is provided, reductions in 4) Water costs. (More than half of a residential building's water demands could be served by rainwater). But not only does the building's owner profit from this initial investment, he also provides value to the city and the community. The benefits to the public include: 1) Easier storm water management; 2) Air pollution mitigation; 3) Carbon dioxide sequestration; 4) Decreased 'heat island effect'; and 5) Contribution to urban biodiversity, since many species such as bats, bees and birds can regain habitats that had disappeared over the past 50 years. The vegetation loss in NYC over this time period has caused the disappearance of many species that were common and useful, for example, in fighting destructive insect plagues.

In certain European countries, green roofs have become mandatory. For example, in Copenhagen, the capital of Denmark, every new building with a less than 30 degree-pitch roof, both private and public, must be vegetated[41]. The focus of the city's policy is not just localized problem-solving: Copenhagen has stated its ambition to become the world's first carbon neutral capital by 2025.

Green roofs are a vital part of this plan. The difficulty of realizing the desire of CO_2 neutrality is that the goal does not promise any immediate financial gain. Although it is obviously the right thing to do for a sustainable future – considering the larger environmental picture and the consequences of global warming – there are still major issues in labeling and justifying these capital expenditures. With regard to resources and emissions, we still operate imperialistically, and have issues with committing to stringent and disciplined resolutions. The lack of authority, and the economic pressure to satisfy economic growth are detrimental to plans to level carbon input and output for global climate control.

While this attitude is perhaps understandable for the private sector, whose main goal is capital growth, governmental leaders must set clear precedents for achieving environmental goals and for creating a support system to realize them. It is often said in American politics that government should only intervene when the private sector cannot do the job. The quest for carbon-neutrality is certainly in that category.

City officials should therefore be less focused on economic feasibility, and more on long-term environmental issues that will not be balanced by private constituencies. In that respect, the city's recent studies that street trees are more cost effective than green roofs, misstate the complex benefit-framework of this issue: urban vegetation cannot be reduced to an 'either-or' when both measures could be used simultaneously and symbiotically.

2.12

2.12 NYC: Current green roof applications. Individual retreats, educational facilities, high-end commercial endeavors, restaurant gardens, roof terraces and urban farming.

What value does a ton of saved carbon emissions have? What is the value of cleaner water? If precise quantitative answers cannot be given, common sense is needed. It is necessary to do more, and not to exacerbate the situation.

If the city's policy making and 'top-down' approaches do not achieve the desired results, what actions can architects take to popularize the idea of green roofs? Why are green roofs not higher priorities in the first place? It is because currently almost nobody sees them. Often, these green spaces, although existent, are not even known to the inhabitants of the buildings they cover. They exist without being used by people – and so they don't exist, at least not for an effective architectural narrative (2.13). The architect's focus should be on developing tangible solutions.

2.13

We decided that our design project has to highlight the roof's surface as a mini-park, which has to become part of the overall architectural experience in order to receive the attention it deserves. Philosophers, behavioral scientists and psychoanalysts have studied how contact to nature supports mental health and alertness. However, despite recent improvements, there are many neighborhoods in New York City that do not have parks for recreation. With a 'roof park' on each flat roof of a building, recreational spaces in proximity to housing would be provided everywhere.

IRRIGATION

Since the upkeep of these roof areas is critical, the system we designed had to be as low-maintenance as possible. There are many reports of green roofs not being as beautiful as promised on the initial drawings. Work is required to maintain the idyll. Semi-public gardens have often suffered from this lack of organization.

A successful strategy might be based on irrigation techniques, which allow vegetation to thrive under relatively minimal oversight. The irrigation should draw on storm water collection, so that the planters would work as storm water overflow buffers, absorbing the maximum amount of rainwater and mitigating peak storm water entrance in to the city's sewage system. Storm water can be put to use instead of creating logistical problems for our under-dimensioned sewage system. In addition, the water comes free of charge.

We looked at drip irrigation systems (2.14) that are economical and controlled by timers. When considering running pipes for irrigation, we also investigated taking advantage of the soil's thermal mass. If we were to place extra pipes, and feed fresh air for the mechanical system through them (2.15), this air could be precooled or preheated before it enters the mechanical room of the building. Since less energy would be needed to condition it, the green roof and the mechanical systems would develop a synergetic effect. However, we did not pursue this idea when refining the design.

WATER COLLECTION

More research remains to be done on the environmental benefits that local storm water catchment would yield. In New York City's combined sewer system, street and roof runoff, domestic sewage and industrial wastewater are collected in the same pipe system, which is often overburdened during rainfall.

When the water volume exceeds the system's capacity, untreated water ends up in our waterways. Even if the system could perform to capacity, treating cleaner rainwater wastes money and energy; keeping it out of the sewage system means lower operation volume. While it is difficult to put a price tag on this issue, it is clear that the growing city must have an interest in improving its infrastructure and in preparing for increasing volume. The roofs of our buildings could, in a decentralized way, mitigate a more regulated storm water flow. While considering the delegation of certain municipal services to the roofs of our buildings, such topics as water treatment and waste management were also discussed. We even considered integrating a 'living machine' into the Roofpod's surroundings to pretreat wastewater locally.

2.14

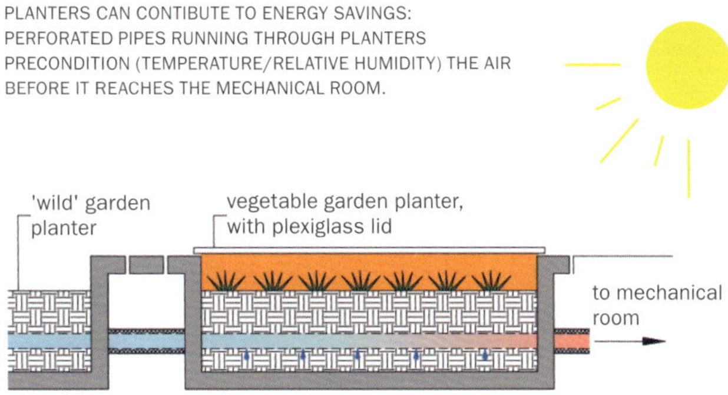

2.15

PLANTING CHOICES

The amount of water anticipated has to be considered when choosing the plants that will thrive under these conditions. We studied many varients, from the reintroduction of native vegetation to full-scale urban farming. The roof garden has sufficient substrate for larger plants due to the given depth of the dunnage system. Since this dunnage system is needed anyway, the weight of the soil is not much of a factor, especially if engineered light-weight soil is being used.

Local fresh food production could not only benefit public health in the form of better nutrition, it could also save on food transportation costs and emissions. It is estimated that a city of 10 million people needs about 6,000 tons of food every day, which has to be brought in from farmland[42]. This heavily impacts traffic and emissions. From an economic standpoint, too, a nearby food source has certain advantages in delivering a fresher product, reducing processing wastes and mitigating logistics costs.

Because of concerns about the suitability of unfiltered urban rainwater to grow edible produce, we also looked into producing biodiesel fuel by planting soy, rapeseed or sunflowers. In fact, there are certain farmers that do this to run their fleet of agricultural machinery. If these plants were processed locally, petroleum energy sources could be replaced with biodiesel as a low carbon alternative for heating oil. Harvesting energy in summer, storing it as biodiesel, and using it as a local complementing fuel source in winter, seemed attractive to our strategy of decentralized local self-sufficiency.

COMPOSTING

We also considered including a composting station on the roof. Here too, keeping logistic streams local would reduce transportation and disposal stresses. If a significant proportion of municipal waste could be processed or pre-processed on site, fertile soil could be produced on the rooftop and used for extending or fertilizing the planter soil. Surplus could be regularly collected, used for public parks or other landscaping purposes, or returned to local farmland to provide nutrients for future crops.

Clearly, the utilization of our roof surfaces would yield many benefits. However, analysis of these benefits is haphazard, since they accrue to many different fields, as described above. If considered separately they might not seem to have an impact worth pursuing, but when considered all together they clearly have the potential to make a huge impact on our urban ecological system.

In order to popularize the use of the rooftop real estate obviously available everywhere for occupancy and vegetation, and to have it under consideration within any future plans, an architect's best strategy to make an argument is to develop a credible visualized project, including a comprehensive strategy. It is even better to realize the project and to collect data to prove that the green roof cover would actually be an efficient way to improve performance for both the building and the environment. The most misunderstood fact about green roofing is that it is often considered a design feature, used to express proximity to nature. We need to work towards an understanding that more explicitly includes performance alongside design.

2.13 NYC: Green Roof Utopia.
2.14 Drip Irrigation system: Inexpensive piping and emitters; Precise control and efficient water use.
2.15 Taking advantage of the planter's thermal mass: extra piping for preconditioning air before it reaches the mechanical room.

6

PROGRAM —MULTI-FUNCTIONAL ADD-ON

It is not only the garden around the Roofpod that is conceived as a multifunctional environment; the entire project was designed to be as flexible as possible. The more programmatic scenarios could potentially be implemented in the Roofpod, the higher the likelihood of a broad approach.

Since the Solar Roofpod was designed for the Department of Energy's 2011 Solar Decathlon exhibit, the required focus was on providing a residential setting. We focused, not without a certain measure of self-interest, on the young entrepreneurs, who, as New York's mayor announced in one of his many speeches, should return to the city and find favorable conditions here to propel the city towards leadership in innovation.

We realized that placing an energy-efficient building onto a structure that itself is up to one hundred years old translated into the need for an approach not be limited to monitoring and improving the Roofpod alone. We also had to be concerned with the host building's performance. Here, we considered the inhabitant of the roof annex to be the 'keeper of the castle', responsible for monitoring everything, from all environmental loads entering the building to all emissions that leave it. The analysis of these in- and out-takes and improvements resulting from that analysis would lead to better operational data of the host building, too, using the Roofpod as an analytical and evaluative device. In the best case, the Roofpod's inhabitant would bring the required expertise to the site. An engineer, functioning as a sophisticated building manager, could advance the city's energy operation toward higher standards. As mentioned, in New York, more than ¾ of total primary energy is consumed in buildings. Performance improvements should happen right at the source; the more people are involved in improvement, the more powerful the action.

After the devastation in the wake of 'Hurricane Sandy' in Fall 2012, the danger of locating building systems machinery at basement level became explicitly clear. Flood waters endanger electrical circuitry and can permanently damage its grid-interdependent infrastructure.

A 'top-down' approach to power and hot water supply is not only advantageous for everyday operation but can also guarantee greater resilience by providing solar emergency back-up. It would also create a viable sustainable alternative to the existing infrastructure.

Moving forward, there are reasons to believe that there will be strong arguments for locating a mechanical room at a building's top and not at its base, where it is vulnerable to flooding. Once this becomes standard practice, there will be also strong arguments for locating the people responsible for the building's maintenance at its top, and thus make the roof a more public space for the building's occupants to visit and experience. The experience would not only be related to the spatial context, which is always interesting on New York's roofs, but also to the performance of both buildings; the power-generating Roofpod, and the host building's operation, too. The roof becomes the place to visit, in order to be informed.

ACCESSIBILITY

Besides its residential program, other plausible options for the Roofpod's use became obvious during the conceptual research. They are based on the following consideration: A residential program always serves only a few people, while a more public function would provide more exposure to the idea that our roofs can be part of an environmental plan. Behavioral change can only be achieved through integration into a broader social and cultural context. Therefore, it would be advantageous if access to this new space were not limited, but if it were common property for the entire host building and beyond.

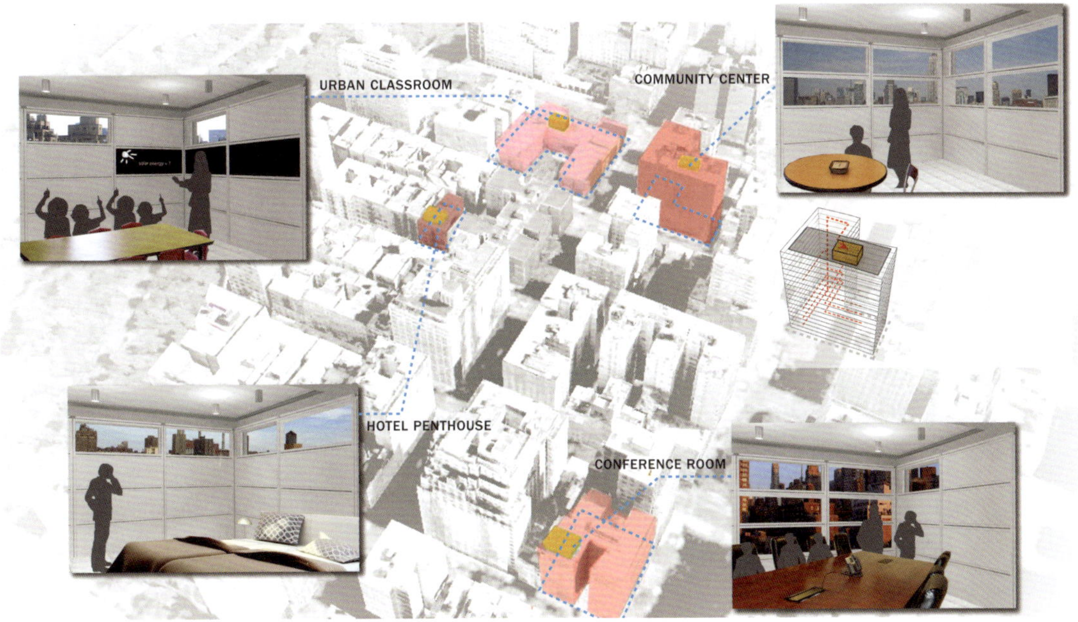

2.16

2.16 Application alternatives in an urban context: functional and contextual flexibility, based on the modularity concept.

Any code advisory committee should consider how important our urban roofs are for this purpose.

There are several public functions that could address the lack of social infrastructure in most of these buildings. The Roofpod could serve as a community room for meetings, parties, and different kinds of small ventures, from yoga sessions to music classes, etc. It also could become an educational hub, such as a library, a space for homework help, or daycare for residents' children.

However, it is clear that investments such as this – construction is expensive, especially in New York – have to be considered in terms of payback. Although this project was not initiated to engage in real estate development speculation, the project's concept should permit uses that could potentially create revenue for the host building. The difficulty is, of course, that a publicly accessible location, which would create a business opportunity, is likely to impinge on the host building residents' privacy. However, it is not unusual in New York for dentists, physiotherapists and other health services to be integrated into residential settings. It might also be worth the try to locate spas, restaurants or other leisure functions on those particular buildings whose residents could see that as an asset. As another conceptual paradigm, some New York apartment buildings at the beginning of the 20th century featured shared kitchens and dining rooms. These functions were considered amenities, and the buildings' residents were attracted by this way of life. Some of these functional mixes have also become very popular in the hotel industry, where roof gardens and cabanas, sky bars, etc. are marketed as places to experience the city from a different perspective and ambience. The popularity of rooftop amenities in that context must mean that there are ample opportunities to implement leisure functions on residential buildings too.

One hurdle to be negotiated with NYC executive leaders may well be occupancy limits restricting the number of people that would be legally allowed to use rooftop spaces without compromising safety and egress. If this issue could be overcome, there are multiple possible programs that would improve existing functional infrastructure. For example, one type of customization could relate to office buildings that need extra space. The Roofpod could work as conference room for a single corporation, or as a shared facility for several tenants. Or, it could simply be used as office space by certain divisions that could be separate from their main offices.

2.17

There are other scenarios worth considering relative to the economics of the Roofpod: New York is a major tourist destination, attracting 50 million visitors from all over the world in 2011. Each New York City household benefits on average by about $1500 in tax savings resulting from tourism, and 300,000 jobs are directly and indirectly connected to it. For a city that advertises its hospitality, an innovative concept for premier guest accommodations would be handy. Roofpods throughout the city could serve as a decentralized "zip"-hotel with rooms all over the city. Tourists could have more flexibility about where to stay, and upon revisiting, could always experience another part of the city. Since this experience would always change, the enticement to revisit would certainly increase.

Also, the fees charged for advanced hotel rooms, especially for those with a unique setting and skyline experience, would indeed justify an investment on such scale.

This last idea reflects significant changes in the way we organize work and leisure. As already noted, there are more and more people working from home. This saves energy and emissions, as well as money and time. The technological advancements in 'tele-commuting' make this possible. However, the ubiquity of work makes it difficult to gain distance from your work life, which is not the most relaxing lifestyle. The commute to and from work could be kept short if the place of work were separated only by an elevator ride to the roof of your building. Several units could even shape a communal workspace in the rooftop pod. What could be better than to imagine having your own architecture office in a setting like that? The space has access to relatively unobstructed daylight conditions, and a view that gives a sense of the surrounding city. Even live-work arrangements, which are becoming increasingly popular, would make great sense in an environment, whose occupiable spaces are adjacent to green rooftop gardens. As has been emphasized in multiple psychological studies: Humans need nature around them!

SPATIAL DESIGN STRATEGY

We decided to design the building as a pavilion that could be interpreted as an observation deck, a refuge – like a Japanese tea pavilion – or as an exhibit

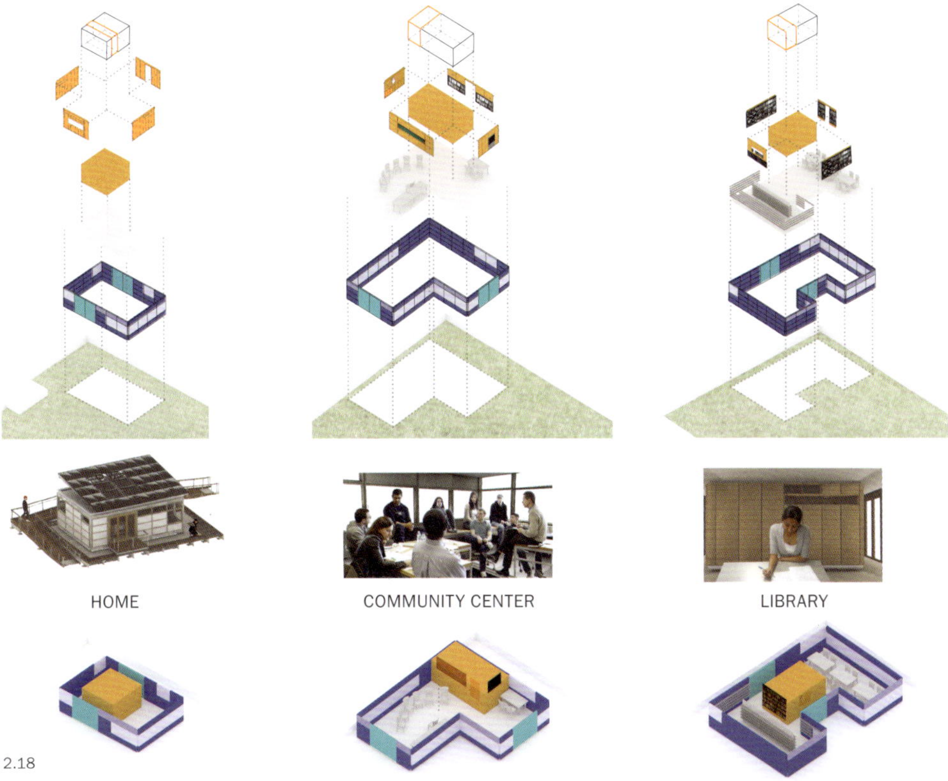

2.18

2.17 Roofpod narrative: first occupants hired: "The keepers of the castle". Young, competent entrepreneurs who will influence the transition of the city into an energy- and cost-efficient setting. The young building managers improve the host building's performance and convert and maintain the roof to an enjoyable outdoor space shared by all the residents.
2.18 Flexibility study: Since the building is composed as a 'kit-of-parts', it can be configured in many ways to allow for diverse functions. These include a home, community center or library; student drawing, communications design.

space. Its exposure, exterior-interior relationships and focuses are therefore quite variable and can be suited to personal preferences. However, what all of these variations on the core architectural typology have in common is their wrap around exposure and their internal organization, which defines the rooms without the use of corridors.

Both the pavilion's freestanding massing and its free plan organization were advantageous to flexible programming, emphasizing a generic and adaptable approach: structurally, functionally, culturally and aesthetically.

THE CORE

We structured the space by means of a single interior element, which functions as 'mega-furniture' in scale to define the spatial zones around it. This element includes all the services needed for occupant comfort: In the case, which assumed residential occupancy, it includes the bathroom, kitchen, murphy bed, home office, home entertainment center and of course, the mechanical room which fosters, among other things, a well-tempered interior environment.

The configuration of this element could be adapted to different uses. For example, a conference room might have to house other items in the central element. These needs can be isolated analytically and interior design strategies developed to best organize these functions and accommodate easy use. For different programs, different optimized cores could be developed (2.18, 2.19).

The result of this design strategy is a more or less continuous, fluid space, maybe related to typical loft space living. The exact position of the center core element defines circulation within the overall space. It also contributes to shaping the interior-exterior relationship and the usable proportion of square footage for the adjacent spaces organized around it. For example, while turning the corner around the core and changing direction, the views towards the exterior were intensified.

We had tested these spatial circumstances in several models. We were also aware of the proportions between the core and the space surrounding it, which contributed to the described experience.

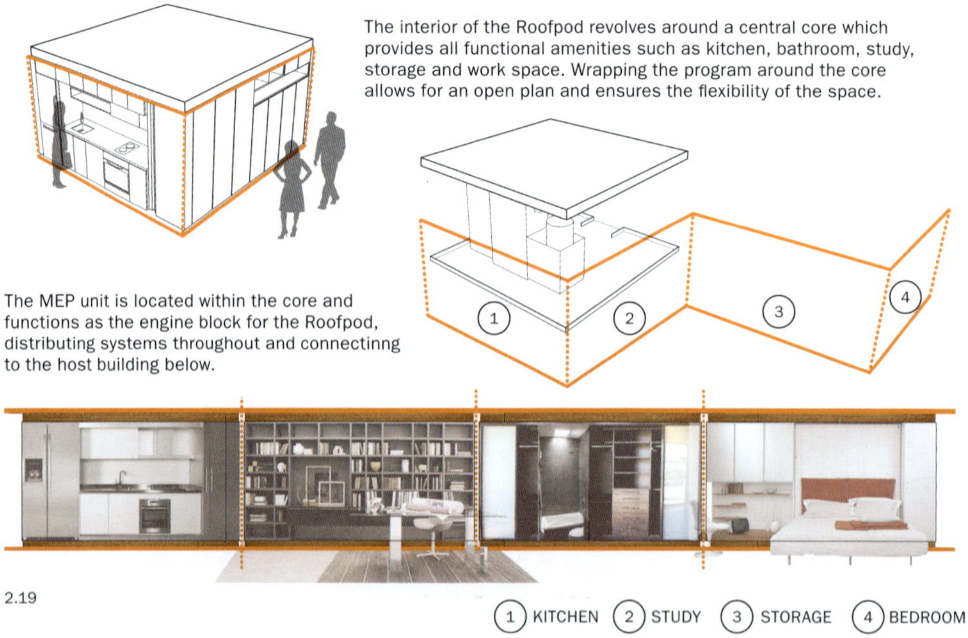

The interior of the Roofpod revolves around a central core which provides all functional amenities such as kitchen, bathroom, study, storage and work space. Wrapping the program around the core allows for an open plan and ensures the flexibility of the space.

The MEP unit is located within the core and functions as the engine block for the Roofpod, distributing systems throughout and connectinng to the host building below.

2.19

THE LOFT

The loft conversion might be a New York invention. It originated in the 60s, when many artists took possession of obsolete industrial buildings, and started converting them into spacious and functionally flexible atelier and living places. This style of living has become very popular, and lofts – especially in the areas where it started (SoHo and Tribeca) – have become a luxury item, the contemporary living space of the rich rather than the resourceful use of neglected building stock by artists.

Our space would not have the ceiling heights that the tradional loft spaces usually feature. Even though the extended height would be beautiful to experience, the additional cubic feet of air to be conditioned would add more loads to the mechanical system. We therefore decided against it.

What is usually considered a deficiency in a living space is the loft's strength: the unpartitioned, continuous space certainly does not offer much privacy. To add functional flexibility in our Roofpod design to the space around the 'mega-furniture' element, we also included full-height sliding doors in a pinwheel configuration. These doors, invisible when not in use, can be used to interrupt contiguous spaces in the interest of acoustic or physical privacy, or for climatizing only part of the interior space, while the rest of the house stays unconditioned. This would again have significant consequences for the energy balance. That is that only the occupied portion of the building requires energy, while the rest of it can be cross-ventilated, cutting the house's cooling loads in half. At night, the building can be naturally ventilated by opening front and back to create a tunnel effect around the core, to efficiently cool the bedroom in summer. These considerations would also change the typical calculations of cooling loads.

The positioning of the mega-furniture thus has spatial significance. It sets up a proportioned configuration of separable space; it establishes relationships between the house's specific areas and the outdoors; and it also can manipulate the way the occupant experiences natural cooling and ventilation.

2.20

TYPOLOGY

The pavilion's organization is the appropriate typological response for the urban rooftop as building site.

We studied several case studies that made us learn about the spatial specificities of a pavilion, but also noticed that the passive responses, which have not had much attention paid to them in recent contemporary architecture, would have great potential. Its all-sided exposure exploits ventilation patterns but also indexes the changing contextual conditions: views to skylines, shading patterns and the dynamically changing interior-exterior relationships.

It also allows for a wide range of programmatic solutions to all different kinds of uses, from residential applications to more public functions.

The pavilion's integrity is also expressed in its aesthetics. It should be identifiable as an add-on to the existing building, and has the potential to become an iconic complement to New York's skyline in many diverse contexts.

2.19 The 'core' of the building, containing the mechanical room, bath, kitchen, and an array of perimeter cabinets, especially, has the potential to customize to the needs of the program; student drawing, communications design studio.
2.20 Schematic design: Rendering of interior space, w/ the core as its main element; student drawing.

7 MATERIAL CONSIDERATIONS

The operational energy which the house requires to generate the level of comfort expected by the occupant, as briefly described in terms of air-conditioning strategies, is only part of the energy fine-tuning involved in planning the Roofpod. A significant amount of energy has already been invested in any house before it is built, in the form of the energy used to produce all the materials needed to materialize it. This energy, which is difficult to quantify, is poignantly called grey energy. In fact, the exact amount of grey energy cannot be determined definitively, since such variables as transportation means and distances, cannot be definitively traced in our global economy. Furthermore, no manufacturers would want to admit that their product is environmentally unfriendly.

Nevertheless, what is the use of building as a very energy-efficient house if the production of its materials required an energy input that will only be amortized much later, if at all? And in cases where the embodied energy does not affect operating energy, such as in interior finishes, shouldn't research on this topic be central to our considerations? Replacing materials doesn't appear very interesting when you're about to build a brand-new building, but the consequences that you (or someone else), and – inevitably – the environment will face, are directly related to the choice you as the architect make originally. For example, different window frames have different life spans, ranging between 20 and 50 years. Or roof membranes, which make up a considerable amount of construction waste, have to be replaced every 10 to 40 years, depending on the material and construction system[43].

Not only must the life cycle of each particular material be considered, but also, since every building is a hybrid assembly, the 'dis-assembility' of the whole has to be considered, too. Otherwise, when replacing the outdated items, you may have to rip out other items that are perfectly fine, just to access the broken one.

We became fascinated by buildings that can be disassembled (i.e. Jean Prouvé, Buckminster Fuller, etc.), and decided to invest in this idea. Not only would maintenance and replacement of deteriorating parts work better, but also the logistics of prefabricating and the assembly on site atop an occupied building suggest a similar approach to assembly. What are the best methods to evaluate 'grey energy', also called 'embodied energy' and multiple other factors such as longevity, energy-savings on operational energy, health aspects (some materials off-gas or drain pollutants), and other environmental consequences created by the production of any particular building material? There are several programs and institutions around that assist in obtaining this kind of data. We, for example, connected with an organization called 'Healthy Building Network' and often accessed their website to use their data. Also, we assigned student research teams to analyze the manufacturing processes of certain materials we favored. Nonetheless, these processes were often unclear, or sometimes even obscured by an industry eager to market its products. The findings are often unsatisfying. Nevertheless, architects should know about the basic production processes of building materials, and their independently evaluated environmental impact. In the end, even once an exact measurement system and nomenclature has been found, the idea of evaluating materials numerically might not be the most satisfying way of communicating that information.

“SOLAR ROOFPOD” SOLAR DECATHLON MATERIAL LIST: COMPONENTS									DATE:	
COMPONENT DESCRIPTION:			# ALT.	COMPONENT LOCATION: (COORDINATES: CENTER LINES; + ELEVATION)					COMPONENT QUANTITY:	
COMPONENT PARTS:	DIV #	QUANTITY	DIMENSIONS (DxWxL; In.)	SQ. FT. (PLANAR)	LIN. FT. (LINEAR)	SP.WEIGHT (LB/...)	WEIGHT (LB)	ENERGY (BTU/LB)	COST/... ($/UNIT)	TOT.COST ($)
REMARKS:							*)	*)		*)

2.21

2.21 Material research sheet: Analysis of weight, cost and embodied energy as a combined overview; all three factors are significant for a design-build project, particularly for a Net-Zero-Energy building located on a rooftop.

For example, if supply and demand of a certain processed building material is factored in, the impact of producing this material will most likely differ when demand is lesser or greater. Often, when a building material becomes ‘hip’, the processing infrastructure to satisfy demands may change, impacting its environmental footprint. Thus, often an exact finding cannot be made. The way architects communicate and address cultural or environmental issues is predominantly qualitative, not quantitative. A project which uses materials that are all renewable or all of recycled content, is a gesture indicative of an approach within a framework of choices. These decisions are linked in a complex manner with spatial and aesthetical intentions. Good architecture doesn’t happen by checking off ‘LEED’ schedules and collecting ‘points’. But it also does not happen if we are ignorant of the greater socio-cultural circumstances within which we move.

MATERIALIZATION STRATEGIES

We decided to apply particular strategies to each area of the project. For the interior, we concentrated on renewable, fast-growing materials such as poplar wood, bamboo paneling, cork flooring and used recycled products, such as glass and plastic paneling. These material choices were the result of a research competition that a team of students initiated (2.22). We also paid attention to the albedo of these materials, since light-colored surfaces reflect daylight better and therefore reduce the need for electrical lighting. For the Roofpod’s exterior envelope, we concentrated on performance-related features. We looked into solutions to conserve energy, to reduce heating and cooling loads, and to provide an array of operable devices that can regulate light, heat and ventilation. At the same time, we chose solutions that add sensibility to the architectural environment.

We developed our modular façade system out of poplar wood, which is relatively fast growing. However, the block system had to be protected. We identified curtain walling as being both flexible enough to achieve this and as the preferred façade technology for urban contexts. For the exterior façade materials, we thus focused on improved glass technology. We researched vacuum-insulated spandrel panels, which can improve thermal properties significantly; light redirecting panels designed to improve day-lighting; glazed elements with integrated blinds to defend against unwanted solar loads; operable units that can be motorized to optimize ventilation. We were also able to obtain newly developed ‘bird-proof’ glass, which had been coated with a pattern invisible to the human eye, but is perceived by birds’ eyes as visual noise. We also preferred glass because it is low-maintenance (except cleaning, but painting, repointing or other façade upkeep is not required); it is waterproof; and it can be recycled to its full extent. The panel’s life expectancy is ca. 25 years. It is also easy to replace the panels when necessary because the mullions can be removed to provide easy access to the panel in need of replacement. We also admired the shiny surface properties of glass, which changed under different light conditions.

For the garden around the house, we focused on economical materials. The plants lend the garden its beauty anyway, whereas the built infrastructure is only an underlying support. For the planters, we used mortar mixing tubs that fit perfectly into our modular grid. The wood framing around the tubs was built out of salvaged cypress wood, which

2.22

is nearly indestructible and plentiful since it is a swamp wood sourced in the southern East Coast States. However, its major drawback was that it is combustible. In New York, there is a 20% area limit on combustible materials on roofs[44]. Therefore, the use of the cypress had been limited to planter trims and the porches, while the walkable material surrounding the house was covered with an industrial galvanized steel plank system. The combination of industrial systems with custom-built was not only aesthetically interesting, – both materials usually look better in this combination – it also makes good use of resources. Where standard solutions are available, they can reduce costs; where these solutions do not suit specific requirements, the architect is in charge of providing a simple and elegant design-build solution. While developing this approach, we often considered weight to be one of the most significant criteria: The lighter the building, the easier it is to assemble its elements, and to bring the elements to the roof. Also, the lighter the materials that are used, the less structure is necessary to support the building and its garden. The lighter structure would reduce the weight and save costs. It would also increase the number of possible sites, since there would be more host buildings able to support extra loads.

However, although the strategy of lightweighting was convincing in theory, when preparing for the construction budget and material specifications, we soon realized that weight savings do not necessarily translate into cost savings. When we investigated, for example, using aluminum instead of steel for certain structural elements, we found that the extra cost would be close to four times what we would pay for steel. Even prefabricated systems, such as the space frame we finally specified for the roof trellis – highly efficient load transfer with a minimum of material – was too expensive to be used to a greater extent, for example to configure the load transfer between the Roofpod and the host building. Even castellated steel beams, which save weight by reducing steel in the web area, were beyond our price range. However, this search for affordable alternatives made us aware of how we might define the market segment and price range of the Roofpod as a product. If the materialization were more strategic and used lightweight systems, it would be truer to the Roofpod's purpose of being placed on top of buildings. At the same time, however, the magnitude of market penetration would diminish because of the higher price.

Balancing time, costs and technology is, however, always a key task in developing a product. Another conflict arises from the fact that the strategy of light-weighting directly impacts the embodied energy in the materials' production. Usually, more refined solutions, which might favor high-tech aesthetics, tend to involve materials with a high content of embodied energy. Aluminum, for example, is among the building materials with the highest embodied energy content of all. Depending on where it was manufactured, its embodied energy content versus that of steel is 10:1. However, if recycled aluminum were used instead of virgin material, the ratio would change dramatically, to about 1:1. In other words, the amount of energy used to produce beams made of recycled aluminum is the same as would be needed to produce beams of recycled steel[45]. Weight savings of 75% would be associated with the use of aluminum. However, we could not find the recycled aluminum beams we needed for

2.23

2.22 Student research competition on sustainable materials.
2.23 Samples obtained by students during the research process. Direct contact with the intended materials is essential to control the imagined design.

our structural framing. We probably would not have been able to afford them either. But if the Roofpod were mass-produced, and the need for sufficient material supply established, industrial manufacture would become more feasible, if not even a business opportunity.

It is apparent that many factors influence the way we specify materials: the economics of supply and demand, the technology used to process products, the amount of recycled content and other environmental considerations, the bulk weight that defines transportation needs, the fuel necessary for processing and logistics, the values associated with popular items, and aspects related to health and well-being – all of these factors can change, and with them, the way we evaluate what we specify. In this respect, material research requires negotiation techniques between all the factors that apply in order to build an inexpensive, beautiful, healthy and environmentally friendly house. Even if the optimal infrastructure to produce certain materials is theoretically known but has not been put in place, there is no value to these considerations from a practical point of view. Often, the processes needed to recycle materials are known, but are not put into practice due to economics. Many building materials that could be recycled still end up in landfills.

There are many ways to improve upon this situation, ranging from entrepreneurship to political agitation for policies that lead to a more responsible use of our finite resources. Stronger socio-political environmental agendas have to be put forward, to build consensus among industry and consumers.

One of the main factors, often overlooked, arises from our sense of being entitled to new things in all shapes and sizes. Usable buildings are demolished before the end of their life span, driven by marketing and consumerism. It is unfortunate that this way of life is still commonly accepted, since the resources used to build buildings must be seen as investments, which are given their value only by sacrificing land, fuel and labor. Only when the investment is returned in the form of regenerated resources does the cycle close and become sustainable. However, we tend instead to exploit all available assets. In the recent past, since the Industrial Revolution began, we have used an unheard amount of anything marketable. The primary objective of modern life is economic growth. There are incentives we give ourselves, related to comfort; because it sells. But, there are no incentives for forced conservation.

The architect is directly involved with a project's longevity, but often at a different scale than expected. It is likely not going to be the grandiose conceptual idea that secures long-term survival of an architectural work, but rather the way it is executed in detail. Details are decisive for the aesthetic quality we mainly perceive. The way smaller motifs are integrated into the greater whole is a challenge in the age of standardization. It is equally decisive that these details are executed to age well and to sustain themselves beautifully over a long period of time. The longer the building can look fresh, the higher the likelihood that people will be interested in preserving and maintaining it. Detailing should encourage and facilitate good upkeep. In the end, materialization and material conservation should also cultivate a social perspective.

2.24

SUSTAINABLE AIR CONDITIONING:
Cooling is provided through heat transfer instead of electricity, which lowers operating costs and emits no greenhouse gases and ozone-depleting refrigerants.

COOLER CITIES:
The hard surfaces of the city amplify heat. Green rooftops reduce this "heat island" effect and mean using less energy for cooling.

MASS TRANSIT:
By using mass transportation to get around the city, Roofpod residents reduce their greenhouse gas emissions in a year by 92 pounds.

DECREASE RUN-OFF:
Stormwater is a major contributor to waterway pollution. Roof gardens retain rainwater and reduce stress on the municipal sewer system.

SAVE ENERGY DOLLARS:
A single Roofpod generates 11,600 Kilowatt Hours per year to save about $2,300 dollars in electric bills.

DECENTRALIZED ECO-GENERATION:
Roofpods form a decentralized power grid generating extra energy for the host building and reducing total demand from the city's conventional infrastructure.

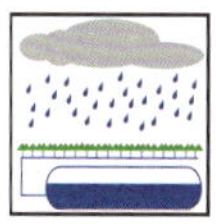

RECYCLE WATER:
The green roof is irrigated with greywater and stormwater. A single Roofpod recycles about 30,000 gallons of water each year, saving 30% on normal water usage.

LOCAL FOOD:
The Roofpod's dunnage garden produces about 190 pounds of vegetables, about half of the residents' annual consumption.

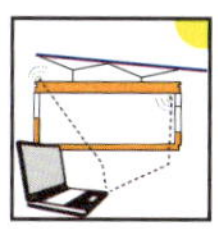

SMART HOUSE, SMART USER:
Meters and sensors communicate to the user information on the Roofpod's real-time energy performance and calibrate its various systems' response to diverse environmental conditions.

DESIGN-BUILD:
The Roofpod was designed to be built by students, to promote imaginative energy-efficient design. Students designed all aspects and constructed every inch of their Roofpod.

8

CODE CONSIDERATIONS AND FEASIBILITY

People generally do want to be environmentally responsible; they are often just not aware of how issues are interconnected. Political outreach is needed to develop the appropriate awareness. However, looking at the bigger picture, the main obstacle to our engagement in more sustainable building practices is our dependence upon economic growth. Usually, there is a reciprocal relationship between economic and ecological conditions. When the economy is strong, people have money. With more money, more goods are produced and consumed, more miles are driven, more waste and pollution occur. People are happier during economic surges since they have something to do. Government is happier, too, since its income from taxes is higher. But waste and pollution are not so positive for the environment. We have not yet figured out how to transform this relationship so that economic and ecological conditions can improve simultaneously. What are the fields of action in which economies and environment can both thrive? The architect is an example. The more buildings can be built, the busier architects will be, but the more resources are put into circulation, the more energy and fuel is consumed, the more waste and pollution result.

It has often been said though that the bulk of new construction was realized during the 20th century, while the 21st century will focus on upgrading, retrofitting and improving buildings. That will change the architect's professional profile: now one has to dedicate time to polishing up buildings invented by someone else some time ago. Is there enough creativity and intellectual reward in this occupation? Any retrofitted building will be more sustainable than new construction.

The question is how to define the problem creatively. Retrofitting is not just wrapping insulation around a building, but developing a creative ambience that makes people see the world differently. This should also be possible with an upgrade of a building, since in fact any architectural intervention is a contextual problem. That is, to look for unrealized potential at a given situation. If so, what are the primary spaces to be conquered, those which have not yet been fully discovered? Where are the spaces in the existing system that could contribute to an improved infrastructure and efficiency, in turn protecting the environment and simultaneously keeping us contented and employed? We discovered the roof as one such place. We realized that the roofscape of our cities is prime real estate, but its potential has been thus far been largely neglected. These areas also receive the highest environmental loads, which nowadays – and more so in the future – translates into marketable energy and environmental improvement. It would be advantageous to start utilizing our urban roofs to improve our city's efficiency.

IMPLICATIONS

The economic advantages of our project are related to improving real estate as well as developing clean energy sources within our urban environment. Both topics are vital for New York's future. The implications of our project are clear, since our proposal both creates a marketplace for a new product and also improves the way we power our cities.

A proposal such as this on a city's agenda would spur economic growth. In Germany in the 1990s, for instance, the economic landscape was transformed by creating green jobs, supported by governmental programs. The impact of these programs determines whether the jobs created are economically sustainable and will stay within the area where they actually improve conditions, or whether the advancement of technology improves a product that can be mass-produced and marketed in a way that it can spread relatively easy. The critique that solar energy solutions are not competitive enough is actually related to what we can install.

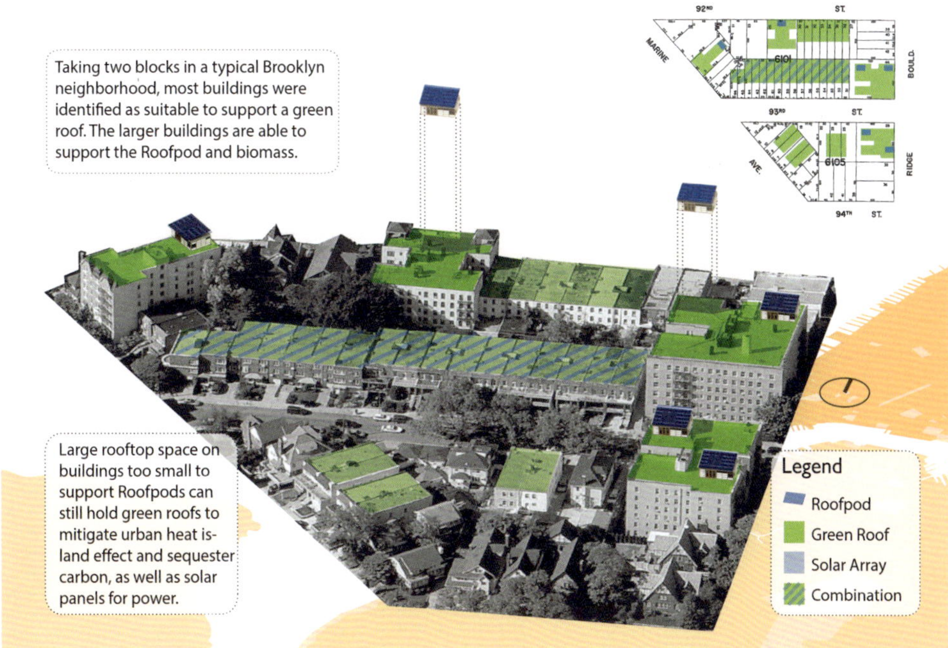

2.25

2.24 (previous spread) Site advantages for the Solar Roofpod, and its related potential for improving urban infrastructure; student drawing, communications design.
2.25 Hypothetical real-estate study in Bay Ridge, Brooklyn: Research on suitability of different rooftop locations and sizes. Most roofs have the potential of being upgradable, either with a Roofpod or at least a green roof system. Implications for the local building economy are promising. Student drawing, communications design.

If the right programs are facilitated, they can make the difference to the economy and ecology of the future. Furthermore, they will have an impact on people's mindsets and cultural adaptation.

LABOR

The Roofpod has the potential to contribute to a localized labor market. Many occupations involved in its assembly are site-specific and cannot be outsourced. In particular, the installation of renewable energy systems, both photovoltaics and solar collectors, would create a market sector that promises growth, especially if the right incentives are given. These incentives are almost mandatory for this development, and are not unusual in the development of new markets. They are justified in order to lessen the emissions impact on conventional energy generation, and to develop a device that both measures the energy intake from the sun and monitors the energy outtake by the host building below. It is clear that the existing building stock, which is currently responsible for a good three-quarter of New York's greenhouse gas emissions, can dramatically improve.

ENERGY POLICIES

The most significant energy draw in most buildings is the roof; especially in dense environments, where the demising walls are often protected by neighboring buildings, the roof is the main contributor to thermal losses. These losses are intensified even more by the convective nature of air movement towards the top. Most of New York's building stock does not comply with current Energy Codes put forward for new construction in order to improve the thermal performance of buildings. To focus on the existing buildings' roofs would not only yield economic advantages, but would also reduce heating and cooling loads, and consequently reduce emissions from powering these systems. While in Europe the thermal improvement of buildings is often mandatory, the enforcement of regulations in the US is not as stringent. New York's City Planning Department just released a proposal to encourage sustainable upgrades, with the goal of achieving high performance standards for the existing building stock (and lower the estimated $15 billion per year it takes to power and heat the one million buildings of New York City). But this proposal depends basically on voluntary effort. The rewards of, for instance, insulating a building's exterior envelope within the property lines is that the area taken by the insulation is exempt from floor area ratio (FAR) density requirements. This does not seem an adequately tempting incentive to getting active, though. In Germany, for example, a building's owner is required to obtain an energy pass when a building changes ownership. This document lists the total energy in kilowatts that the building is expected to consume per year (kWh/m^2a). It is a record of how well the building is doing in relationship to the environmental challenges. There has already been political discussion about whether property taxes should be adjusted to this stated energy value.

The approaches are quite different. While in Europe political pressure is applied on the owners of buildings to adjust to high performance standards, our approach to increasing the efficiency of buildings' envelopes is instead voluntary. A debate about the advantages and disadvantages of these different approaches would probably remain undecided, but the objective must be to move things forward. In our studies, we did not focus on the politics of implementing higher performance standards since our direct influence was limited in that area. Rather, we were interested in setting out the architectural problem: how could retrofitting be articulated in a way that would also support measures for improved performance. We assumed that policy change could only be addressed if a concrete solution were presented. The underlying thesis is that socio-cultural responses are more likely to change if there is access to direct sensual experience that communicates the meaning of our topic. We were hoping that if we were able to present a convincing solution, we would attract the interest of developers and policy makers.

We discovered that there were many different scales involved in improving the performance of the city, the host building, and the roof pavilion itself, in terms of active and passive measures. The solar-powered equipment, designed to decentralize power supply by providing extra power from 'top-down,' is the primary active component. The proposal, however, complements this supply with a series of passive approaches, improving performance, none of them as high-tech as a solar panel array, but effective nonetheless, usually with less capital investment.

One of the obvious measures, especially when penetrating the roof membrane to install the dunnage connection to the pavilion, would be to insulate the roof of the host building sufficiently. Most existing roofs are not yet sufficiently insulated against heat loss, and would have to be updated anyway. This intervention would be far more effective than the city's 'cool roof' initiative, which only involves increasing the albedo on the roof surface. Additional insulation would instead block the transfer of thermal energy from the building's interior, through all seasons, and with a relatively quick break-even point. The light-colored surfaces supported by the 'cool roof' initiative only manipulate the roof's surface properties of reflectance and emittance, sometimes, i.e. in winter, even to its disadvantage. The insulated roof saves money and avoids CO_2 emissions at an entirely different magnitude to the 'cool roof' initiative; it is much more effective.

An additional green roof, which means another layer of protection against UV radiation, will also protect the roof membrane and prolong its life to a greater extent than any 'cool roof' could. Besides that, in urban environments, light-colored materials soil quickly anyway. Without proper maintenance to keep the material clean, the energy savings created by the lighter-colored surface will diminish over time as albedo degrades and soils. This example demonstrates that one needs to differentiate between catchphrases of greenwashing, and initiatives that have real potential to transform the way we retrofit our buildings. Again, surveys must be conducted to create concrete strategies and policies.

ZONING

It was clear to us that we could not change the current building codes and zoning regulations, but that there is leverage to adjust these regulations over time.

For this reason, we tried to understand, how we could best work backwards: our strategy was to stay close to current regulations, but to indicate the advantages of changes to whatever requirements impeded the ingenuity of our approach. Our thinking was that if the city can adjust building mass and height (FAR) to allow additional square footage for example for stores selling fresh food in neighborhoods with a high incidence of obesity, then there must be similar mechanisms to facilitate additional square footage for powering our city with renewable energy. Increased density in the right places would provide more efficiency. It would improve the quality of life and the environmental outlook.

An urban environment is, of course, a complex organism: building density and ownership conflicts create a challenging physical and legal environment. Over the past century, the equality intended by zoning regulations has deteriorated. Air-right trading and other less controllable practices have created very heterogeneous environments in some parts of the city, creating shading patterns that are disadvantageous to solar energy generation. However, most of New York City is still quite homogeneous, especially the residential neighborhoods which were built earlier under less capital pressure than the city's commercial sectors.

The more homogeneous areas in Manhattan can be found north of 57th street, where Central Park and the residential quarters begin. For our strategy, which at least in part relies on widespread application of photovoltaic energy generation, equal access to sunlight is obligatory and would in the future have to be better protected by zoning regulations. We chose to operate in areas with relatively even building heights, such as the Upper Westside of Manhattan, and Harlem, where our school, City College of New York, is located. Most of the residential buildings in these areas have 5 to 10 storeys, with additional FAR available.

HOST BUILDING COMPATIBILITY

We looked at several building typologies and construction types characteristics of apartment buildings. Construction types are defined by building codes and are one of the deciding factors for the building's maximum bulk. We understood that if we want to complement a building with our Roofpod, its construction must be compatible with that of the building beneath it. Construction types protect public safety interests, focused on fire resistance and structural frame. Since most buildings in New York are non-combustible, most of their materials are fire-rated. The Roofpod itself must thus comply with this standard. We decided to realize the Roofpod's frame in steel, which can resist collapse caused by fire better than wood. It facilitates compliance with a number of construction types. The steel frame guarantees that this addition can be placed on thousands of buildings with similar fire-ratings.

BULK / DENSITY

Massing of buildings in urban environments is regulated by Floor Area Ratio (FAR). FAR is the total square footage of a building divided by the total square footage of the lot on which the building is located. Usually, developers try to maximize the square footage to the last square inch in order to maximize return on their investment. But, this practice has not always been so intense. Buildings built in the prewar period in particular often do not reach the limit defined by local ordinances. Over time, too, mechanical systems have been updated and take up less space than originally allocated. Mechanical rooms do not, however, count towards FAR.

There are countless buildings with potential additional square footage. We did not have the opportunity to verify exactly how many buildings could currently be considered, but we also chose not to limit ourselves on that basis, anticipating solar energy-motivated rezoning. The city could also adjust FAR to support the public interest by using the so-called density bonus to permit Roofpod construction. A plan to power our city with less CO_2 emissions, to reduce the heat island effect, to revegetate, and to provide mixed use opportunities are all facets of this concept; the Roofpod deserves the additional new floor area required.

Incentive zoning can only be adopted after considerable dialogue with communities and developers. We have talked to community boards and developers, and both groups considered the Roofpod idea worth promoting. It was recommended to us that we define where the implementation of this concept would be most effective. Zoning incentives are often tied to overlay districts, or mapped areas in which a specific set of additional regulations supersede the usual zoning classification regulations. Success with the implementation of overlay districts is heavily dependent on identifying those existing conditions that can foster favorable development.

EGRESS

A critical challenge is that often the roofs that could serve as a building site do not have the infrastructure for legal occupancy. The law requires that any space that can be occupied must have two ways to escape in case of a fire; this is called means of egress. While a single staircase extending to the roof is quite common, this condition is not adequate to meet current codes. This deficiency might immediately eliminate some of the functions suggested for the Roofpod, especially the ones with high traffic, but it is worth considering whether one staircase could be sufficient, or how a second escape route could be generated. There are examples of higher traffic buildings, such as some schools, which are already permitted to use their roof spaces as part of their everyday operation, although there is only one means of roof egress. One argument in favor of this practice could be that the flow of people from the top floor is considerably less than on lower floors. If this is not acceptable to fire departments and code regulators, then guidelines on types of function and occupancy could define how to move forward.

There is another alternative that conforms to the code as it now exists: if the Roofpod is simply an extension of the host building's top-floor apartment, connected to it by an interior stair, the egress rule for the upper floor no longer applies. The resulting duplex is already tied to the two means of egress on its lower floor, which satisfies the current building codes. There are several conceptual arguments against this solution, however. First of all, the Roofpod, despite appearing to be an independent pavilion, would actually only be an annex of the apartment below. This approach is aesthetically misleading, since it creates a form that belies its content. More importantly, we dreamt of creating a space that would be publically accessible and engage public participation. We prioritize public access because it lends the Roofpod the ability to influence the public's perception and emotional investment in thinking about how we run our cities. This is not only necessary to influence long-term change, but also in recent years, the power of architectural environments to influence social practice has been increasingly acknowledged. In our approach, architecture is not seen only as a formal exercise: it has to consider and reflect the people who are using it. Architecture is assigned the role of an agent of socio-cultural change.

Therefore, in order to implement this project's crucial social component, there are two ways forward, one weak and one strong. Either the costs of extending a second staircase are calculated into the budget, which would be probably so costly that half of all potential customers would be lost; or the regulatory obstacles must be addressed by finding a code solution that defines the maximum of people that can be served safely by one staircase.

ACCESSIBILITY (ADA)

When President George H. W. Bush signed the Americans with Disabilities Act (ADA) in 1990, his intention was to insure that no American would ever again be deprived of access to "life, liberty, and the pursuit of happiness." But he had no idea what problems he would cause for our project. The 1990 standard states that no individual may be discriminated against on the basis of disability in seeking access to goods, services, facilities, or other offerings of any place of public accommodation by any person who owns, or leases that place. Public accommodations include most places of lodging, recreation, transportation, education, and dining, along with stores, care providers, and places of public displays, etc. As expected, similarly to the problems arising from the means of egress issue, the programs possible would diminish greatly if accessibility rules were to dictate them: in order to conform, an elevator would need to access the roof. However, this is not the case on most New York City roofs. The elevator bulkheads common in rooftop panoramas only house the hoisting engine above the cabin, and do not offer the means to ride the elevator all the way up to the rooftop. To extend an elevator costs several tens of thousands of dollars, a cost which does not positively influence the implementation of our idea. It again separates the project's potential uses into two different groups: public accommodation and private dwelling. While the former is legally obliged to be handicapped accessible, the latter, according to our interpretation, needs only to meet the standard of handicapped adaptability. The difference is that the adaptability must not be implemented immediately, but has only to offer proof that it can be upgraded to the legal standard at a later date.

Therefore, the group of residential dwelling is, at least initially, the preferred target market to establish and proliferate the concept. At the same time, we hope that enough semi-private events will take place in those Roofpods as are needed to cultivate people's greater awareness of how attractive it is to enjoy time within New York's roofscape. As a second step, perhaps it will become worth the investment to make public functions with true handicapped accessible access possible.

LIGHTWEIGHTING

When looking for ways to reduce the weight added by the pod to the building's roof, one frequent consideration was the choice of the most suitable structural material. Actually, according to the code, there are not as many options as one would imagine. Lightweight materials, common in many other fields, cannot be used in architectural applications because there are not yet appropriate testing certificates, whose generation is a complex endeavor that can take decades. This class of materials includes fiber-reinforced materials, Kevlar, glass fiber, and carbon fiber, most of which are usually bound with different sorts of resin. These resins are, like wood, combustible, so although they might have a superior strength-to-weight ratio, they do not comply to code fire-ratings. There are also new developments in fiber-reinforced concrete, which can be cast or extruded but neither are these materials certified for structural applications yet.

The point is, however, that material research is constantly ongoing. The decisions we made are not just aesthetic, but based on legal and economic constraints, reflected in our choice of structural steel. The development of a lightweight prototype needs both time and powerful partners in the building industry to achieve approval for a product that is slimmer and more efficient. The idea of a mass-produced pavilion – similar to car prototyping – certainly would help to develop high-tech lightweight alternatives.

ROOF GARDEN / IRRIGATION

The idea of having a beautiful mini-park on the roof requires thought about irrigation strategies, too. While excess water from precipitation can favorably be absorbed by the planting substrate, the need for water in dry seasons is more problematic, for financial reasons, too. The city only gives green roof tax credit incentives to roof gardens that prohibit irrigation. The assumption is that irrigation would make the roof garden less able to retain stormwater. However, in order to create a garden beautiful enough to attract people, either the irrigation policy needs to be reconsidered or the amount of moisture-absorbing matter has to be increased. The increase of moisture-retaining matter, however, contradicts the goal of making the structure lightweight.

Scientific projects must address the quantification of the environmental benefits of irrigated roofs in order to reevaluate the city's limitations on its incentive. This would also open the door to urban farming with the potential to reduce carbon footprints and produce healthier nutrition. The technical advancement of grey water catchment and irrigation control could significantly contribute to the goal of only irrigating with grey water, so that green roof irrigation restrictions could be rendered obsolete.

9

'SMART HOUSE' —TECHNOLOGY

COMMUNICATION

The contemporary house has the capacity to communicate the efficiency of its operations to its users. We can learn how this works from automobiles, which are ahead of the curve in comparison to buildings. Many current car models have some sort of a dashboard computer, which tells the driver, among other things, the car's fuel efficiency, how much fuel is available, and how far the driver can still go under the current circumstances. In a house, although we use houses more intensively and frequently than cars, we are blind: comparable data on energy consumption is seldom available.

MONITORING + CONTROL FUNCTIONS

The need for real time monitoring and communication is fundamental to propagating greater energy efficiency. When receiving electricity bills after a month, one will not be able to adjust spontaneously to the occurring energy consumption. We should know more, and have easier access to data, about the state of the systems running the house: their consumption, and especially in our case, their generation of energy. However, nobody wants a dashboard at home. It is not very relaxing to be confronted constantly with this data. The aesthetics of how this information is communicated thus also requires further definition. Most current approaches are heavily based on the model of a computer interface. If that is the future, computer displays need to be better integrated into our everyday surroundings without impeding our need to relax at home. One discreet and obvious solution can be found in the growing number of apps for smart phones.

Monitoring and even remote systems control via phone is possible and relatively easy. However, electronics of this sort offer little benefit to architectural expression, nor do they have any presence within the space. Instead, we were looking for something analogous to the old-fashioned counterweighted 'grandfather clock', a common accessory in our great grandparents' living rooms to indicate the time; not only an accessory, but a tangible part of a house's interior life. In those times, the clock was seen as a cultured item.

Of course we weren't looking for anything old-fashioned, but rather for a device that could have a similar presence, as part of the furnishings or interior fit-out. We experimented with color-morph materials, which change color when they receive a small electronic impulse. We chose to integrate these indicators into everyday activities, for example, into the kitchen backsplash and the bathroom mirror, in order to communicate current power and water conditions. These indicators were part of our monitoring and control system, which had background computer control.

Beyond tracking and communicating water and energy, the system also operated the blinds to shade the windows, the heating and cooling systems, the irrigation of the roof landscape, electric lighting, the collection of solar power and the inhabitants' overall energy consumption. The program was designed to analyze these systems' diverse impacts holistically (2.26) in order to operate the house with the minimum of energy while maximizing comfort. Due to the extensive data volume required, a contemporary house – like a car – has to be computerized.

PHOTOVOLTAIC SYSTEM

It was particularly important that the photovoltaic array, a system component that is usually viewed critically because of the slow return on investment (ROI), albeit under conservative calculations, had to function optimally. In reality, ROI analysis may not be the greatest influence on consumer uptake, since many consumers are less swayed by a pessimistic investment evaluation and just want to feel good about their environmental footprint.

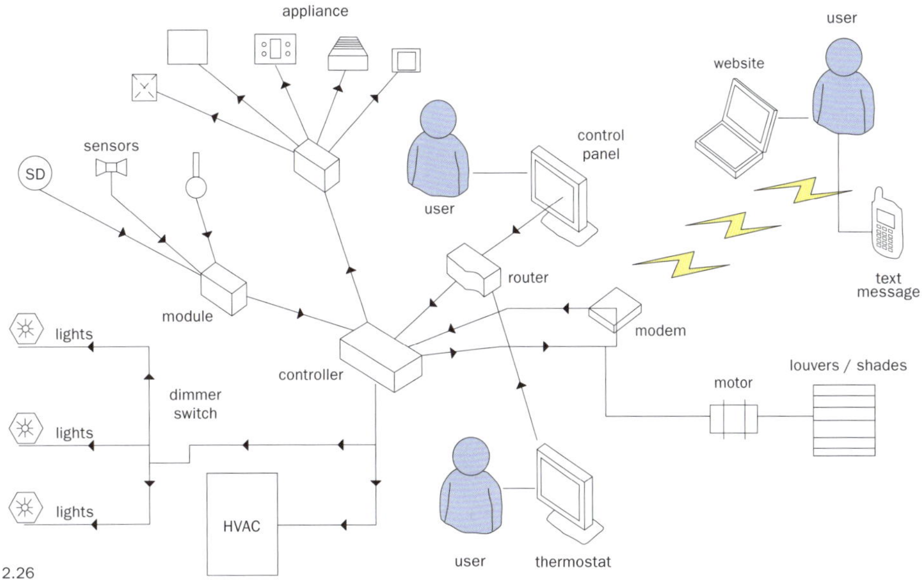

2.26

2.26 Electronic 'Smart House' concept: The performance of all energy related functions of the house are monitored by sensors, and optimized by a control system, which at any time can be accessed by the user, to become familiar with the houses consumption; schematic student drawing.

Although these emotionally motivated desires are commendable, they are often not rewarded by the agencies that promote the transition to renewable energies. The New York State Energy Research and Development Authority (NYSERDA), for instance, only approves incentives for these arrays if their angle to the sun promises optimal power generation. For example, if the house you intend to retrofit, which might have been built decades ago, unfortunately has an unfavorable roof orientation, the State of New York will not reward your step towards producing emission-free power.

The message is clear: the money spent has to have respectable yields. We developed several design moves to insure optimal PV yield.

Flexibility of the trellis: The original idea of the trellis was to facilitate the retrofit of the system as technology progresses, and also to keep it well-ventilated, which improves the performance of the solar-cells by lowering their operating temperature. At its base, we also integrated adjustable jacks that could be used to redirect the trellis in different directions and angles.

Because of the diverse requirements of orientation, street patterns and specific roof conditions, this feature makes it possible to fine-tune the trellis' orientation to the sun. Although initially considered, this system was not intended to track the sun dynamically by moving the panels, as has been tried in other projects before. Instead, the significant effort to move the panels as well as the additional maintenance necessary for moveable parts convinced us to keep it simple.

We also had to consider wind loads, which are much higher on the top of buildings compared to their bases. Actually, our trellis had to stay relatively shallow so that the wind loads, with wind gusts up to 100 m/h, would not catch the trellis like a sail. The trellis became symbolic of the whole intervention, especially its solar-power attributes. The technological flexibility of the photovoltaic installation derives from a relatively new approach to interconnecting the panels. Whereas conventionally installations would have had a series of panels connected to each other before the wiring went to an inverter, transforming the voltage-to-grid connections, our arrays took advantage of micro-inverters, one of which was dedicated to each of the 40 photovoltaic panels. The advantage is reduced power loss. Low voltage is susceptible to higher transmission losses; by avoiding these losses, efficiency can be increased by 10 to 15%. In complex shading pattern situations, such as those in urban environments where certain panels would be shaded from time to time while others would be exposed to the sun, a conventionally wired array would cause all the panels in the same series of a shaded panel to lose efficiency. With a micro-inverter network, only the shaded panel would be affected. The advantages of this technological innovation are convincing. Using 40 panels, we were able to produce roughly 10 kW, which is usually good enough to power about three to four mid-sized households, depending on people's use patterns.

That means that the surplus not needed for our modest 800 s.f. Roofpod would go right into the host building's power supply to provide extra solar power, for example for corridor lighting, mechanical equipment, the laundry room and other functions.

SOLAR COOLING AND HEATING

While this achievement translates into greater clean energy produced self-sufficiently on site, we discovered that a much more efficient solution can be achieved for certain items that are particular drains on power.

It is important to know that the efficiency of photovoltaic panels is not great. Only about 20% of the energy from the sun can be transformed into electrical power by the highest-yielding photovoltaic panels on the market[46]. However, the transformation from thermal energy into electrical power is not necessary for heating requirements, including both space heating and domestic hot water. It is less well known that thermal energy can also directly generate cooling mainly through evaporation and condensation using hot water as the driver, and in our case, silica gel adsorbers as refrigerants. The process is called adsorptive cooling. Taking advantage of these thermodynamic processes promises much higher efficiency than the conversion of solar energy into electrical power. While solar cells have an efficiency of up to 20%, the efficiency of thermal collectors, producing heat of about 180°F, can be four times that[47]. It is therefore clear that whatever equipment can be run directly with hot water (2.27) rather than electricity will be more energy-efficient. Furthermore, any surplus of this production can be used to serve the hot water needs of the host building, and provide clean, self-sufficiently produced energy.

One of the greatest benefits derived from this project is the ability to control the performance of the host building and, by extension, the city's existing building stock.

When most of our buildings were built, control mechanisms were primitive, which has led to obsolete energy operations. We envisioned the 'Roofpod' as a sort of smoke-stack-information-filter, which could sense and analyze the host building's performance, and subsequently improve its operation. When asked who the first 'honorary' inhabitant of our Roofpod should be, we decided to make the building manager the keeper of the castle (2.17). He or she would bear the responsibility of transitioning the existing building by becoming the main observer and analyst of necessary changes. We as architects, always interested in expanding our knowledge of analytic skills, have to take on a similar role, understanding our responsibility, and anticipating building management and performance.

STRATEGY CONCLUSIONS

Clearly, any successful approach to urban renewal requires knowledge accrued in many different fields. This is because the inherent nature of cities is complex and full of different scales of operation. Economic and legal constraints create ectoscopic factors that contribute to the logic of solving the problem, even if they are not part of the original conceptual thread of architectural thought.

We know that living in dense urban environments is one of the key paths to sustainability, since its potential to offer an infrastructure for many, not just for a few, means less resource intensity to sustain everyday life.

Architects have yet to discover that they are responsible for entire cities, not just buildings, at all scales from the small detail to the big picture. This means that architects have to adopt a methodology of problem-solving in which a concept is not developed from one single strong viewpoint alone, but from an amalgamation of viewpoints that has the power to contribute significantly to problem solving. In our Solar Roofpod project, these viewpoints – occupying and harvesting from the roof; relying on renewable, decentralized and self-sufficient energy production; considering the existing urban infrastructure as part of the project; providing an anticipa tory system of modular construction and different programmatic applications; integrating a green roof strategy; using environmentally-friendly

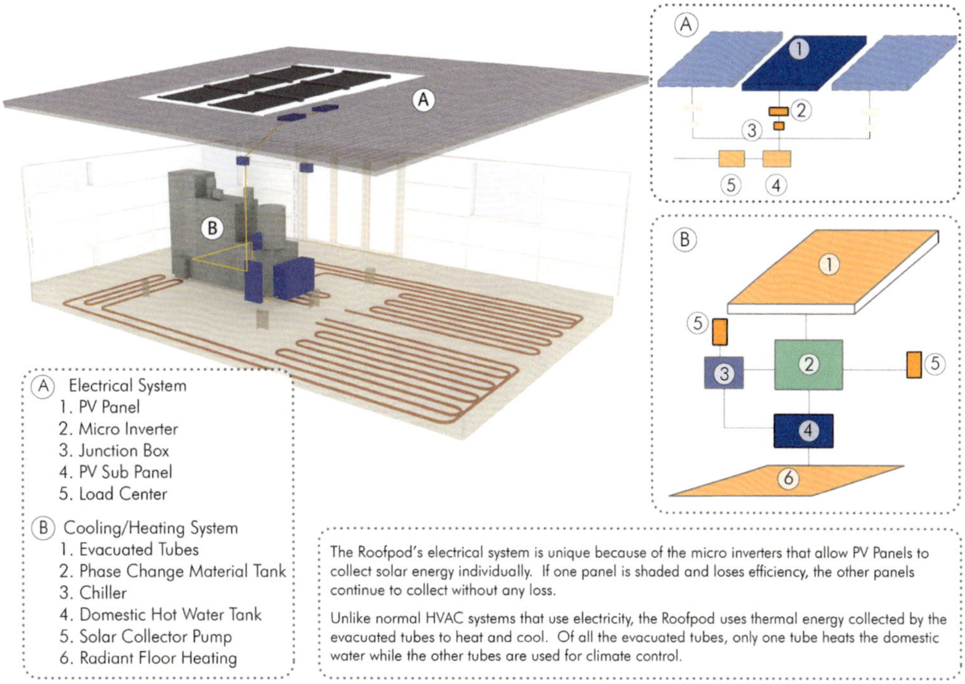

2.27

2.27: Solar cooling and heating system: While photovoltaic power generation has an efficiency of less than 20%, cooling with heat can develop a much higher capacity. Student drawing, communications design.

materials and operation modes; and utilizing computer technology to monitor and control operations – are all strategies that are instrumental to distilling a reasonable approach which considers aesthetic, practical, legal, technical, cultural and fiscal responsibility.

The architect's job is to put things in order, and find their appropriate place in the configuration of the whole. Often, topics that seem to dictate an approach intellectually are not as expressive visually or performatively, and sometimes other items that did not seem as important become important features in the developing planning process. The architect has to direct this process.

In that sense, the architect's role is similar to a product designer's: to develop an attractive product, which is favorably adapted to the environment and practical for the user. In order to design this product, the architect has to be intellectually literate but at the same time savvy and crafty. He or she has to be what Colin Rowe called a "bricoleur", with the ability to make use of what is at hand and combine it masterfully. Simultaneously the architect has to be something of a scientist, who knows how to achieve the best performance results, and how to fine-tune the functionality of the design product.

Beyond the realm of the designed product itself, the architect has to be able to have a broad understanding of the environment in which he or she is building, and the preferences of the users for whom the design is intended. Even more broadly, politics, law, philosophy, psychology and sociology are all involved in the process of design.

The environment is not limited to site conditions and climatic influences, but encompasses all conditions which impact the project extrinsically. In that sense, it also comprises the infrastructure of water, gas, electricity and other media connected to the city. It includes typical – and extraordinary – manufacturing and assembly methods as well.

Over the course of the past decades, the architect has often been depicted as a sole form-maker, but this is far too superficial. The architect should undoubtedly have the skills to create beautiful things, but also, to fit them into a physical and intellectual context. Without this context, things become profane, uninspired and meaningless.

What the architect seems to have lost is the ability to materialize the projects he or she dreams. While the project is untouchable in its drawn form, control over a design is often lost as soon as the design enters into the conditions of real life.

2.28

Contractors, developers and mechanical, electrical and structural engineers take over and sometimes dictate, to the detriment of the product, and the outcome of the built artifact.

However, what excites us is the beauty and elegance of something that works!

In order to gain better control of our architectural design, we have to relearn how to be master builders, able to choose proper materials and methods with confidence and care, and amalgamate them with the complex extrinsic framework that determines our lives. The expected experiential qualities can only be found, if the realization processes were controlled adequately beforehand. This project was a quest for a superior product, but even more so for a superior process of design and for an improved skillset for us as its designers.

Not least of all, this project has been an educational project. It hopes to develop awareness for the permanent learning progress, which the comprehensive process of prototyping architecture demands of us.

2.28 'Solar Roofpod': Rendering of a still uncommon life style, taking advantage of the forgotten, but architecturally tangible urban spaces. What skills are necessary to materialize the imagined?

COMPREHENSIVE INFLUENCES ON ARCHITECTURAL PROTOTYPING

ENVIRONMENTAL INTEGRATION

Climate analysis, weather data, hours of sunshine, orientation, wind patterns + speeds, thermodynamic laws, sensible + latent heat, ambient temperature, peak temperatures, day-night differential, seasonal differential, moisture, precipitation patterns, snowfall, condensation, daylight hours, daylight intensity, UV impact, albedo, soil moisture saturation, thermal mass, passive gain strategies, longevity, optimized integration of renewables, heat sinks, storage; environmental impact of production methods...

URBAN INFRASTRUCTURE

Power: grid distribution, interconnection, power generation, peak loads; carbon dioxide emissions per sector. Traffic: public transportation, car traffic, walkability, vehicle sharing. Heating: heat generation methods, district heating, heating losses. Cooling: cool generation, heat island mitigation. Waste management: collection methods, processing, recycling. Water: fresh water infrastructure, distribution, storage. Waste water: sewer / canalization distribution, water treatment, stormwater management...

ECONOMIC CONDITION

Return on investment (ROI), financing, mortgage market, interest rates, construction budget, break-even points, land costs, building costs, property taxes, sales taxes, subsidies and tax rebates; value engineering. Cost analysis: design alternatives, scope reduction; initial costs – operational costs comparison, maintenance + repairs, design anticipation, longevity; material costs – labor costs; budget-related detailing + specifications; resale value, real estate market performance, attractiveness index, zoning plot ratio, time = money!...

LAW – CODE COMPLIANCE

Zoning regulations, bulk restrictions, height limits, set-backs, zoning incentive rules, policymaking to encourage renewables. Building codes: FAR, construction class, occupancy group, energy code compliance check, means of egress, handicapped accessibility (ADA), building permit approval; fire-resistance ratings, separation distances, mixed construction. Inspections: structural stability, welding, bolts, sprinkler, fire protection, mep systems, curtain wall; construction safety (OSHA); certificate of occupancy...

HISTORY – CULTURE

"Unfinished city", cultural landscape, urban design, industrialization; references. Ensemble of typologies: skyscrapers, office buildings, apartment buildings, water towers, bridges, below/above grade infrastructure. Systems (not just buildings): transportation, services, social, ecological, educational; lots + blocks parcellation, street grid, network, skyline, panorama, views; concentration of power, money, energy, activity, trading, commerce. Modernism: roof garden, communal spaces; healthy living: light, air + sun...

FABRICATION METHODS

Industrialization: one-off or mass production, customization, manufacturing waste. Material: raw materials, pre-industrial recycled, post-consumer recycled; finishes, treatments, prefab, 'styles'; artisanal methods, per-unit-cost. Performance data and specifications: thermal, sound, moisture, UV, fire-resistive characteristics. Processes: glue, lamination, extrusion, annealing, tempering, welding, milling, routing, cutting, coating, sanding, polishing, vacuum forming, 3D printing, CNC fabrication, life cycle assessment...

ASSEMBLY/CONSTRUCTION

Schedule and assembly sequence, material procurement, tools, staging, storage, waste. Connecting: bolting, welding, screws, nails, glue, curing, sealing, fire-proofing, painting; anticipation of tolerances; level + plumb; shimming; rigging. Hoisting: fitting of preinstalled units, 'location finder'; mock-ups; caulking, fire-safing, gasketing; construction layers, avoidance of thermal bridging, electrochemical corrosion, leakage (water+air), UV aging; longevity, retrofittability, repair access; structural/mep integration...

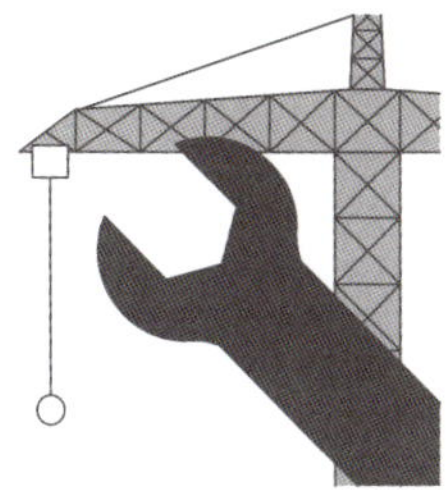

SCIENCE – PERFORMANCE

Material properties, material science, composites: material engineering, performance data + analysis; software simulation programs. Monitoring + control systems: daylighting, air-conditioning, shading, irrigation, ventilation management; real-time data and indicators; weather-forecasting; advanced systems commissioning; cooling + heating load predictions. Energy loss survey: blowerdoor testing, air flow rate. Photovoltaic technology: energy-efficient micro-inverters; phase change materials, latent heat storage...

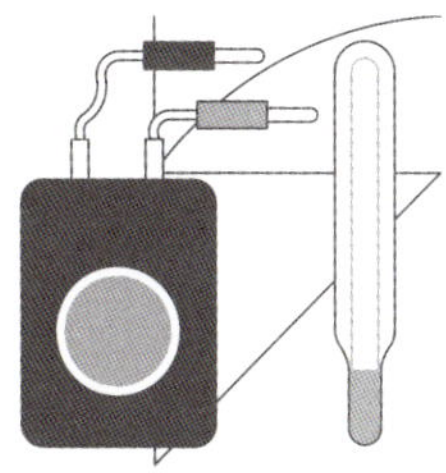

LANDSCAPE – VEGETATION

Urban green index, greenhouse gas sequestration, heat island mitigation, stormwater management, rain- and greywater usage. Biodiversity: flora + fauna, air quality, asthma rates, suspended particulate matter removal. Green roof advantages: energy saving effect, roof membrane longevity; possible urban farming, organic fuel source generation, composting: possible reduction of municipal waste, "mini-park" and retreat: space for social development, environmental education and recreation...

DESIGN – REPRESENTATION

Conceptualization, spatial thinking, project narrative; proportions, sketching + rendering, technical construction documentation; form, light and shadow, spatial sequence and experience, functional interrelations; detailing: anticipation of sequence and tolerances, joinery and joint lines, reveals; structural coordination, mep systems integration; dimensions, distances + orientation; anticipation of legal constraints, code compliance, budgetary + time constraints; design alternatives; operational diagrams, ...

PROTOTYPING –MANUFACTURING

0

SECOND APPROACH: "HOW?"—'CRAFT'

Design is very much about anticipation and judgment. It is not the control of a singular thread of thought, but instead the integration of many factors, and the comprehensive, iterative process of working through them towards a satisfying result. In the process, one has to eliminate the solutions that cannot work; one has to keep – although with skepticism – the ones that might be plausible. Finally, one has to work hard on the solutions that seem to have the most potential for realization. However, since judgment is subjective, the ability to evaluate results draws on one's intellect, and which hat one is wearing at that moment; the artist, the developer, the inventor, or the activist, to name a few. The tools used in the process are diverse, changing to address each specific design challenge.

HOW?

When designing, one has to combine the 'ectoscopic' framework laid out in the first two chapters with the 'endoscopic' factors that apply to the inner materialization process. Design decisions have an imperative to relate to manufacturing strategies and concrete inventions. How are materials put together? How can they be connected? What is the sequence of an assembly? What is the visual effect of a finished surface?

However, these design tasks are not isolated from one another; they conflict and have to be balanced out. That means that details have to be looked at in different contexts, zooming in-and-out to process these findings (3.1). The approach can sometimes be deep, very focused and isolated, but then has to reconnect to the bigger picture again. It is thus necessary to separate design problems into components and subsystems before reconnecting their altered meaning again.

Organizational capability is essential to the materialization of architectural components; it depends on a repetitive sequence of the same tasks: 1) Identification of all important factors that have the potential to shape the component; 2) Hypothetical prioritization of these factors; 3) Evaluation of the factors' interrelationships, whether symbiotic, neutral or in conflict with each other, in order to develop a value system as an instrument for the design process.

The broader interrelationships, physical and intellectual, are so complex that understanding is often developed using hypothetical situations. Assumptions have to be tested in sketches and other abstract simulations, and thereafter, critically analyzed. What are the advantages and disadvantages of certain solutions? What follow-up hypotheses can be generated? Often, this design process goes through countless iterations because discoveries emerge in creative ideas that shift priorities, changing the design values of what was seen favorably or more critically.

The ability to follow earlier thoughts or to retrace other paths is a strategic tool for realistic conceptual development.

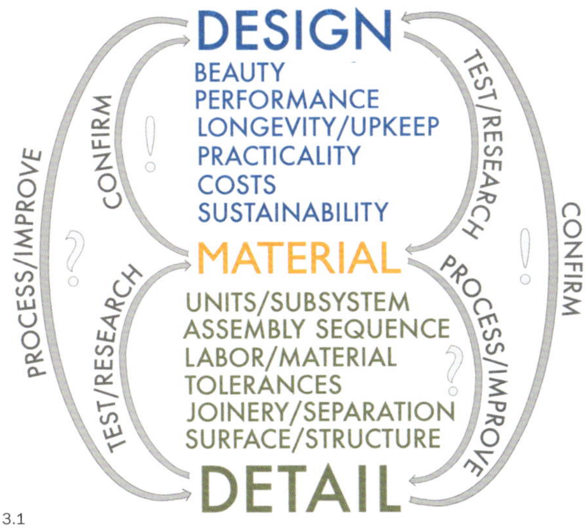

3.1

3.1 Practical design thinking: If one is to conceive an abstract conceptual idea that can also be materialized in a concrete manner, there must be an iterative negotiation process between the initial design criteria and the specific, detailed, composition. The prototype must develop alongside a feasible, justified method for its realization.

This iterative process happens to an even greater extent when specialists, for example engineers and craftsmen insert their own value scale, and influence the design through their subjective expertise and priorities. This is most likely to be the case when design moves closer and closer to its realization phase, when design thinking has to be coordinated with technology and material processing, moving a design towards a realizable artifact.

This process is a sophisticated negotiation between financial feasibility, time constraints, practical means, realistic benefits, infrastructural, political, environmental and legal factors, and of course the criterion of beauty; an elegant appearance, which is rich and consentaneous with experiences. Moreover, the resulting artifact should not just be useful but also convey an intellectual input, on the cultural level.

One should not forget that architecture, above all, creates human-scale enclosures for social interaction. Although difficult to predict, these enclosures affect human behavior in the long-term, for a life time. An understanding of the particularities of fundamental social patterns, and suitable spatial configurations, is thus also mandatory. Architecture is a comprehensive discipline: particular relationships articulate the full set of design values, and thus generate form and space.

Do the technical means of construction have something to do with humanistic culture, or with a big picture way of looking at architecture? Both technical and cultural aspects are related to intellectual stimulation and comfort: overall, architecture focuses on experience and cultural content.

For example, the house as a machine was a common paradigm in early Modernism. Because of changing manufacturing conditions within industrialization, the house's experience was supposed to relate to those technical processes and amenities which had been influenced by the machine age to change people's lifestyles. How can this thought be expressed architecturally? Modernism reacted by simplifying details and emphasizing industrial methods of production and connection. The way in which this was materialized started with the integration of industrial *objets trouvés* and included the simulation of mass-produced building systems, even though many of the elements at stake were only one-offs.

The articulation of an idea thus often depends on the detail. Furthermore, that particular detail is not the common detail, which is readily available in industrial catalogues, but an interpretation, or even an exaggerated version of a condition. It is "cultural communication in the form of a functional unit."[48]

If this communication is transmitted materially, which seems to be the case particularly in Modernism, then the understanding of construction must be paramount to expressing the bigger picture.

1

PROTOTYPE – PRODUCT

3.2

PRODUCT

Why emphasize the act of 'prototyping'? Because architecture, even though initially conceptualized through hypothetical means, doesn't exist on paper! Imagined architecture needs to move into materialized form and space, since its origins are in spatial and haptic experience. Plans, sections, renderings and other forms of representation are thus utilitarian tools to understand the materialization of an experience. But these tools are not suitable to convey the means of materialization. Skills that combine the idea with its execution are needed to make architecture. This is represented as the juncture of design and technology. The tools of transfer, or technique, have to be understood to produce something tangible. In that respect, architecture is tasked to develop an attractive product, similar to processes in industrial design, where factors, such as customer desirability, economic viability and technical feasibility have to be balanced to develop an artifact (3.3).

There are other, less technical, artisanal activities that have similar relationships between the idea and its execution. For example, a recipe does not produce food without actually being cooked; music does not become a melody without being played. Architecture must be materialized in order to be experienced. A proper evaluation is not possible without experiencing the product.

Spinoza's famous aphorism that "the concept of a dog does not bark"[49] asserts that the idea has to be connected with the 'ideatum', or the thing itself. Therefore, factors directly related to materialization must be emphasized. They decide in the end whether a product, as the synthesis of idea and realization, is a holistic success.

PERFORMANCE

In its quest to become a 'high-performance' house, our Roofpod, for example, had to have an envelope that would keep the interior 'climate' as constant as possible, with enough continuous insulation to avoid conductive heat transfer and without air leakage from one side to the other. The structure had to be detailed and sequentially constructed to support this goal. Certain methods of materialization were thus excluded. The façade's final appearance was a direct consequence of this research.

Beyond the thermal quality of the envelope assembly, it also had to be user friendly: the more effective ventilation, shading and daylighting regulation were in achieving user comfort, the more attractive the building would be. This, too, had consequences for its appearance. Performance is one of the defining factors for a product; competence and anticipation are needed, particularly in the realization process, to achieve that goal.

Drawings and simulations do not perform: they are virtual, and can sometimes be misleading, often only flattering their author. On the other hand, shortcomings are obvious in a realized work, revealing whether real-life performance was considered during the design process. The right tools must be used to anticipate correctly; monitoring is required to confirm the predictions. Performance is, at least at first glance, not visible.

VIABILITY

The competence to evaluate, which technical solution would be most economical, cannot be extracted from drawings. The evaluation needs to be either based on experience, or on trust, by relying on statements of experienced professionals.

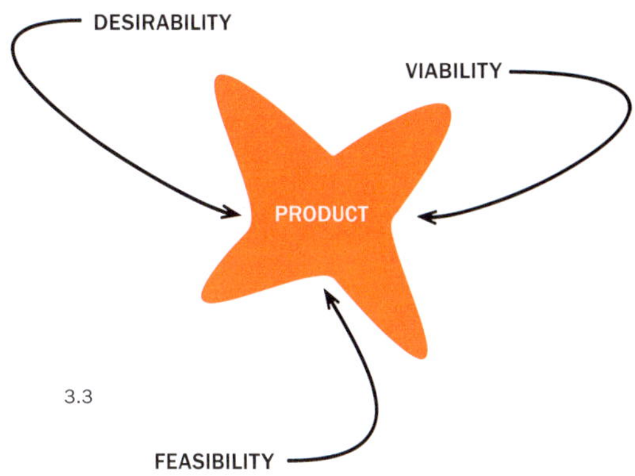

3.3

3.2 "Ceci n'est pas une pipe", or "The Treachery of Images" René Magritte, 1929: No matter how naturalistically we depict an object, we never capture the item itself.
3.3 Integrative process: Balancing desirability with technical feasibility and economic viability. The ability to exploit opposing ideas and opposing constraints to create new solutions. Redrawn and adapted: Tim Brown, "Design Thinking", 2009.

Material costs can be quickly researched. In case of the Roofpod design, many material decisions were actually based on this research: steel is much less expensive than aluminum, poplar much less costly than cedar. The resultant question was always related to whether the cheaper material can still meet the performance criteria. However, in industrialized countries, labor costs are usually more significant for a budget than material costs. To extrapolate the amount of labor from drawings is not possible, if not experienced before in a physical way, another argument for design-build learning.

FEASIBILITY

Technical feasibility is also related to haptic experience. Drawings are not subject to gravity, so that any physical connection defined on paper will not reveal its practicality in the real world. Also, drawings are not sequential. Any assembly can be drawn, while the access to fasteners usually defines the logical order of construction. The complexity of this sequence generates the effort, therefore the labor costs. Technical feasibility is thus a main contributor to the economic conditions of a product.

DESIRABILITY

This factor is in fact generally defined by visual conditions, which designers are usually most familiar with. What is deemed attractive is of course also related to trends within a socio-cultural context. The means of drawing can be easily utilized to evaluate a customer basis's interest. However, whether the drawn object is as attractive as the real artifact is not really proven yet (3.2).

LONGEVITY

There is great significance to whether a product can sustain its performance for a long time. In fact, longevity of an artifact is not only important in the consumers' commitment to and bond with a certain product; it is one of the eminent factors in improving sustainability within our society, and in the smart utilization of finite resources. This is not limited to the building's thermal qualities only, but to all elements in the house with which the user is in contact, from hardware to paint and wall surfaces, to appliances and fixtures. In its material composition, a house is a complex organism, involving many components and systems which must be understood interdependently and fine-tuned in relation to one another from the design phase to the punch-list and beyond.

It is clear that the development of a successful product is dependent on an integrative process of often opposing conditions. The optimal negotiation between the applicable factors is the creative part of the architect's profession. Many of the criteria do not reveal themselves visually. They must be discovered with experience and anticipation. Visual skills are thus not sufficient to comprehend an artifact fully. The prototype does not lie! It thus should be utilized in the design process.

2

PRODUCTION METHODS

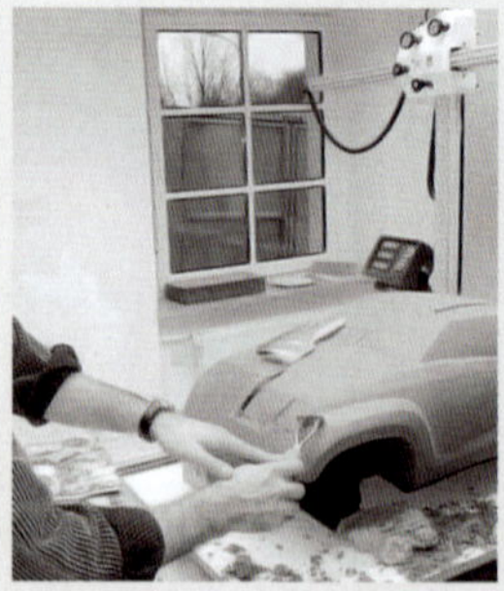

3.4

CAR DESIGN

The comparison with luxury car marketing insinuates an understanding of product development: Certain car brands have a reputation – and often a record – of being constructed in a way that makes them more reliable, user-friendly and elegant than others. The ability in product development to stay in front of competitors and improve product attractiveness enough to insure sales is an economic prerequisite that should apply to the field of fabricated architecture, too. Architects are inventors: they have to control the attractiveness of their products.

However, while car companies have the chance to improve a project throughout a series of realizations, creating an improved version for every annual development phase of the newest model, architectural projects almost always seem to be one-offs. That suggests that architects are always at the development stage of the 'Model T'. While car companies can develop their improvement agenda by analyzing the product and surveying customers, an architect doesn't have this feedback available and will therefore work somewhat more intuitively.

Also, detailing in automobile design is less restricted in component sourcing than in architecture. Due to high unit quantities, car manufacturers can often dictate to a greater degree how components should look and perform. Architects usually have to approach this problem differently. Because it has no economies of scale, architecture has to start with what is at hand, either producing 'from scratch' using raw materials, by adapting industrially manufactured items, which at least circumvents the logistical problem of coordinating a custom fabrication process. But the use of mass-produced items has the disadvantage of being unspecific, since such products are usually designed for a broader and thus unspecific market.

ARTISANAL PRODUCTION

To develop an architectural component from a raw material based upon a design idea requires craftsmanship. In other words, the tools and techniques to shape the particular material must be mastered. Furthermore, the ability to decide its suitability depend upon first-hand haptic experience of that material's properties and the proper ways of working it. The haptic experience is thus generative to judging design suitability, not only knowledge of fabrication techniques. Contact with and exploration of the world of materials is the only way for this knowledge to grow.

BRICOLAGE PRODUCTION

For industrial products, too, a knowledge base of materials and methods is helpful, although the relationship is different. The essential skill, knowing how to use an unspecific semi-product as a conceptual component, deploys a bricolage logic, since no influence on the production process is possible: it is dependent on knowing how to combine the available parts with ingenuity as part of a whole, and how to use them to satisfy all applicable requirements. Therefore, this way of working often requires greater conceptual flexibility.

It also means that often, the materials finally specified were not those originally intended, but others that were adapted physically and intellectually. Inventiveness and flexibility is key to developing a design solution using industrial products.

3.5

3.6

3.4 Mock-Up: Clay modeling of car body for wind tunnel testing; craft-related.
3.5 Mock-up: Combined craftsmanship and serial fabrication, combined to generate an artifact. Carrozzeria Bertone, for Alfa Romeo, Italy 1954.
3.6 X-ray drawing of the same Alfa Romeo sports car, but now focusing on systems integration within the overall form.

The skills must be focused on adaptation to the specific purpose. Surprisingly, these skills are quite close to the artisanal abilities described above, since the detailing of prefabricated items must be reconceptualized to make them suitable to the design idea. In this process, hands-on skills learned through craft practice are generative, and should be integrated into design thinking. For both production strategies, therefore, the negotiation between the conceptual design intent and the knowledge of viable detailing (components, assembly, joinery) should generate the overall materialization strategy (3.1).

MATERIAL PROCESSING

Often, manufacturers take advantage of both production strategies (artisanal as well as industrial) in the course of product development. This underlines the fact that both skill sets are needed to advance a product, even in an industrial culture.

For instance, it was common in the car industry to produce full-scale mock-ups to understand how to design the tools and equipment needed for the production run to come (3.5). Other artisanal skills are needed to carve scaled clay models for wind tunnel tests (3.4). This method is still vital for the process of performance control, even with current advanced digital modelling techniques. It appears that, especially during the crucial form finding process, the development and control of proportions, surfaces and transitions, artisanal, tangible and physical techniques lead to a better result.

The difference between mock-ups and real prototypes is grounded in the fact that the mock-up concentrates on particular qualities and criteria in isolation, whereas the prototype addresses the product as a functional whole. According to this definition, the architectural rendering has mostly mock-up qualities, since it focuses on isolated, specific qualities, for example proportions, overall appearance and context, while structural and systems considerations are often intentionally toned down, since the medium is not best suited to expressing those thoughts.

There are exceptions to this definition, since drawings can be manipulated in a way that takes advantage of their very weightless nature, so that the object can be exploded or x-rayed to reveal its structure and composition (3.6). In that case, the drawing serves as an expression for system thinking. The same can also be achieved to a certain degree by using large-scale detail sections, to control transitions between elements.

MOCK-UPS

A mock-up is a large-scale working model. It is used for design testing, demonstration and development. It thus provides haptic feedback on design ideas. The term is quite young: It can be traced back to the 20s, when it was originally used in airplane design. It is thus immediately related to an industrialized design process, integrating traditional craft and mass-production techniques within the development phase. Since architecture is influenced by both techniques, at least since the machine age, its relationship to the mock-up is essential: mock-ups are suitable for investigating contemporary architectural production.

3

MOCK-UP PROCESSING 'BUILDING BLOCKS'

The investigation by means of mock-ups was already manifested in the conceptual idea of the Roofpod. To build the entire envelope of modular 'building blocks', which were transportable in an elevator, was easier said than done.

Mock-ups were absolutely necessary to develop an understanding of the technical means – the 'endoscopic' design factors – that do not reveal themselves in drawings. There were at least four large-scale mock-ups involved in the design process, in order to test diverse factors of the intended design concept. These tests influenced the construction and appearance of the final building.

PHASE 1

The first full-scale mock up – built in cardboard and foamcore instead of wood – was used to determine what scale of blocks would be easiest to deal with, based on the idea of transportation by elevator but also based on several other criteria, derived from K.Wachsmann's essay 'Modular Coordination'[50]: Material, performance, geometry, handling, structural capacity, joinery, installation and planning module.

The size of this mock-up was at that time 3'-0" x 6'-0", based on the initial plan to use standard photovoltaic panels for the exterior cladding surfaces of these modules. The dimensions of common photovoltaic panels were close to this. Also the space planning seemed to work well for the interior space (3.9). We were aware that this was a preliminary dimension, an approximation that had to be confirmed during design development.

This module was in fact transported in our school's elevator. Even though the mock-up was built in cardboard, and thus the joints were just glued together, there were concerns already regarding the labor needed to produce more than a hundred of these elements. We realized that the chosen module was small in size, which had certain advantages for handling (3.7) and the overall geometry related to the aesthetics of the house, but also would cause more effort in terms of joinery and installation.

At that early stage of the design, we were confident that a rationalized fabrication approach, potentially taking advantage of digital fabrication methods, would ease the problem. Repetitiveness of the block elements was therefore considered important.

PHASE 2

The second mock-up was built of wood and metal fasteners, at half-scale to save money on the wood. It was related to the structural concept of the envelope system. In the early design phase, both walls and ceilings were conceived in the same way (see section, 3.9); however, while this is possible in the gravity-less world of drawing, the real-life application had its issues. Stacking the units was of course easier than spanning them.

The use as ceiling panels had to favor a different approach: the connection had to configure a beam, made of several, jointed building blocks. It didn't take long until a structural engineer discouraged this aspiration by stating that it was crazy to cut material into pieces only to then reconnect it again; it was twice the effort. The comment was justified in spite of its cynicism since it focused on the typical engineer's structural and material efficiency viewpoint.

However, the students bounced back; for them, the systematic idea of elevator transportability, not considered in the structural engineer's assessment, was so essential that they didn't want to give it up easily. They decided to drill holes into the block elements to integrate tension cables for ceiling applications.

They completed the mock-up, and were able to prove that they had come up with a solution to the problem, integrating an efficient structural connection. The students presented the mock-up at the following 'advisory board review', and although

3.7

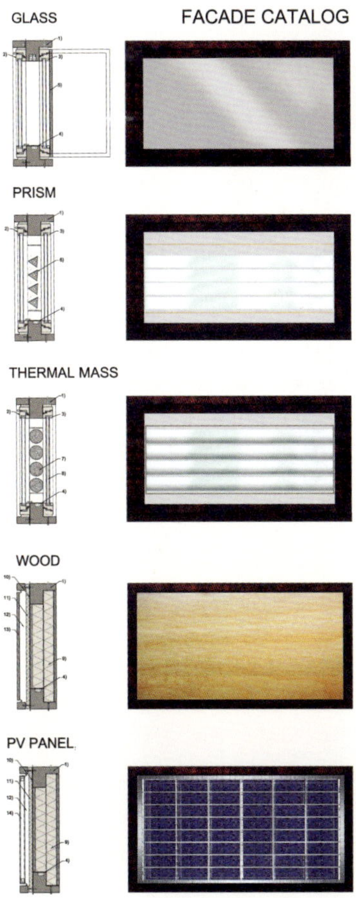

3.8

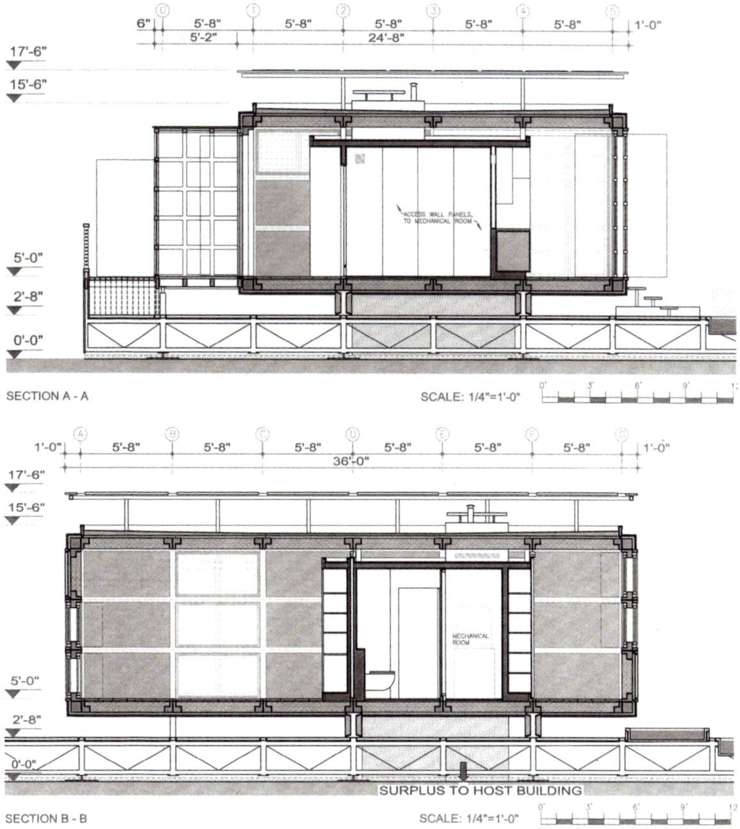

3.9

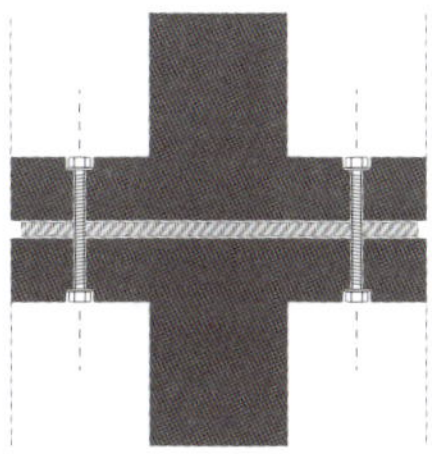

3.10

3.7 Roofpod: Mock-Up 1. First full-scale verification of transportability + modular coordination.
3.8 Roofpod: Schematic design 'building block' catalogue; different options addressing passive design and renewables.
3.9 Roofpod: Early sections. Wall and ceiling blocks were conceived as identical modular elements, although their structural nature is completely different.
3.10 Roofpod: Schematic 'building block' connection.

3.11

3.12

they were euphoric about the solution developed and materialized, the specialists' concerns grew. Post-tensioning cable installations is dangerous, and also needs expensive equipment. The uncommon system would have to go through testing agencies to obtain approval, which would take months or even years.

Our General Contractor, also an alumnus, who is passionate about prefabrication, voiced his concerns about blocking a residential building's elevator for days for delivery purposes, while simple hoisting with outriggers from the roof's parapet would be cheaper, more flexible and lead to fewer complaints. When he also explained his view of the relationship between prefabricated module sizing and the development grade of industrialized countries – usually, the more developed, the bigger the construction modules – he advised going with a solution that considers simple hoisting, and removing the focus from the elevator scale.

Another concern was related to the combustibility of these blocks: The blocks, though easy to fabricate in wood, would not achieve a construction class necessary for New York's roofscapes. It would be easier to use the blocks for space configuration purposes only, while structural concerns would be satisfied by a separate steel structure.

It took a while for the students to adjust to the new design requirements derived from several diverse criteria. The steel structure was developed with their structures professor, in such a way that the building blocks could be inserted into the structure, like cartridges[51].

This was the plan for our exterior walls. Floor and ceiling were configured with other cartridges in the form of prefabricated structural insulated floor and ceiling panels, to reduce the workload on the students. Instead of our tabled post-tensioning method, these elements could be inserted in one piece, making us deal with element sizes of approximately 4 x 10 feet. Since these now defined the modular coordination, we also revisited the way the wall blocks were to be produced. The structural engineer's comment had come up in many discussions with the students; they also started to comprehend that smaller pieces are more difficult to assemble, since irregularities add up that have to be taken account for by figuring out how to compensate for them.

We therefore discussed whether it would be wise to fabricate singular units, at that stage 'shrunk' to a dimension of 2'-0" x 5'4", which we had discovered was more efficient and less wasteful in terms of material use. Or, would we produce 'preinstalled', combined units, which would have the advantage of being able to be stacked four-high to match the height of the super-units at 8'-0" x 5'-4", providing greater accuracy, less work and faster progress. For example, the jambs could be used as one continuous element and wouldn't be separated into four pieces any more. The discussions revolved around the question: What do we consider to be modular?

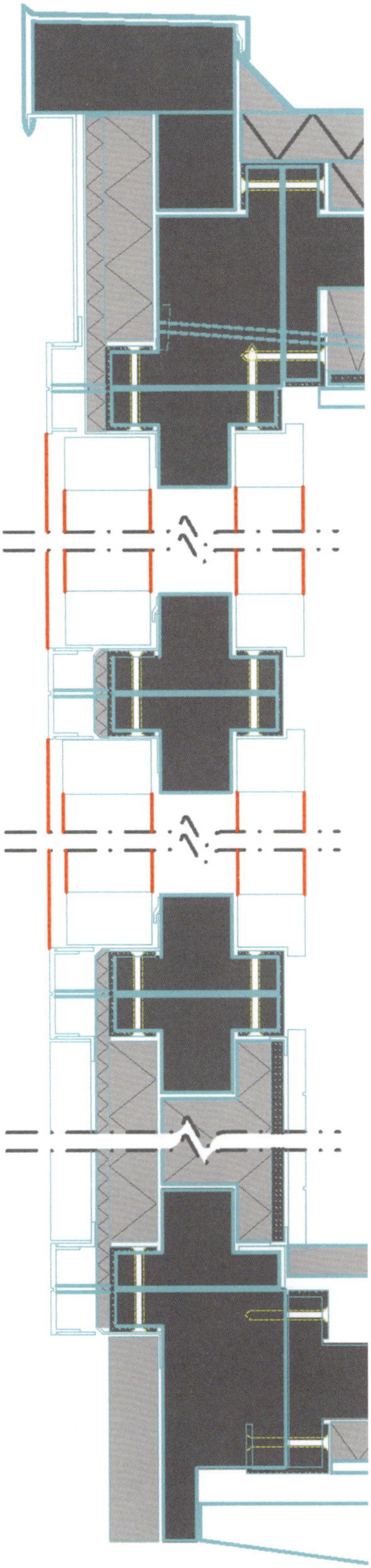

3.13

3.14

3.15

3.11 Roofpod: Mock-Up 2. Half-scale; the wall 'building blocks' are clamped w/ aluminum channels.
3.12 Detail, block connection.
3.13 Schematic wall section, addressing structural and insulation issues.
3.14 The roof 'building block is now structurally differentiated; tensioning rods are inserted.
3.15 Test of post-tensioned assembly.

3.16

3.17

3.18

3.19

3.20

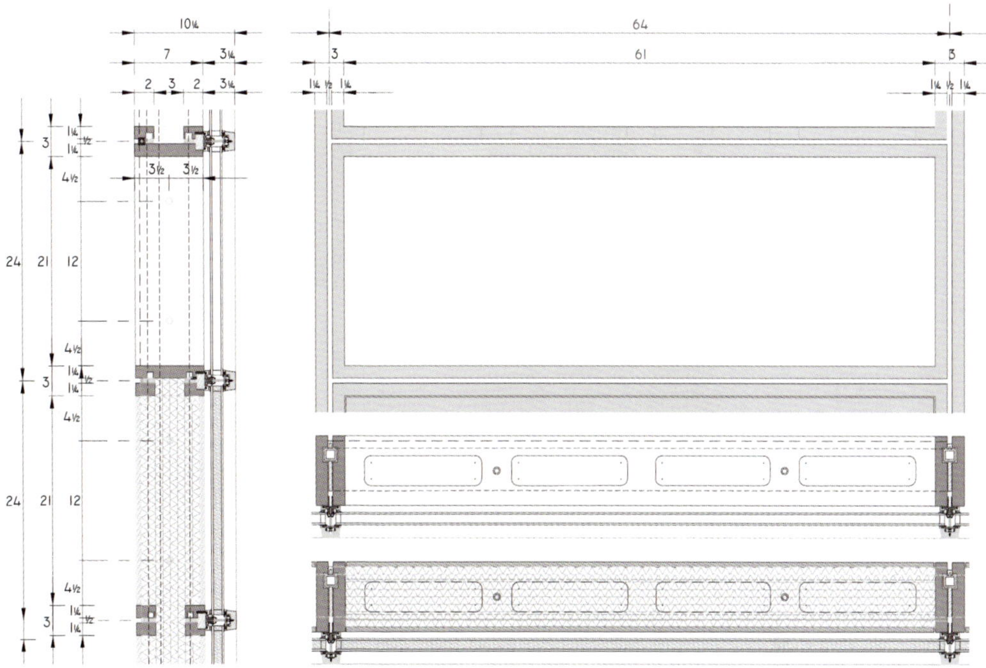

3.16 LEGO: Do 'building blocks' have to be created separately, or are 'preinstalled' units OK?
3.17 Roofpod: Mock-Up 3. Detail, block connection.
3.18 Stacking the 'building block'; note the frame cells to reduce frame conduction.
3.19 Stacked 'building block'; installed curtain wall sample track to test joinery.
3.20 Preparatory drawing for fabrication.

As a constant reference, we used the comparison with LEGO blocks. The traditional blocks came only in one regular height, about 3/8", while width and depth differed; more contemporary LEGOs, however, add blocks as multiples of the standard height (3.16). While the modular system remains intact – everything can still be combined due to the modular coordination – the material is used more efficiently (less plastic for joinery of elements), and less labor is necessary for installing the assembly. Height 'tolerances' of block assemblies are also controlled more easily.

Conceptually, however, the decrease of 'possibilities' reduced the 'playfulness' and some of the supposed creative potential of the system. Again, comparisons with customized products (cars, electronics, etc.) were undertaken. The conclusion: Options should only be available if they sell the product (desirability), or improve performance. Otherwise, they should be taken off the catalogue. It became clear that only between 10 to 15 compositions of four stacked blocks made really sense, due to daylighting, ventilation and view criteria. A super-unit catalogue would still allow for a sufficient amount of options, while production efforts would slim down.

Since the students were split in their opinion, we decided to build mock-ups for both options: The singular 'building block', with its stacking connection, and also the other, preassembled super-unit, which would even be fully glazed to integrate the issue of attaching the exterior façade layers.

PHASE 3

The singular mock-up (3.17, 3.18, 3.19) was built in full-scale, and besides the stacking mechanism also considered improving energy-efficiency by reducing thermal conduction within the frame. Also, the integration of the steel frame fastening 'fins' between blocks was made possible, by configuring a groove in the jamb to conceal the fin, and equally to lock in the blocks by means of the fin.

During the process, the students gained an increasing understanding of how difficult the production of 64 equal 'building blocks' was, if they decided to go with the singular stacking connection. This amount of units was necessary to enclose the building. The utilization of digital fabrication was not yet in our reach, so we depended on conventional tools. The haptic experience of fabricating the precision grooves and corner joints conveyed to us the level of difficulty and real scope of work.

PHASE 4

It became clear that the super-units would be more economical to produce. For this mock-up, the last in the series (3.20 ff.), we prepared precise shop drawings for a bay of four stacked blocks, and also

3.21

3.22

integrated curtain wall components to complete the assembly. Our conclusion after this try: It is easier and more exact to produce the blocks in a stacked way, especially with regard to producing continuous jambs. However, the aesthetic idea of emphasizing the blocks' appearance in the repetitive planning module weakened with it. The continuous jamb had to be manipulated to express the modular strategy. By fabricating the last mock-up, however, we learned how to manipulate the configuration so that the initial strategy could be articulated. A simple groove in the jamb, applied after assembly between block segments, achieved the intended effect. With this method, the dimensional accuracy of the continuous jamb could be maintained, while the expression of the modular unit was possible, too.

Aesthetics and buildability were realigned. The idea of the repetitive unit was readable while the means of construction allowed for an efficient assembly.

During the process of the four mock-ups, the students excelled in their application of their skills with materials. Although at first the judgment on materialization processes was solely based on drawing, – what can be drawn should also be possible to build – this attitude changed. Drawing was instead used as an anticipatory tool.

The development of this essential building component was, however, not completed. The upcoming construction would reveal more information on the practicality of this element during its repetitive assembly.

The production of the 'building block' prototype was based on artisanal fabrication. The experiences gained were based on one-off strategies. For a mass-produced building block, tools would have to be utilized that would facilitate an easier fabrication, especially in terms of precision control.

If CNC equipment had been available for our production line, certain decisions we made would likely have been different.

However, the fact that we were able to produce an artifact that could serve as a basis for discussing the refinement of these methods, was exactly what prototyping in other sectors is about. It is the start of tangible communication: the intellectual exchange with specialists in the field is more easily accomplished with haptic input. It is usually more effective in streamlining the design than abstract drawings.

3.23

3.24

3.21 Roofpod: Mock-Up 4. Studio discussion, interior elevation: finetuning the 'building block' module appearance.
3.22 Attachment of curtain wall carrier profile.
3.23 Studio discussion, interior elevation: finetuning the 'building block' module appearance.
3.24 Learning to use proprietary tools to install curtain wall gaskets.
3.25 Assembling 4-block 'super-unit' to minimize jamb tolerances.

3.25

3.26

3.27

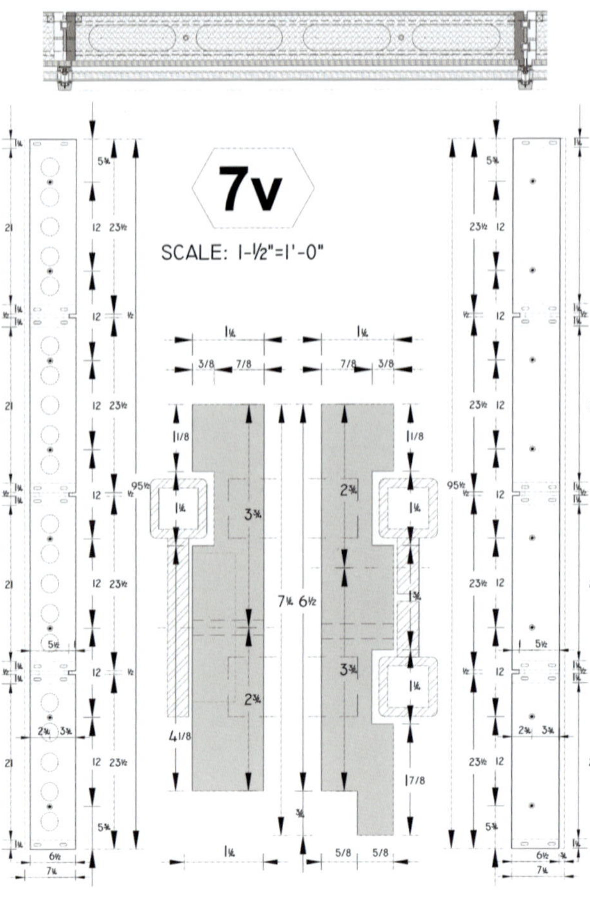

3.28

3.29

3.30

3.31

3.26 'Building Blocks', batch-production: fabricating components.
3.27 Shop drawing catalogue; a necessary guide throughout the fabrication process.
3.28 Units prepared for spray-insulation.
3.29 Stain and finishing preparations.
3.30 Spray insulation for opaque units; windows and ventilation units were covered.
3.31 Assembly of 'building block' super-units.

4

SYSTEM THINKING

SUBSYSTEMS

The series of 'building block' mock-ups concentrated on structural qualities, specific modularity, tolerances, joinery and interconnection. The testing and the subsequent comprehensive analysis would not have been possible with the drawings alone. The results were more conclusive, drawn from the experience of fabrication. This decision making process was essential in arriving at a solution that, in the end, was realizable. The articulation of a subsystem emerged, integrating concepts of appearance, performance, assembly, costs and labor, among others.

However, the development of a building component is only significant if it can be reconnected to its overall design context.

The building block design alone would not be able to solve all the problems addressed in the overall design approach. How could the Roofpod be connected to the host building?; How could mechanical, plumbing and electrical systems be integrated?; How would storage occur?; How would functions such as bathrooms or kitchens be integrated?

Even though the particular component could respond to multiple design functions, it is likely that other subsystems will emerge that are more suitable to addressing specific key issues.

These subsystems have to be adjusted, to respond to each other so that they become an internally consistent assembly. The articulation of their relationships, and the hierarchy that they establish in their configuration, are essential for the architectural expression; they are again multi-layered on many different scales.

The skillful combination of these subsystems, so that they are subordered towards the bigger idea, but still express their individual character, is thus the design challenge.

SYSTEM THINKING

The articulation and organization of a system is what often determines contemporary design.

A system is reproducible. In an industrialized context, within a consumer market, architecture has to be concerned not only with production and fabrication, but also with the conditions of its reproduction and marketing.

The effort of reproduction is smaller than that for the original attempt; thus, the efficiency and magnitude of reproducible design is greater.

In our project this was essential, since we conceived our house on the basis of an urban analysis. The design was therefore already related to a market, and to the industrial approach of a kit-of-parts. The conceptual idea of the Roofpod as infrastructure in an urban environment demanded that its materialization strategies integrate mass production. We had to become 'prototypers' to develop a customizable, but systematically reproducible system.

Also, in order to process a design efficiently, especially as a team effort, subsystems have to be identified that reduce the number of participants involved in the particular design activity. The number of parts to be controlled is reduced by this method, too. It is necessary to comprehend logically where subsystems can be articulated, and where the borderlines between them are situated. This definition is dependent on materials and trades, and the intended construction sequence. The articulation of a subsystem is one of the typical zoom-in methods to process a design in detail.

In our project, this method was essential. An efficient teamwork organization was key. We had to understand what subsystems were feasible, for creating strategies for a coordinated materialized design (3.32). We identified five building subsystems:

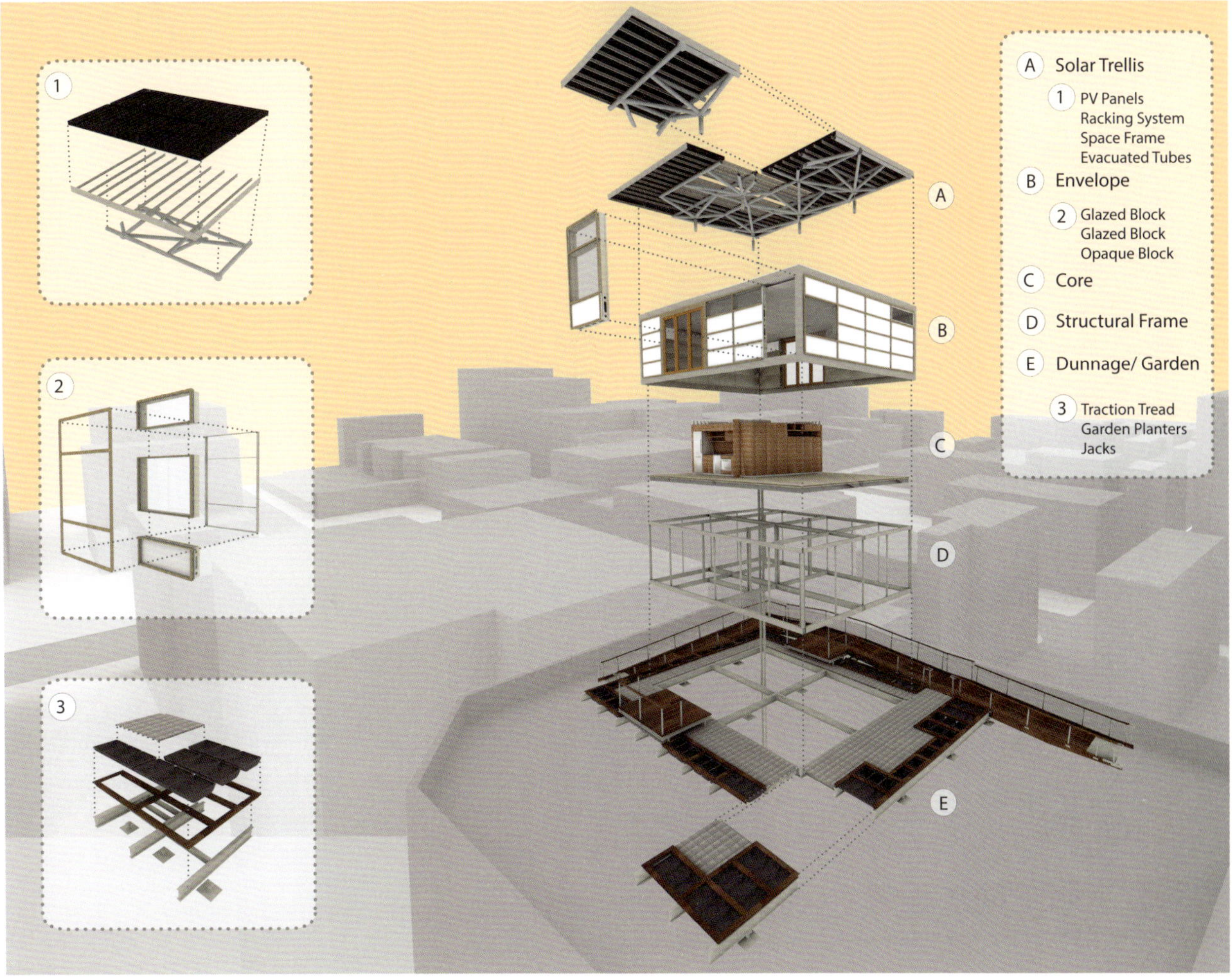

3.32

Solar trellis, envelope, core, structural frame and dunnage garden. These subsystems were then again subdivided. For example; the envelope into roof and floor panels, 'building blocks', curtain wall, glazing, doors, etc. With this fragmentation, we were able to develop materialization strategies, which were all based on the primary module dimension of 5'-4". Therefore, we were sure that they would fit back together again.

SYSTEM PROCESSING

We constantly had to virtually reconfigure these subsystems, to interrelate them to form a whole. Zooming-in meant working separately, while zooming-out was necessary to reconnect the work as a whole. Any interrelated issues between systems had to be addressed by all the parties inolved. For example, the running of the electrical systems, with lighting fixtures, outlets and switches and all kinds of sensors to control the house's performance, had to be coordinated throughout the subsystems.

This adaptation – the to be able to adjust the design to its specific and buildable conditions – happens on different preparatory 'platforms' before construction starts. These platforms include large-scale drawings of details which may reference a zoomed-in element that was part of the original schematic design drawings.

In our process, we often executed details in very large scales (3"=1'-0", 6"=1'-0", or even full scale) to develop an understanding of the real dimensions of the building elements. In that respect, large scale detailing connects 2-D drawings to the world of 3-D mock-ups.

3.32: Exploded perspectival rendering, demarcation of subsystems. The identification of these systems worked as wan effective transition between overall design thinking and detailing for construction.

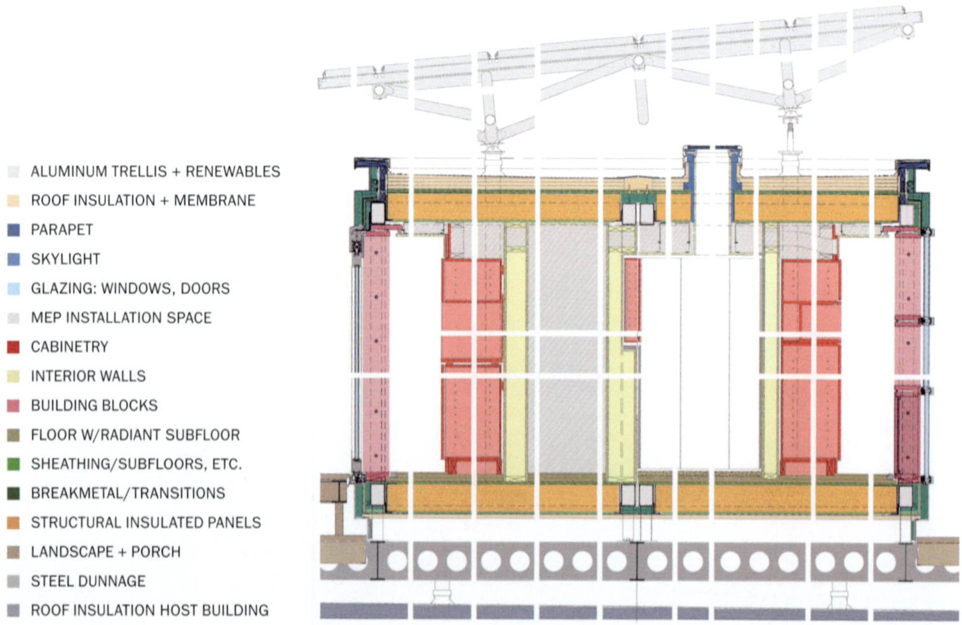

3.33

These drawings are used to simulate materialization, since physical characteristics of the abstracted assemblies, such as joints, connections, sequences and the management of tolerances, are revealed. This intellectual simulation is, however, only possible if the experience with mock-ups and haptic construction is prevalent.

The experience in full scale 3-D construction reveals the real limitations to the possibilities of materials.

It is arguable whether in the early stages of design education the abstract drawing can convey fabrication limits, because the student feels empowered to invent everything graphically, without having the complementary fabrication knowledge. A real feasibility confirmation, however, cannot be made. In translation from 2-D drawing to 3-D object, therefore, hands-on practice must lead the decision process.

Both working methods, from 2-D to 3-D and from 3-D to 2-D, depend on each other. While the detail drawing often departs from three-dimensional geometric relations and the overall design picture, the mock-up instead emphasizes material production methods, connections and joinery. The drawing, especially when computer generated, bears the imprint of the zoom tool, moving constantly between the whole and its parts, while the mock-ups relate to the production process and have all the qualities of the unzoomable future artifact.

With practice, the detail becomes the connection between design and fabrication, the generator of materialization. According to Semper[52], the knot as the generator of an assembly is "the oldest technical symbol"; it is the "prime motif of human tekhne, because a structural necessity becomes an aesthetic, meaningful image."[53]

Both these knot design problems, and the related meaningful systems integration was therefore our main planning task. This is not uncommon to the architectural profession; however, it is not common in an academic context, where these adaptation skills are in tension with the typical conceptual training. While conceptual thinking is usually solely processed in small models, sketches and drawings, the act of prototyping needs large-scale three-dimensional support.

3.33 Roofpod: Wall section, addressing subsystems and components; monitoring the transition of the separately processed elements back to a whole assembly was crucial for the design development.

ARCHITECTURAL FABRICATION

Whether in an industrial or craft-based context, the detail defines much of an architectural design's overall appearance. It is a processing tool to translate the idea into a real-life object.

Since architectural design is linked to market forces, especially when there is the intention of mass-production, the effort that can be spent to generate these details is not unlimited. Whether invention-based prototyping is really necessary has to be evaluated on a case by case basis. If, in our industrialized world, suitable substitutes exist that could be fitted into the overall design without any detrimental effect, wouldn't it be better to make use of them?

With standardized materialization, there is a preconception that our architectural environment will look more and more repetitive and uninspiring. Is there an inevitability to embracing these conditions?

Mass-produced items have contributed to an unhealthy dependence among architects that – to return to the cooking metaphor – compares to processed food. In fact, if one were totally dependent on pre-prepared food without practicing the act of cooking, one would soon forget all about this skill, and even lose the ability to understand techniques for preparing and combining ingredients anymore. A vital part of our culture and a source of creativity would disappear.

At the same time, one cannot deny the world we live in. There is an advantage to Taylorized prefabrication that is difficult to compete with.

Multiple strategies are important: if you cannot compete with an industrialized market, you have to provide a product that is superior. That means that you have to twist the conditions (3.3). Cook with better ingredients (materials), use more sophisticated recipes (material techniques), plate a more beautiful product and know where you are likely to find your customers. There will still be economic pressure on a product that will probably cost more time and money. Improving it means to look for ways to navigate around these flaws. Can prefabricated elements be utilized that do not take away from the product's quality? Can techniques be altered to save time? One should take advantage of prefabricated methods whenever they make sense. Constant assessment skills and advanced creativity are needed to make the best product possible.

In architecture, practicing the detail and investigating control of its fabrication has the highest significance. It is needed to understand which processes are groundbreaking.

One has to be interested in all possible, more or less craft-related fabrication methods.

One cannot just turn back the clock and deny industrial fabrication. Industrial processes are a significant part of our culture, in terms of both costs and appearances. However, they have to be taken advantage of for the purposes of advanced fabrication. One has to learn the conditions that apply in order to improve them. The exercise of practicing fabrication does, in fact, not deny or fight against this industrial setting, but develops the abilities to use it better. It is a necessary part of training in order to understand assemblies and to alter semi-products competently for use in a particular design.

In the process of our Roofpod design, we also tried to learn from case studies that comprised some of the stringent issues.

One aspect was the design integration of prefabricated elements, and the effects on the building's appearance; another was related to the requirements of handling and transportation (in our case to Washington DC).

We were also interested in the change in the way prefabricated design ideas were processed from the high-times of prefabrication in the 1950s to the present day. The changes that have occurred in the past 60 years, related to changes in conceptual architectural approaches.

CASE STUDIES

The house as a kit of parts was the point of departure. Therefore, we did not concentrate on one specific case study building, but on an architectural approach, which reiterated a series principle.

Two particular cases came to our attention: One conceived in the early 50s by Jean Prouvé's "standard house" studies, the other by Kieran-Timberlake

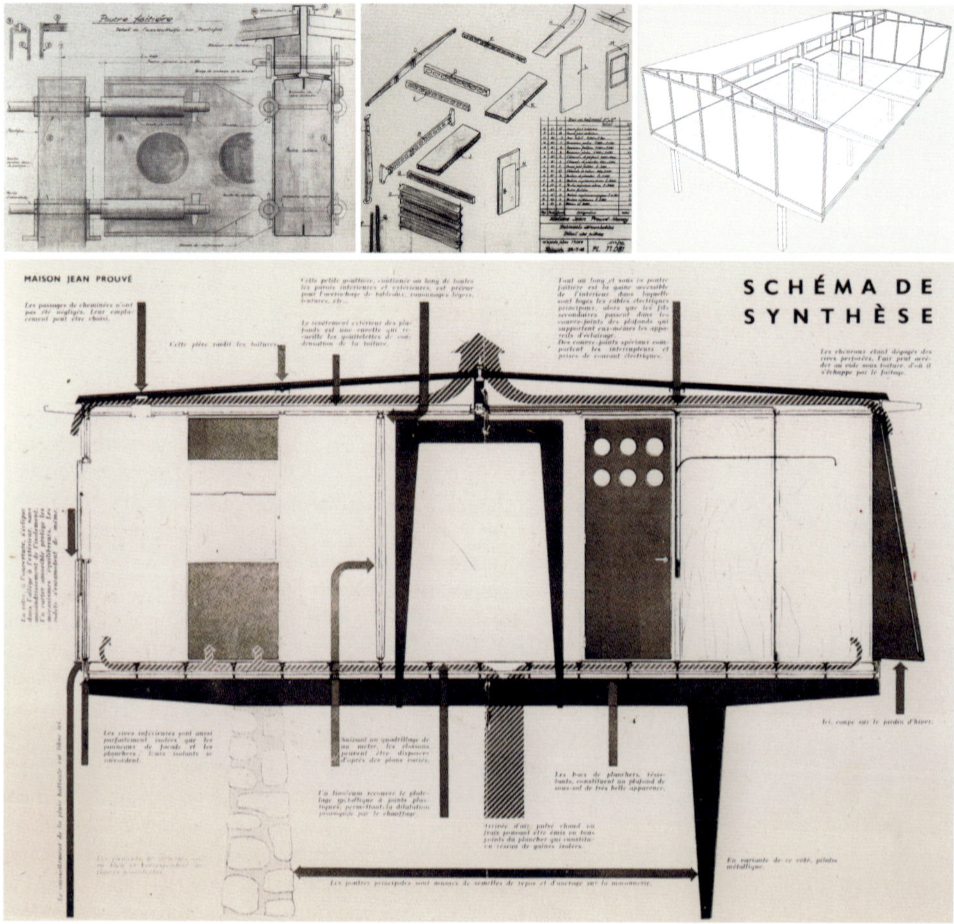

3.34

Architects's aluminum-framed dwellings, manifested in the Loblolly House and the Cellophane House.

Both of them were mainly prefabricated off-site, thus transportable, and made of subsystems that could be easily identified. Both were demountable; they were also focused on passive design strategies.

The "standard house" was conceived almost like an automobile, where all its parts were fabricated in Jean Prouvé's own metal shop to the requirements conceptually identified in the design process. All the connections between parts were also designed and fabricated from scratch. Prouvé started as a metal worker by trade, and knew how to do these prototypes. These skills allowed, for instance, for custom beams that would be maximally perforated to pass through ventilation. Thus it was possible to optimally intertwine structural and operational concepts to manipulate the interior climate.

All parts were based on a one-meter (3'-4")planning grid[54], with elements being multiples of this grid. This system is cleverly resolved in its geometry, but it is strict. There are issues related to the responsiveness towards context or orientation, since the grid inserts don't leave much space for reacting to particular site conditions, solar orientation or views. The planning grid, which positively influences the constructability, since all joint intersections are resolved and the instances of these connections minimized, restricts the flexibility to respond to specific conditions or any occupant preferences. Many of these houses were therefore altered afterwards by the owners. Also, some of the original owners, who despite falling in love with their extraordinary houses, are now concerned about finding replacement parts, since the one-meter elements came from one proprietary source only, and are not produced anymore.

Even though the intention of the design was to be truly industrial, it is essentially an artisanal artifact, based on the design-build know-how of the fabricator.

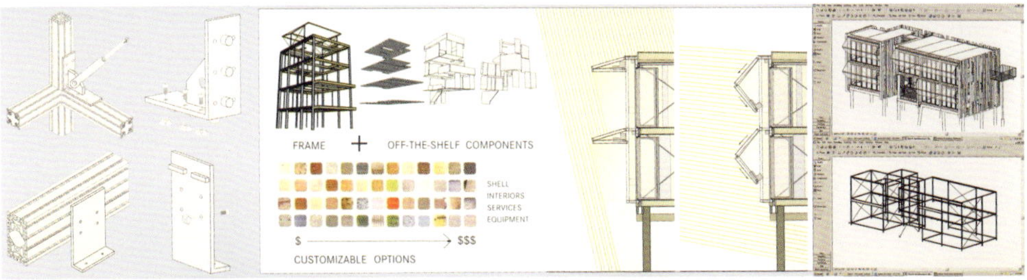

3.35

3.34 "Standard House", 1947 - 1951, Ateliers Jean Pouvé: Prototypical building system; among them were the realized "Maisons Meudon" and "Maisons Tropicale", 1947 -1951. The houses rely heavily on the manufacturing processes set up in Prouvé's factory.
3.35 "Loblolly House" and "Cellophane House", 2005 - 2008, Kieran-Timberlake Architects: The building system relies on combining industrial off-the-shelf subsystems, adapting them or complementing them with custom solutions when necessary. The combination of subsystems is coordinated by means of Building Information Modelling (BIM), compensating for less geometrical rigor.

The system developed by Kieran-Timberlake Architects responds to the limits of standardization. It utilizes off-the-shelf components for typical space enclosures, while interiors, services, and especially elements related to establishing contextual or orientatial conditions are developed in a customized manner.

For instance, the south façade's shading and light manipulation system is built from scratch, while the overall framing and the floor and roof panels are sourced from industrial fabricators. The structural framing, which is actually a proprietary system designed for industrial shelving, is emphasized by developing custom solutions for the joints. By adding these features, the system is adapted to its structural functions, and its industrial origin can be exposed as a symbolical gesture.

The building system integrates many subsystems that are industrially available, with the option of customizing them to the particular requirements. The flood of information caused by this approach is processed by digital means. A virtual Building Information Model can coordinate the different building elements, with the advantage of being less tied to a strict geometry.

Prouvé's product is an artisanal house, manufactured by his own shop in the vocabulary of standardized industrialization. Kieran Timberlake's product is in this respect more industrial, since it incorporates off-the-shelf products. It utilizes industrial prefabrication when advantageous, while developing customized approaches for adjusting the artifact to its specific conditions.

Both strategies were incorporated into our Roofpod design. The decisions were based on the specifics of the particular subsystems.

5

THE ART OF COMPROMISE —'DESIGN-BUILD'

3.36

The methods of prototyping in architecture are concerned with feasibility and buildability. Balancing the negotiation between different design factors most likely leads to formative compromises, since the economic, aesthetic, legal, environmental and practical circumstances only rarely align to their mutual advantage, but instead create conflicts. The more or less elegant resolution of these conflicts results from effective planning and processing skills. Design optimizations do not occur by ignoring the holistic nature of a problem. Accurate decision-making is based on the art of compromise. To find the best solution within all factors that shape the design problem; to balance the priorities in such a way that architecture – as a realized artifact – can prevail. Trial and error, and over time the force of experience and anticipation, will lead to a productive design process. This process depends on developing the most positive consensus between the multiple interests and forces derived from concrete criteria. It is a negotiation between the many sides of one's own expertise: both artist and businessman, scientist and occupant, builder or manufacturer or fabricator or engineer or developer.

It is also related to interaction with others, to communication and learning skills in a social context.

COMMUNICATION KNOW-HOW

It is apparent that the know-how necessary for fabrication must be broad. An architect needs to know the techniques of the trades, in order to navigate between common practice and invention. But, you cannot learn everything about the many trades usually involved in common construction. Basic knowledge of haptic techniques is, however, necessary. The architect is the middleman between fabrication and a finished product.

What is needed in this function are clear communication skills, a good oversight of the processes and a strong vision of the final product.

It is comparable to the profession of an orchestra conductor; many of the world's leading conductors know how to play many instruments and how to produce the techniques necessary to generate a certain tonality and musical cadences. But they don't have the ability to play all the instruments as perfectly as the musicians do. Like the conductor, the architect has to be holistically knowledgable. Also, affection for the detail is needed to develop a superior product. The good conductors – to stay within this metaphor – separate themselves from the rest through their ability to interpret written music in a more beautiful and sensitive way. They observe the smallest details and integrate ideas and techniques into the given overall scenography. Both the know-how of techniques and the oversight and intellectual focus on the final result are mandatory in shaping the work.

The architect is the conductor of product negotiation. He is the coordinator of all the different influences that shape the final artifact. Therefore, the architect has to have knowledge in all fields that are associated with the disciplines of design and building.

The profession is inherently based on both broad knowledge and specialized knowledge. The broad knowledge is related to understanding the big picture and responding to it intellectually. It is necessary, as a reaction, to come up with a creative visionary response, the articulation of an idea.

The specific knowledge is related to the implementation of this idea. The specific knowledge is thus related to craft.

Know-how can never stay broad; it has to accrue depth. This requires understanding ones partners in

3.37

3.36 Balance, manipulated.
3.37 Material–Know-How: Dynamic interrelationship between processing knowledge and possible material expressions.

the typical design-build processes, and the critical values at stake.

Furthermore, especially if certain values conflict with the architect's aims – i.e.: architects are typically interested in a beautiful artifact, while developers are interested in inexpensive production – it is vital to develop methods to satisfy both requirements and to keep them under control. Therefore, an efficient strategy to develop interdisciplinary consensus is paramount for design-build projects.

MATERIAL KNOW-HOW

The traditional yin-yang symbol can be appropriated to express a deeper understanding of material processing techniques, showing detailing as a dynamic balance of interconnected forces (3.37). Materials can only be understood by tangibly gaining know-how of how they are processed, including the appearance that results. The transformation of a natural resource into a manufactured element for a specific application is part of the magic of architecture. Knowing how materials can be refined will influence the understanding of matter; it will change our perception of a particular material and with it, of all materials. This deeper understanding can give rise to more advanced techniques which in turn offer better control over the final product's appearance. Know-how and material expression are inseparably linked.

In other words, without advancing know-how, creative development of materials could not occur. Architectural materialization would be limited to a preexisting vocabulary. Evolving know-how is thus crucial to gaining freedom as a designer. Access at an intellectual level to material processing techniques has to improve constantly throughout one's professional life. One has to be actively involved in acquiring knowledge of how materials are processed, using multiple, diverse techniques and making intellectual connections with the finished artifacts surrounding us in our everyday life; one has to build one's own palate and an intellectual research basis in order to understand what can be done to achieve a desired material appearance.

This skill of obtaining material know-how is currently underdeveloped in the architectural profession. Architecture is considered a white collar job; however, in order to advance in ones material techniques, contact with the manual blue collar workforce is needed. Again, the architect has to wear both colors. Messy work is necessary to create extraordinary artifacts.

With time and advanced experience in the tangible part of the aforementioned material techniques research, it will be possible to overtake some of the responsibilities that conventional fabricators usually take on.

It becomes conceivable to develop architectural detail based on individual research, which is uncommon to the typical fabricator.

Furthermore, techniques could be improved in collaboration with those fabricators, which often feel passionate about their tools and equipment.

However, it is likely that the architect is the partner in a team, who is most inclined to experiment. Interests diverge: the architect is interested in intellectual creativity, while the fabricator has to sustain economic conditions, too. If financially feasible, architects could set up their own shops to develop a particular kind of detail that would otherwise be difficult to realize. Currently, this is more and more possible in digital fabrication, in which architects can bring in their expertise of processing digital shapes (3.38). If the architect could play a stronger role in fabrication, using emerging or conventional technology, their control over the product would increase.

3.38 CNC fabrication, based on architectural drawing: The direct transfer from planning to fabrication contributes to better control of the product. (The CNC router was installed after the Roofpod's construction. Therefore, we weren't in the position to utilize it.)
3.39 The use of tools and any hands-on installation activities are vital to competently control architectural assemblies; the obtained insights can be easily transferred into 2-D detailing activities. (Photographs depicting students during the construction of the Solar Roofpod.)

3.38

Again, a similar tendency can be observed in the car-manufacturing sector, where car companies are able to develop prototypical parts through their own research and development shops. The link between fabrication and research is crucial. The challenges within architecturally advanced building are similar: strong ties to contractor, fabricator and engineer, allowing the architect to better understand and possibly to execute some of their tasks would be advantageous to greater control, and to developing the product's appearance and performance.

The architect cannot sit behind the computer and think that the world can be controlled from there. It is necessary to be proactive in acquiring skills and knowledge in the fields of construction and fabrication, science and engineering, and all fields related to manufacturing. This objective must be encouraged in education, so that the delineation of thinking and doing disappears in architectural culture.

Competency in Design-Build must develop through hands-on experience, and can most likely not be learned efficiently from books or online courses. Fabrication is inherent material culture, with many different ways of not-doing. Literacy in common fabrication methods is necessary, as is curiosity about new developments in the field of product development. Contemporary architects have to be more involved in this culture.

LEARNING FROM MISTAKES

Mistakes in realized projects are much more visible than in drawn representations. In fact, the nature of materialization is that mistakes are irreversible (unlike drawings, where there is always another way out), and they are therefore intimidating. However, errors are – at least initially – unavoidable when learning, since decisions are based on predictions, which draw upon abstract representations; experience teaches how to deal with the margin of error.

Nobody wants to make mistakes. However, these mistakes are often more productive and intellectually formative than hiding in the abstract realm of paper representations. These physical mistakes build experience: while craftsmen's apprentices learn concrete materialization skills from their masters, in architecture, the transfer of tangible knowledge about materialization often does not occur. Schools don't teach fabrication, and architecture offices are often specialized in plan production, making site visits for certain staff unlikely. How can one develop the skills to act in the world of materialization while sitting in an ivory tower?

TOOLS

The use of tools is mandatory for architectural design training. Haptic material research has to be encouraged early on in one's education, through furniture design, mock-ups and possibly the work on

3.39

whole buildings. It is the use of tools, utilized for full-scale material processing, that conveys the means of making. Even though the first tool handling and machine operation might not lead to perfect results, it will communicate issues that can be incorporated into 2-D detailing activities. For the same reason, construction site and shop visits are mandatory so that discussions can be held with experts.

Prototyping, in a do-it-youself manner, should be considered research on tangible materialization. Research depends on originality: errors – and the anticipation of how to avoid them – are part of the risk-taking training necessary to develop control over the fabrication process. If a method can be found that communicates common errors and teaches the anticipatory thinking to prevent them, as was done in traditional trades, the architect could again develop the skills to negotiate the process of building. In Germany, before the Industrial Revolution, the word for architect was Baumeister = Master of Building. The profession was based on realization, not just planning skills. In an industrial culture, only know-how of sophisticated fabrication can ensure products that are out of the ordinary. The aspiring architect has to engage in this kind of apprenticeship again.

TIME MANAGEMENT

Another piece of information that cannot be experienced in drawings is the factor of time. How long does it takes to fabricate an idea, articulated in a 2-D detail? This information is crucial for designing a product based on economic viability.

It is astonishing to realize the different amounts of effort needed to describe a detail in words, drawings and built form. While it will only take minutes to describe it verbally – with all the associated ambiguities – it might take hours or even days to document it in technical drawings. And, if drawn without close consideration, a single line can create work that translates into weeks of labor and potential conflicts with the craftsmen who have to pay for the lack of anticipation. The phases through which a construction detail passes from an idea to physical reality, although interdependent, imply very different scales (3.40). To realize that the detail had been ignoring feasible realization conditions, is unsatisfying and can compromise the intended execution of the design.

In our project, students were surprised how different the time scales of drawing and building were. Many of them, having lived so far only in a world of drawings, had no initial experience for this essential estimate. Careful drawing as a tool for directing construction has thus to be learned empirically. If both worlds, drawn and built, are experienced together, accurate conclusions can be settled upon for proper time coordination with the craftspeople.

In order to be able to lead the materialization process, more than social coordination skills are needed.

One has to be familiar with the actual means of fabrication; for instance with ways to account for time-related dimensional tolerance adjustments, or for general preparations that do not exist in drawings. Without the knowledge of tools, processes and appropriate material techniques, it is hardly possible to maintain credibility as a conductor, in charge of a project's outcome and success.

How should one develop these skills? Again, close collaboration with contractors, fabricators and engineers is generative in learning time estimation; shop internships should be mandatory. It is also important to engage in individual experiments in materialization, i.e. design-build furniture. Curiosity is key! With the ambition to complete the 3-D object with the same precision as the perfect 2-D drawing, one will probably find that accuracy in 3-D objects is more difficult to achieve than in the virtual world.

This activity will establish a relationship with experts in these fields. It also establishes know-how related to the particular professional expectations encountered when designing in a negotiative process. And, it creates the opportunity to compare aesthetic goals with time- and labor-related conditions, in the best case, to advance both of these goals.

It is well-known that one of prefabricated design's main principles is the reduction of elements and the overall number of pieces. This is related to labor costs, transportation and the logistics of assembly. The consequence is not necessarily a less beautiful product, but the ability to streamline architectural production in a comprehensive way when it is already in its 2-D design phase, either customized or serial.

CONCLUSION

Successful architecture is dependent on countless factors, but in the end is – or should be – read as a single artifact: consistent, differentiated and comprehensible. Coordination, especially in the translation between idea and execution, is one of its most urgent issues. A beautiful artifact is more than the sum of its part: a set of drawings can still give rise to dozens of different interpretative possibilities. We have to be responsible to limit its interpretation in order to bring consistency and harmony to the product's final appearance.

In order to negotiate between the conceptual idea and the possibilities of material realization, it is necessary to wear different hats: within the task of prototyping, the hat of the manufacturer, or of the fabricator is a very important one. The architect has to be part contractor to anticipate the means of construction. At the same time, his eyes must be those of the vigilant entrepreneur, able to pursue both the big picture and the small detail simultaneously. Different methods, from drawings to mock-ups, must be utilized to conceive architectural form. The information that does not inherently exist in drawings has to be anticipated nonetheless. It includes time, sequence, tolerances, weight, performance criteria, energy consumption, and also upkeep and aging. Only then can the design product prove itself during construction and beyond, when it is finally adapted to its environment. The anticipation of all conditions in order to create a superior artifact is the professional task.

Therefore, the focus on materialization has also to reconnect to the cultural conditions and vision of how we can live better and more sustainably. The political 'hat', tied to policies determining rules and regulations in our urban environment, was thus also significant for our Roofpod project. Amidst the obvious shortcomings of how we currently frame our lives (see Chapter 1), the architect must be a vital contributor to renewing our environment.

3.40

3.40 Time-labor relationship, graphically explained: "Galaxies of design". Different time scales of production processes have to anticipate the difficulty of construction.

The creative challenge is to address how to make the necessary changes tangible, so that people relate to and identify with them, but make them viable and feasible, too. The proposals, to develop our city and make any addition to its infrastructure meaningful, need to be seen with fresh eyes as well. The developer's hat needs creative refashioning.

The architect must have the ability to connect design factors, working at many different scales. The architect is not specialized in broad-based thinking, but rather in the ability to work competently among different scales and factors, representing those contrary forces which must be cross-referenced in order to form a whole: money and beauty, big and small, nature and built environment, the measured and the emotional, future and past, the abstract and the concrete.

It is clear that the process of inventing and ordering, of imagining and prioritizing, has to be conducted extremely carefully and professionally, since there usually is only a single chance to materialize a project. After the project is done, everything else returns to the realm of abstraction. "What if" becomes irrelevant. Architectural design has to be spot-on. Even if planning occurs slowly, it has to be right before it goes into production.

Creativity is based on experimentation. Prototypes are an expression of materialized creativity, of originality and innovation. In architecture, where product development is dependent on ingenuity, – one-offs are predominant – influences that foster creativity have to be supported, while tendencies that undermine this ability have to be fought. Ingenuity in production and material refinement belongs to the realm of architectural creativity. In fact, it is one of its foundations.

Sigmund Freud quoted the 18th century German poet Friedrich Schiller's letter to a friend. The friend complained about his lack of creative power. Schiller replied with this allegory: "In isolation, an idea may be quite insignificant, and venturesome in the extreme, but it may acquire importance from an idea which follows it... In the case of a creative mind, it seems to me, the intellect has withdrawn its watcher from the gate, and the ideas rush in pell-mell, and only then does it review and inspect the multitude. You are ashamed or afraid of the momentary and passing madness which is found in all real creators, the longer or shorter duration of which distinguishes the thinking artist from the dreamer... You reject too soon and discriminate too severely."[55]

Both are needed: Momentary madness to "review and inspect the multitude", as well as the skills to realize what has been imagined. Good architecture contains both – in many different scales – it cannot exist without both. While creative madness is not hard to find, the common skills needed to materialize sophisticated architecture are underdeveloped in today's industrialized world.

	TOPIC	IN CHARGE	INDICATOR
URBAN ENVIRONMENT	CONTEXT APPLICATION SITE CONDITIONS	POLICY MAKER DEVELOPER	1 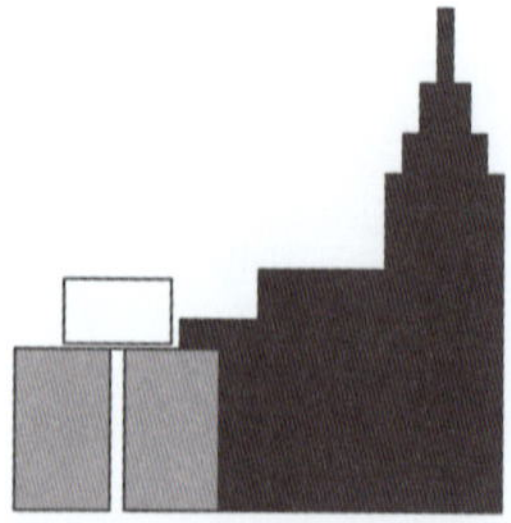
DESIGN PROCESSING	PLANNING, PROTOTYPING SCIENCE AESTHETICS	ARCHITECT ENGINEER	2 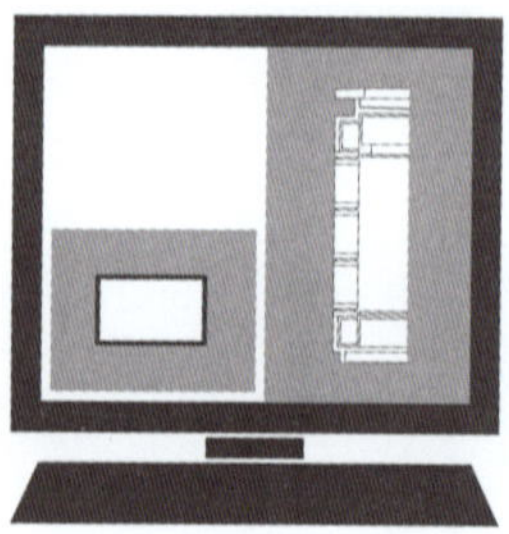
MANUFACTURE/CRAFT	MATERIALS, METHODS PRODUCTION CONDITIONS	MANUFACTURER FABRICATOR CONTRACTOR	3 

'DESIGN-BUILD' KNOW-HOW: PROBLEM SOLVING

RELATED

- ORIENTATION, SHADING, VIEWS
- ADAPTATION: STRUCTURAL, MECHANICAL, ELECTRICAL
- CODES: EGRESS, FAR, FIRE RATINGS, SET-BACKS
- SITE ASSEMBLY: SEQUENCING, LOGISTICS, HOISTING
- MAINTENANCE, UPGRADING, RETROFITTING, REPAIR
- POLITICS: ZONING REGULATIONS, ENERGY, INCENTIVES
- GRID INTERCONNECTION, ENERGY DISTRIBUTION
- REAL ESTATE: PROFITABILITY, RESALE VALUE

- BEAUTY, PROPORTIONS, FORMAL + SPATIAL QUALITIES
- PERFORMANCE: THERMAL BRIDGING, LEAKAGE, GAIN
- MATERIAL KNOW-HOW, MATERIALS INTEGRATION
- ASSEMBLY CONCEPT: SEQUENCE, LABOR, MAINTENANCE
- REPAIRABILITY, FLEXIBILITY, RETROFITABILITY
- BUDGET, COST CONTROL
- ENVIRONMENTAL FOOTPRINT AND PEDAGOGY
- SYSTEMS INTEGRATION

- STANDARDS: PERFORMANCE, CODES, SAFETY
- PRODUCTION: VOLUME, MARKET
- ECONOMIC OPTIMIZATION: COST-PER-UNIT, MODULARITY
- PRODUCTION METHOD: LABOR, EQUIPMENT
- ENVIRONMENT: ENERGY, LONGEVITY, WASTE
- RESISTANCE: FIRE, MOISTURE, STRESS, HEAT/COLD
- ASSEMBLY METHODS, FINISHES, TREATMENTS
- TRANSPORTABILITY

REALIZING THE 'SOLAR ROOFPOD': HANDS-ON DESIGN NEGOTIATION

1
5
6
2
7
3
4

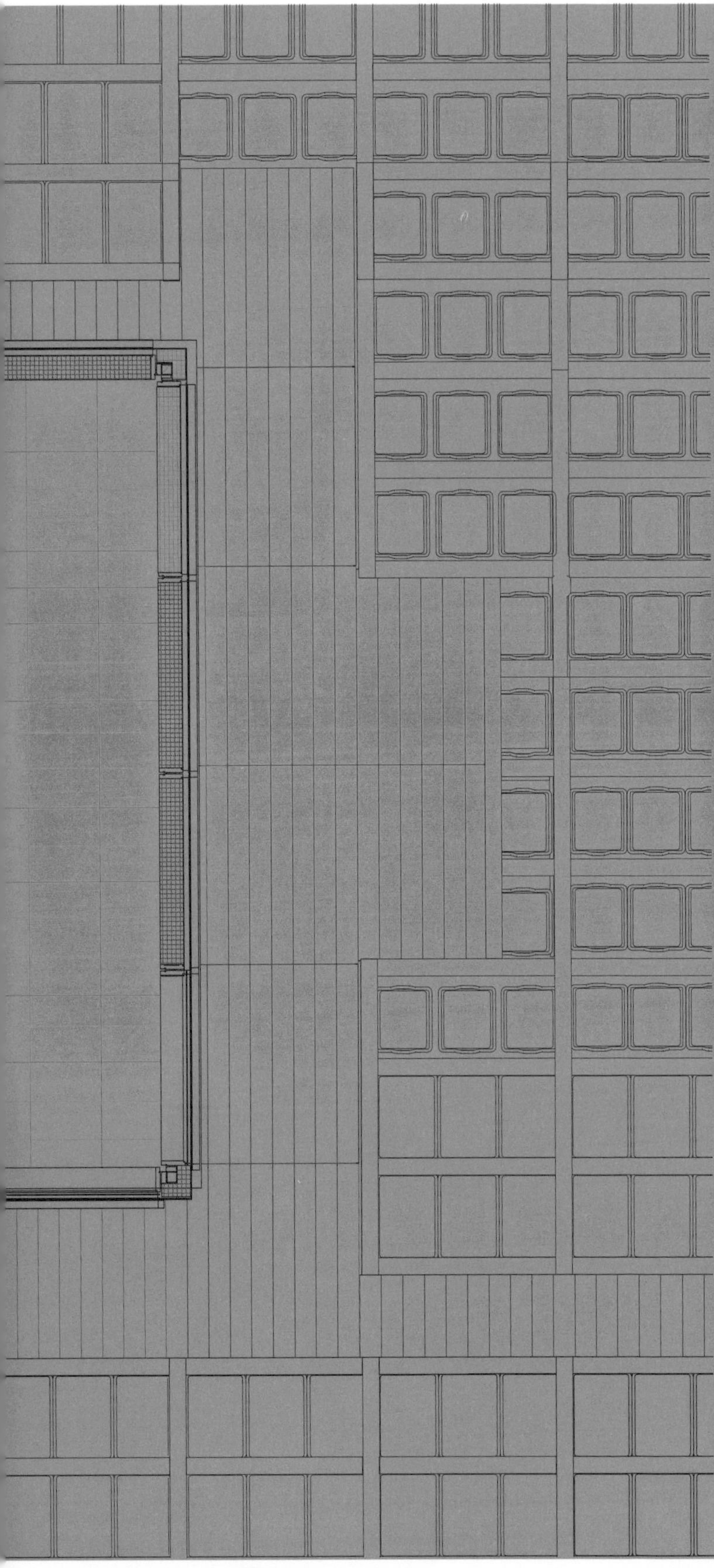

PLAN

1 ENTRANCE

2 LIVING ROOM

3 KITCHEN AREA

4 SOUTH PORCH / GARDEN ACCESS

5 INTERGRATED SLEEPING AREA / MURPHY BED

6 BATHROOM

7 MECHANICAL ROOM

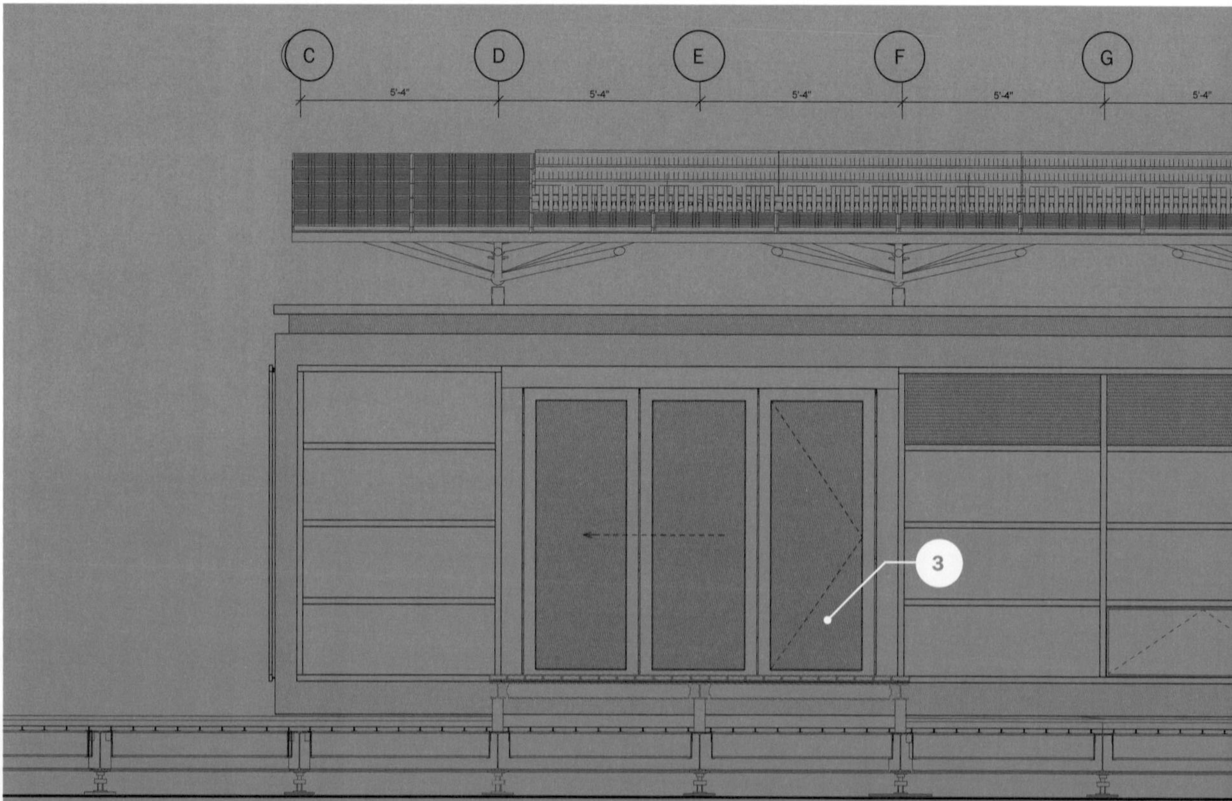
C
D
E
F
G
5'-4"
5'-4"
5'-4"
5'-4"
5'-4"
3

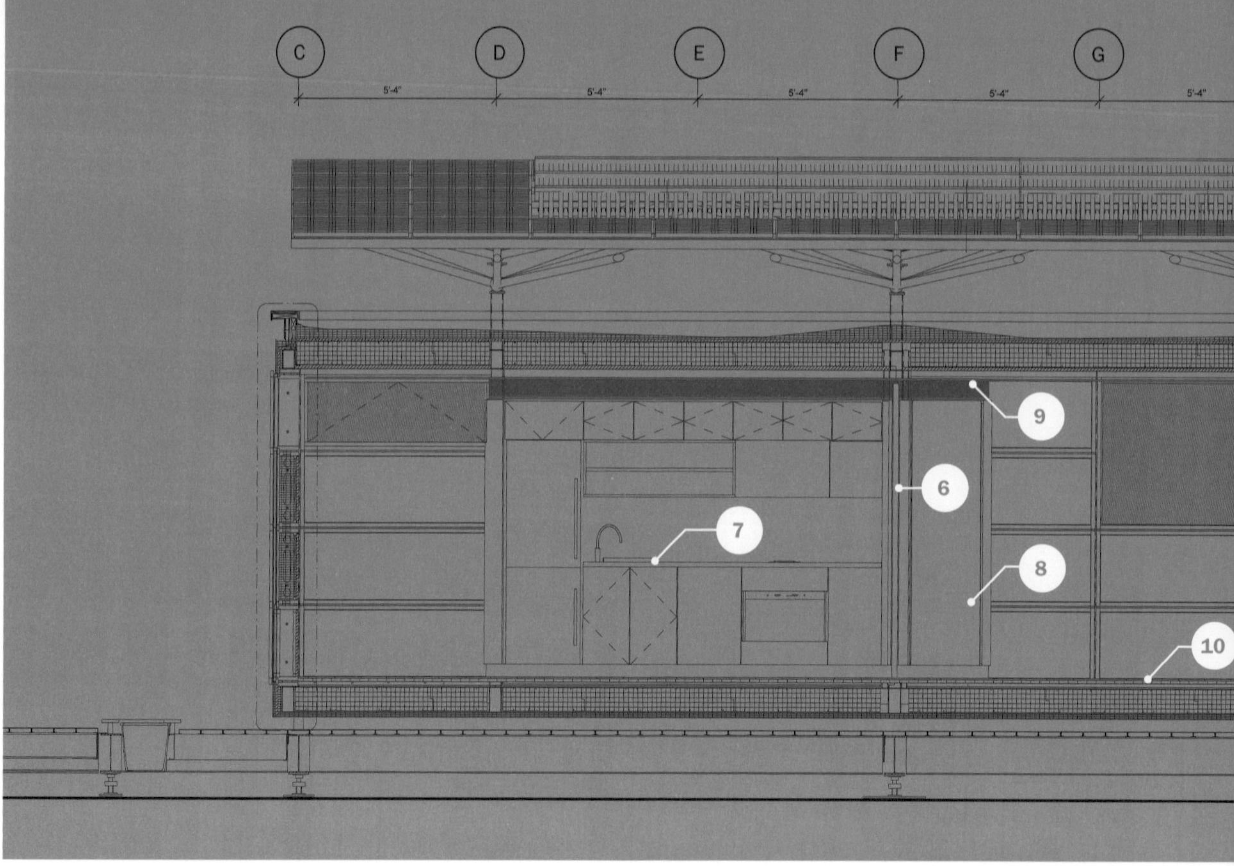
C
D
E
F
G
5'-4"
5'-4"
5'-4"
5'-4"
5'-4"
9
6
7
8
10

SOUTH ELEVATION +SECTION

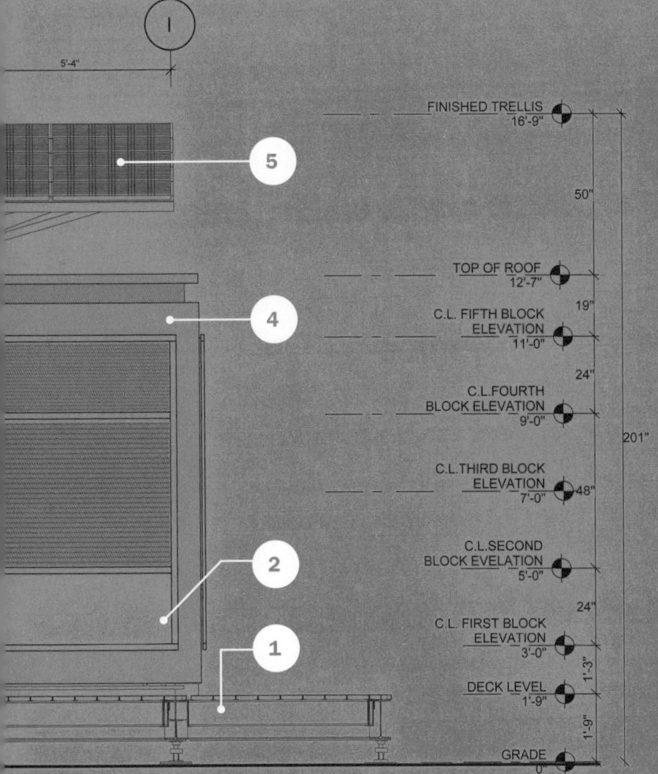

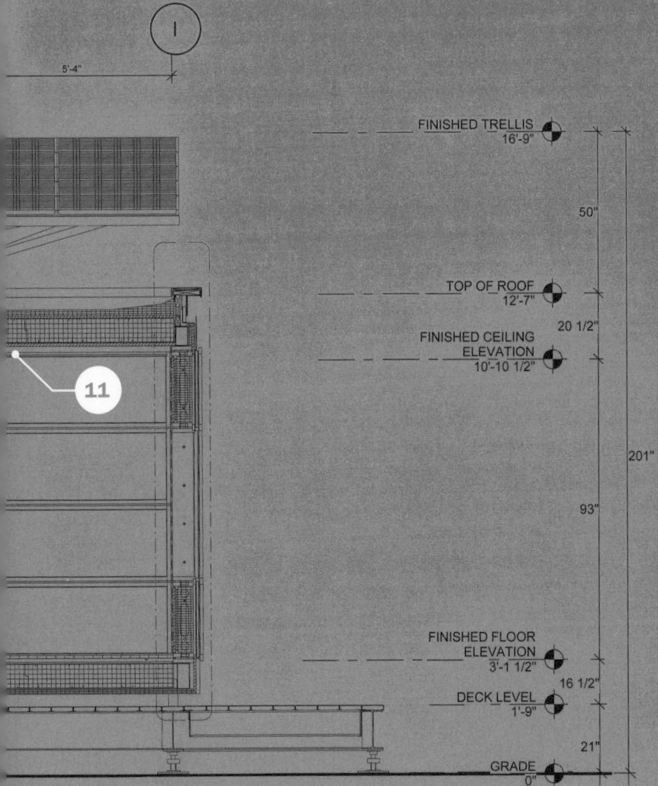

1. DUNNAGE SYSTEM
2. 'BUILDING BLOCK'
3. DOOR UNITS
4. BREAK METAL TRANSITION
5. SOLAR TRELLIS
6. SLIDING DOORS, SPACE DIVIDERS
7. KITCHEN CABINETS, COUNTERTOP + APPLIANCES.
8. LIVING ROOM CABINETS
9. CORE GRILLE: VENTILATION + TOP EDGE
10. CORK FLOORING, TONGUE AND GROOVE SYSTEM
11. SUSPENDED DRYWALL CEILING, W/COVE LIGHTING PERIMETER.

4.1

0

INTRODUCTION

In architecture, comprehensive design thinking has to drive the act of making. After the thoughts of a design are organized, and a strategy of materialization is articulated on paper, the architect's involvement in the tangible realization of his/her product is often limited. The relationship between materialization skills and product in the profession of architecture is a peculiar one. While a cabinetmaker achieves a concrete output of handiwork in the form of his cabinets, or a pianist demonstrates his/her skills immediately in the performance, the architect's output in construction, the labor that leads to the product of a building, is for the most part invisible, and usually not done with their own crafting abilities.

It is quite easy to explain what a cabinetmaker or pianist does professionally. However, when asked what an architect does, people often have issues with defining the actual activities of the profession in a way that relates to building. "The architect produces drawings", is what is often said. Nevertheless, the conclusive subject of the architect's work has to be the final product: the building. The drawing is nothing more than a clever device of communication that is used as a tool to involve many others in the complex act of building itself. Thus, this tool is only effective if it is based on construction skills. Yet, in school we learn to draw without exposure to building, meaning that control of this tool remains undeveloped. In order to compensate for these deficiencies, we have to start building as an educational program, and make manufacturing part of the training necessary to realize what has been visualized in an effective way.

This chapter combines planning and construction activities, because they are inherently interconnected. The way of learning construction as a strategy of materialization is documented based on hands-on prototyping. Two overlapping approaches are taken:

A) In the subsequent overview drawings (1-12), aspects of a Roofpod application on a generic NYC rooftop are explained phase by phase, with multiple design criteria vignettes leading to the concrete materialization. Here, the balancing act of diverse design requirements that negotiate between conceptual preferences and realistic construction is described, based on specific categories of construction subsystems.

B) The final materialization strategy is documented in photographs and details, related to the act of construction on campus with the students. These photographs also document the separate phases of construction related to the material subsystems.

The main objective is to look behind the surface and train future professionals to comprehend the depth of design within construction.

This construction is the basis for the product of the building, which the architect claims as his. It would not exist without the architect's knowledge of its composition (as is the case in cabinetmaking or piano playing). The connection between planning and materialization was best expressed by one of the students, when he saw one half of the building being lifted in the air by a crane to load it on a trailer for transport to the Solar Decathlon exhibit in Washington DC. "Look, my section!" the student shouted, full of excitement.

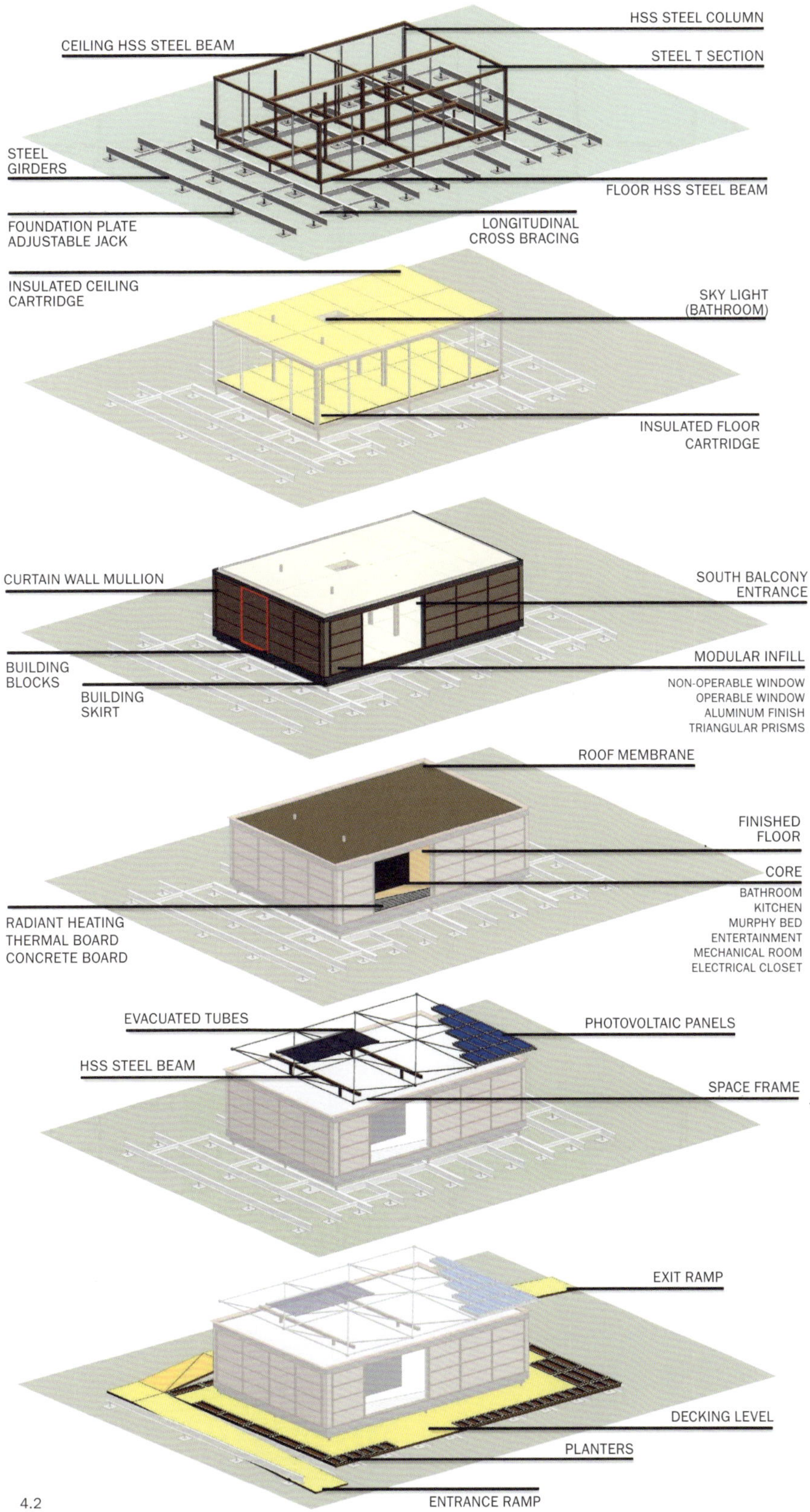

4.2

4.1 SD11, DC: Rigging of sectional modules.
4.2 CCNY: Understanding the project in construction phases. By identifying crucial design elements and their related sequence of assembly, students were able to identify separate subordinate design problems. Specific groups managed the materialization of significant subsystems within the 'kit of parts' strategy. (Design Studio, 11/2010)

The urban environment is a system in itself, with countless subsystems. It is a manifestation of our culture, of how we think and how we adjust to and address changes. The water towers, an intrinsic and distinctive element of New York's skyline, are actually an example of the adjustment of a subsystem's functionality. They were built when the water supply pressure had to be regulated to serve the increasingly diversifying building stock. The add-on of the 'Roofpod' to a residential building wants to achieve something similar: to become part of the urban infrastructure, and make a contribution to a more efficient city. At the same time, it connects with one of the most signicant assets of New York; the views towards an extremely urban environment: the skyline, experienced from within.

An abundance of environmental loads is threatening the longevity of the roofs of our buildings. Sun, wind and rain are affecting the roof membrane and interior climate in a negative way. The effect of UV light, high heat, temperature-related contraction and expansion, moisture penetration, etc. make it necessary to repair roofs frequently. The strategy of warding off these loads might be just wrong. The approach demonstrated instead shows how to take advantage of them, in order to develop a more energy-efficient and environmentally friendly way of powering our buildings, and with it the entire city. These impacts of nature are not a threat, but a potential form of income through harvesting sun, wind and water and responding appropriately to seasonal and diurnal fluctuations.

Usually the first expense for a building project is dedicated to the purchase of the ground; this site is for free. The roof already belongs to the Coop, and can be instantly surveyed for best usage. The most suitable building system must be highly adaptable for different roof sizes, orientations, views and functions. The 'Roofpod' must also be able to transition easily to the structure and MEP systems of the host building. To be competitive in the construction and real estate market, it must be able to allow for fast assembly and versatile marketing. A customizable building kit of parts, mainly transportable to the roof by elevator or hoisting, has to be able to accommodate personal preferences, cutting edge in its technology and superior in its green energy efficiency goals.

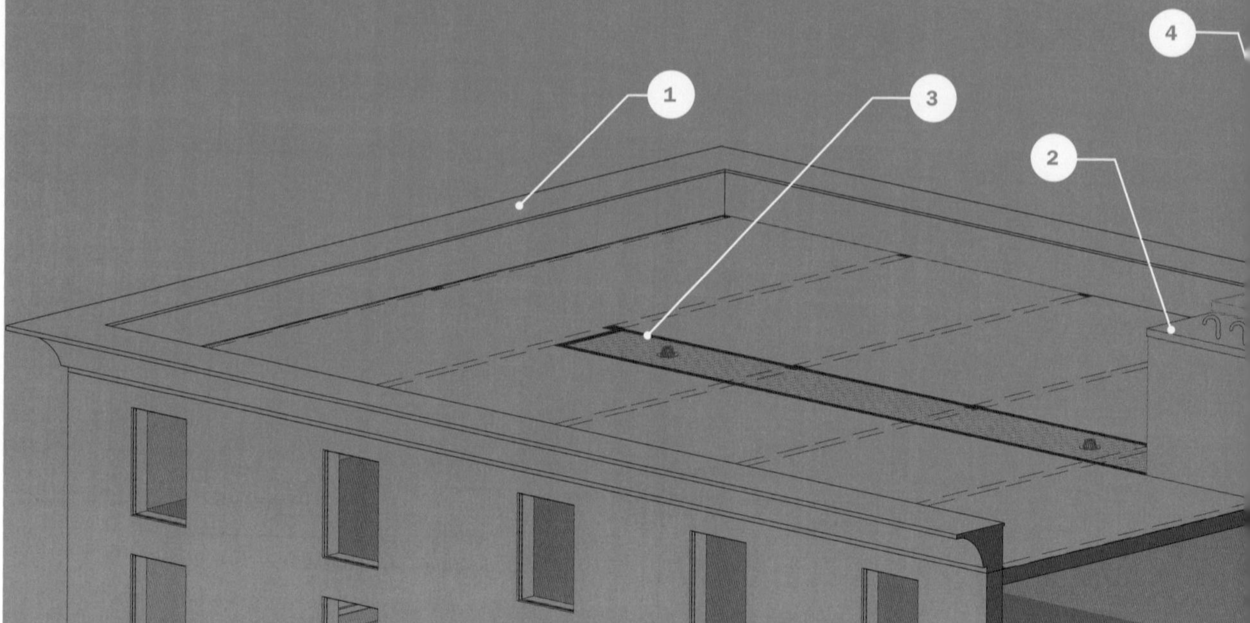

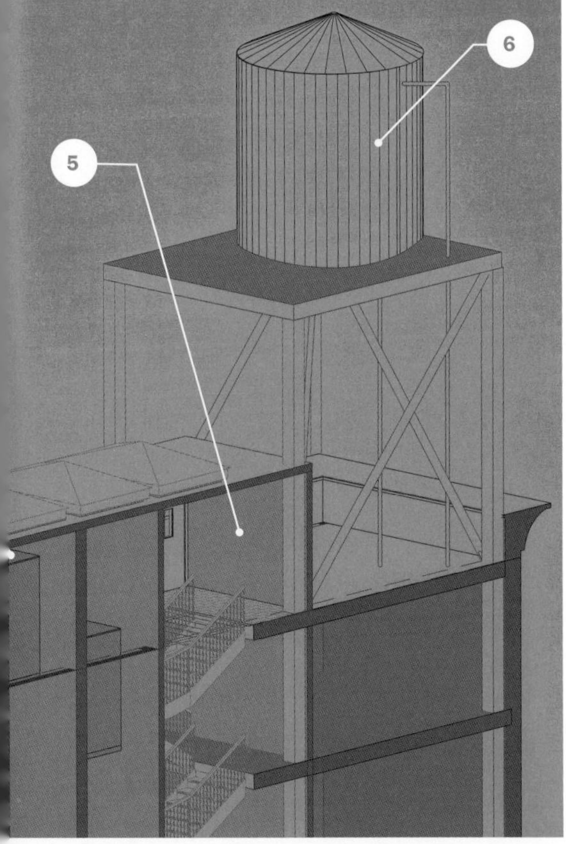

SITE

1

1 BUILDING PARAPET / CORNICE

2 RISER SHAFT AND MEP DISTRIBUTION

3 DRAINS, DISTRIBUTED TO RISER

4 ELEVATOR SHAFT BULKHEAD

5 STAIRCASE (MEANS OF EGRESS)

6 WATER TOWER (“SKYLINE BEACON”)

1

SYSTEM 1: SITE – HOST BUILDING'S INFRASTRUCTURE

4.3

Many of New York's apartment buildings are based on similar building systems, since they were predominantly constructed within the same time period. Additionally, common systems were employed because construction of large-scale residential buildings represented a substantial cost and thus the building industry was heavily involved in the development of such construction systems. Building codes also heavily influenced the way residential buildings were constructed, specifically fire-protection related regulations and certificates complying with these requirements. In New York, many of these buildings are constructed as steel frames, with masonry shells to provide fire resistance. Steel construction was the preferred material choice at the time, since the USA had been one of the strongest steel manufacturers worldwide since the beginning of the industrial revolution. Steel has several other advantages in that it is relatively inexpensive, and the assembly of a steel frame is one of the faster ways to configure a building when considering the timeframe of construction.

We investigated the apartment buildings on the extended Upper West Side, close to our school, predominantly built around the early decades of the twentieth century, which were rendered accessible with the opening of the subway system in 1904. The advent of the subway not only revolutionized public transportation, but also changed the way people lived in the city, when this building type superseded the once common townhouse. At that point, the buildings that began dominating the city were about 10 stories high, and contained multiple apartments and common amenities. Most of the buildings were built to the same regulations and with the same builders, resulting in an appearance with similar detailing and a consistent cornice height. It was this homogeneous building height that was important for our study, because any shading of the Roofpods installed on these roofs from surrounding buildings would be detrimental to generating solar power.

This building type also has the potential to be intrinsically energy-efficient, since the façade and roof surfaces are relatively small compared to the confined volume (A/V ratio). Thus, any energy losses due to the compactness of the building could be minimized. However, these exterior walls and roofs are not well insulated, since at the time of construction, fuel was inexpensive and global warming was unknown[56]. It goes without saying that most of the exterior envelopes of our urban buildings are in need of substantial improvement.

The roof itself is one of the most important areas and a practical method to improve the thermal resistance of the apartment building's envelope. This area is relatively easy to improve, since there are no façade details, nor potential transgressions beyond the property line, which would complicate the same undertaking for the façades. Also, applying an additional layer of insulation is a straightforward approach, since the currently typical roof membranes have to be fixed or replaced every ten to twenty years anyway: an upgrade is possible within the regular maintenance plan of the building. Better yet, this solution addresses convection losses, or the loss of heat rising through the roof, which is substantial. We thus assumed that this upgrade will take place as a prerequisite to our intervention.

4.3 Upper Westside: Predominantly apartment buildings. A new building typology accessed by new means of transportation.
4.4 Typical floor plan of NYC Upper Westside apartment building; simplification of its structure and circulation system, to develop a 'starting point' for a generalized approach.

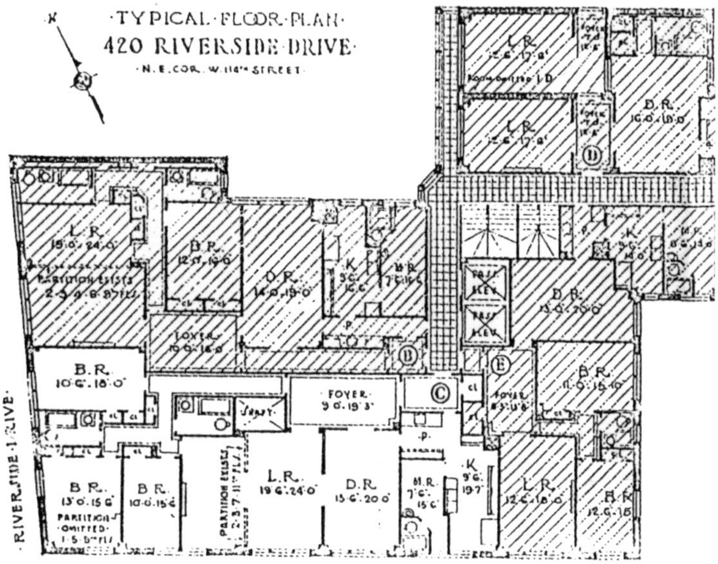

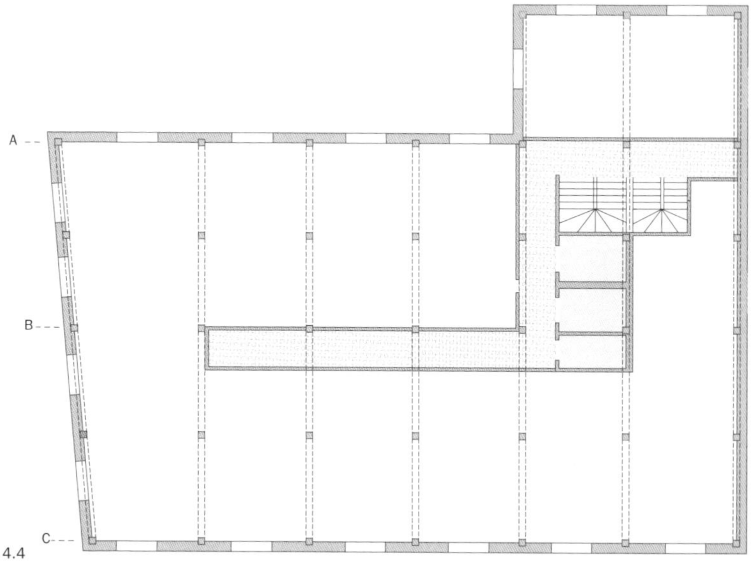

4.4

We chose a building on Riverside Drive that represents this building type (4.4). Typically, the interior circulation is organized in a double-loaded corridor configuration, with apartment units paired on both sides. The plan spreads out into different wings, and concentrates at the elevator, staircase and plumbing riser core. The structural system was commonly organized in a rationalized grid. In such cases, the columns are located at one of the two corridor walls in the center of the building, and at the exterior perimeter walls of the building (4.4: **A**,**B**,**C**), so that they do not interfere with the desired apartment layouts. Even though the structural frame is not always regular, there are certain typological characteristics that can be generalized when inventing a system that works well on the majority of these buildings. Of particular difficulty was the task of relating to the dominant modular characteristics of steel framing.

However, many of the traditional methods eventually faded and there are now countless different influences determining the frame geometry of a building. We decided therefore that our modular system would not relate as much to the host building, but rather to the practicality of installing the assembly, and the coordination of available elements involved in this configuration.

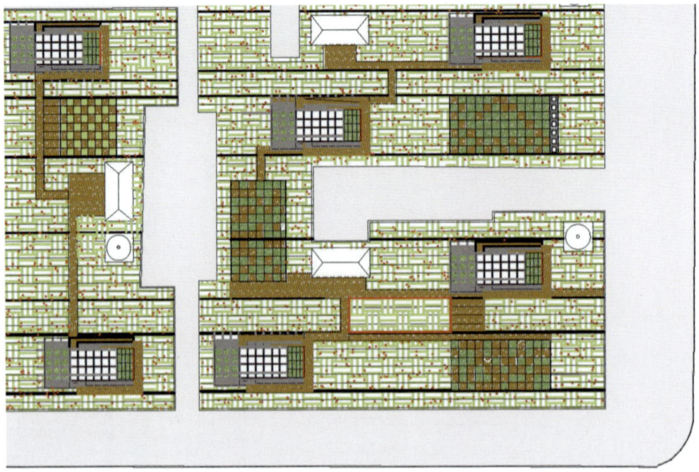

4.5

The simplification of the structural system shown in image 4.4 suggests many ways of adjusting to the existing frame of the building after the structure is known. A structural engineer would have to design this system anyway, and verify load capacity and particular detailing. This same process happened, for instance, when the water towers in New York were built. Feasible strategies to carry additional loads were addressed, and compared to each other to assess their advantages and disadvantages.

In order to engage in these assessments for our intervention, it was important to presenting a case that is specific, and can also be superimposed for a more vast and general application. It thus constitutes a prototypical test case!

Of course there are obstacles that have to be overcome in order to make the idea compliant with the current building codes. However, codes are, and should be, somewhat flexible, and depend on both policies and cultural development. The benefits of such an addition to a building suggest a discussion in two directions: What has to be changed within the building design to make the addition of a Roofpod a safe feature for its built base? And, as is the case for other benefits desired by the city, a concept has to be developed to facilitate codes responding to the addition that promote Roofpod applications at a larger scale.

The main topics to be reviewed in this respect relate to the following issues:

1. Means of egress: The NY Building code requires two means of egress for every apartment. There must be two independent ways of escape from any place of occupancy. Since nobody has yet lived on the roofs of these buildings, often only one staircase leads to it, and is in place for purposes of maintenance and general access. This is a shortfall for most of our potential rooftop building sites. However, since the way of escape always runs from upper to lower floors, the volume of people required by code for a protected escape from the roof is much smaller than from that of the lower floors. The Roofpod has room for approximately two occupants in its 800 s.f. of living space. One might then ask: how many people at maximum would spend time in or around this small annex? Or: how many staircases would be necessary to safely serve this quantity of people? In our opinion, the reduced number of possible escapees would justify a less stringent way of egress.

2. Handicapped accessibility (ADA): If handicapped accessibility were compulsory, direct access to the roof by elevator would be required. Since the last floor typically accessed by an elevator is the top floor, the entire elevator shaft would have to be extended to the roof. Therefore, the machine that pulls the elevator up and down would also have to be moved up a floor. The additional budget necessary would be significant.

But, there might be exceptions for certain applications, based on the interpretation of major life activities. There are currently new apartment buildings being approved, wherein only a limited number of apartments are required to be handicapped accessible. Since a residence is not a public place, one could argue that access to the roof is not mandatory. If, however, the elevator has to be extended, a developer would only be interested if the rentable square footage on the roof justified the additional expense. The singular Roofpod would have to become one of a series in a dwelling system in and of itself to satisfy this change in strategy (4.5).

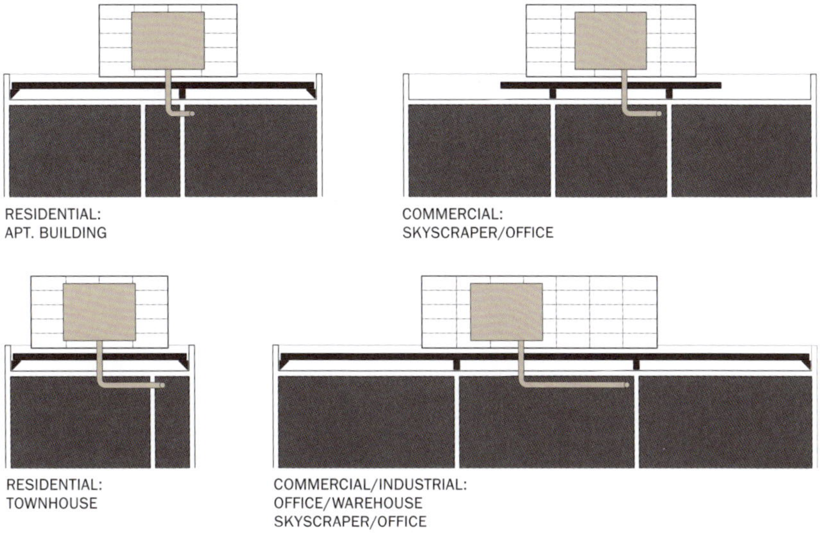

4.6

4.5 Application of a dwelling system: Many Roofpods on the same building could be accommodated, if particularly egress and structurally related issues were resolved.
4.6 Adaptation to structural system and mechanical circulation, related to different building typologies.

3. Floor Area Ratio (FAR): The same strategy was important in addressing whether additional floor space was available for adding to an existing building. Many buildings are "maxed out" in their FAR, due to high capital pressure in the real estate market.

However, by studying the history of how FAR has been used as a political instrument encouraging developers to focus on improving the city, one realizes that many different kinds of policies were promoted in the process. Therefore, it doesn't seem too speculative to promote a policy that would supply additional clean energy to the city, which is much in need of additional sources. All the other positive side effects, such as life cycle extension of roof membranes, storm water management, reduction of cooling loads, etc., would only strengthen the case. Setback rules and other regulations could easily be integrated, since the Roofpod would not fill the entire area, and its modular system is flexible to accommodate variety.

The main fact is indisputable: The environmental loads that our roofs are exposed to have detrimental effects on the buildings' materials, and the city in general. The typical response of warding off these loads is not sophisticated, and relates to the fossil-fuel era, where cheap fuel regulated any thermal issues or power deficits with the push of a button. Much more could be accomplished if sun, wind and water were considered as efficiently managed resources. And thus, these adapted roofs would become 'learning labs', in order to experience and understand how to most efficiently take advantage of these loads.

The requisite investment is minimal, in the sense that the 'ground' is already purchased, and the distribution system within the host building, made up of plumbing risers extending from basement to roof, is in place. This is the case for all buildings, since the ventilation lines that ensure air can escape when displaced by water in the plumbing system, are required by code. Therefore, the direct connection of any additional roof infrastructure can easily be tied into the building. On the other end, the connection of the host building to the city's below ground infrastructure, from sewer to power grid, is also already in place.

It is well known that one of the most important considerations in sustainability is the utilization of existing resources and infrastructure, and the most efficient integration of a new device into the overall system[57]. If the groundwork is right, the new device will have exponentially superior functionality. By tying the Roofpod into the existing urban infrastructure as well as the infrastructure of the host building, the prospect of creating a 'device' with a better performance is much more likely. For example, any surplus of power generated in the Roofpod would be fed directly into the building, therefore avoiding transmission losses. The system doesn't work for itself, but in connection with a greater demand, it serves the 'host building'.

The Roofpod as a 'device' inherently becomes part of the infrastructure and can be referenced in the skyline, very much like the water towers of New York City. It becomes a cultural fixture of the city, a symbol of infrastructural reorganization. It connects with one of the most significant assets, the views towards an urban environment: the skyline, experienced from within.

In order to assemble time and material efficiently, the transfer beams – configuring the dunnage system – must be sized in an optimized way. They must be continuous enough to avoid unnecessary connections but also small enough, to allow for simple rigging and assembly. Before this part of the installation can start, the building connections have to be established: roof penetrations for structural purposes; column extensions and parapet brackets are necessary to make a connection. Also here the same rule counts: the fewer elements, the more efficient the installation. In this example, the two parapet walls and loadbearing columns adjacent to the center corridor are thus utilized.

Countless buildings in New York City have additional floor area available, defined in the zoning regulations. One of the best ways to extend a building in a dense environment is probably the roof. Not only is there plenty of space available, one can also take advantage of the existing 'infrastructure' of the host building underneath, and – by upgrading the roof insulation and membrane when preparing for the dunnage installation –, one can profit from several incentive programs related to improving energy efficiency and environmental impact. However, there are also several code obstacles to be overcome, which would need the attention of policy makers.

Roof penetrations are expensive. The puncturing of the roof makes extensive repair work of the membrane necessary. The connection points should therefore be reduced to the minimum amount possible. In the early design stages of the project,we considered using a 'space frame' which would be much lighter in weight, and able to distribute the loads with a minimum of support points. However, the space frame, and other design alternatives (aluminum, or cellular beams), were too expensive for the budget. The economic attractiveness of a prototype is dependent on the right choice of materials. In a mass-produced setting, 'unit cost' and design possibilities influence each other; fine-tuning is key.

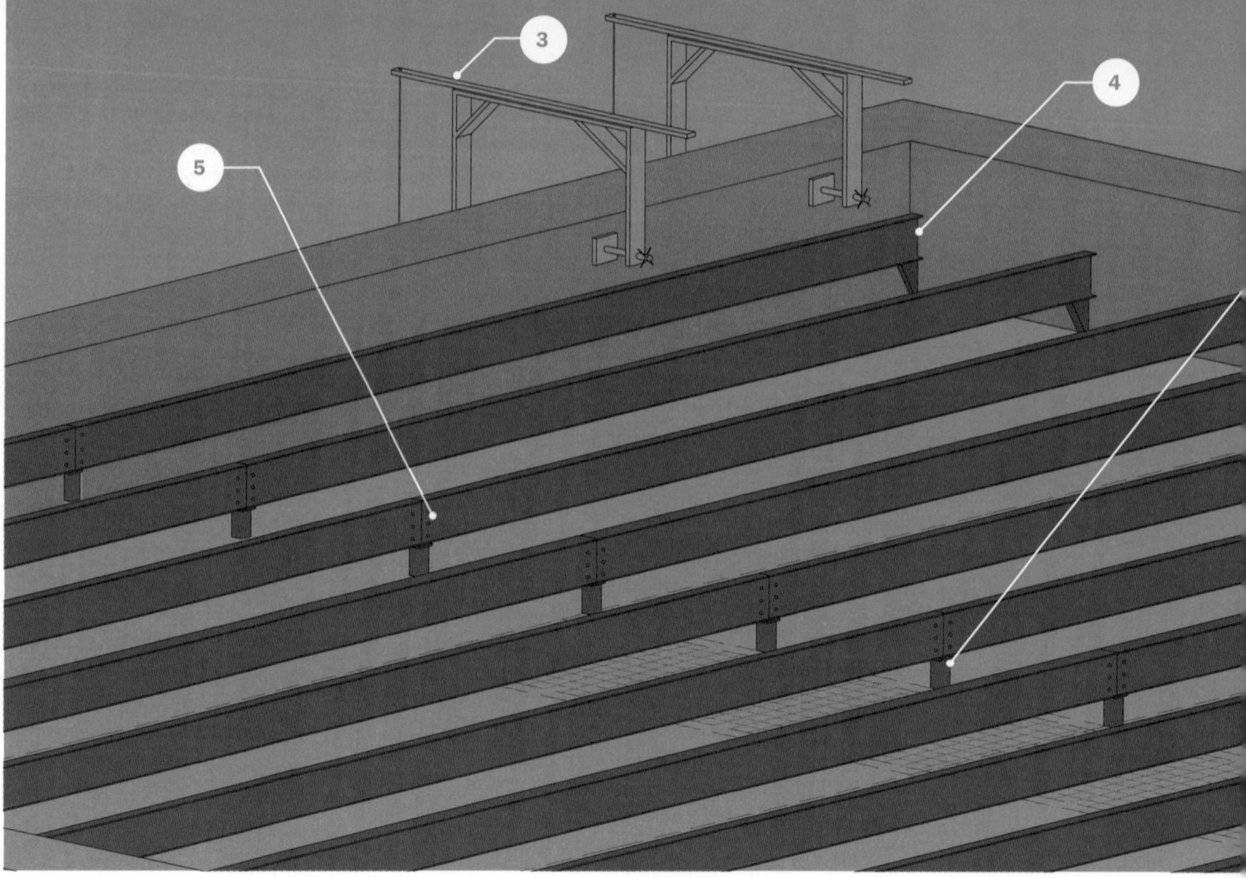

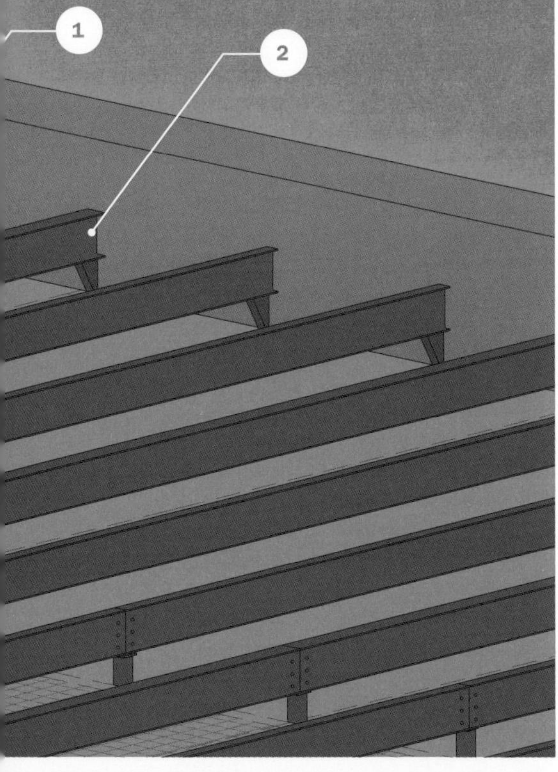

DUNNAGE: LOAD TRANSFER

2

1. ROOF PENETRATION, WITH STEEL STANCHION TO EXTEND COLUMN
2. PARAPET CONSOLE, WITH STEEL BEARING PLATE
3. HOISTING: DEMOUNTABLE PARAPET OUTRIGGER
4. DUNNAGE: STEEL WIDEFLANGE BEAMS (W 12 X 22, TYP.)
5. SPLICE: STEEL BEAM CONNECTION CLOSE TO CENTER STANCHION

2

SYSTEM 2: DUNNAGE –LOAD TRANSFER

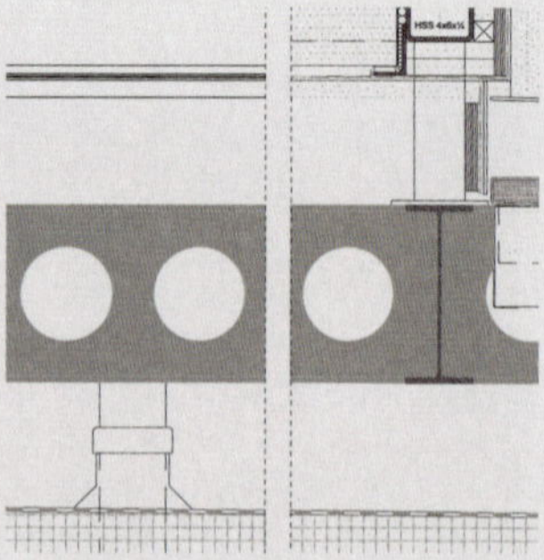

4.7

The structural transfer system between the Roofpod and the anticipated host building underneath was a complex design problem. On the one hand, the design had to be conceived to be as generic as possible, so that the structural system could adapt to diverse flat roof shapes and types. On the other hand, the prototype had to be designed more specifically to accommodate travel to Washington DC for the Solar Decathlon exhibit. Furthermore, the geometry of this transfer system, called dunnage, would establish the dimensional grid for the project's kits of parts and all subsequent structure and design. Thus, the structural optimization of the dunnage material, and the simplification of all related components had to be coordinated carefully. The results of this investigation weren't necessarily consistent. For instance, the sizing of steel sections relates to structural efficiency, but the sizing of inserted units is subject to handling and transport as well as to coordination with standard building material dimensions. While the former is more economic with longer building elements, the latter profits from shorter and lighter units. And of course, both investigations were related to budget constraints.

One of the early exemplar houses we discovered was based on a modular grid system. This was the Almere House by Benthem Crouwel Architects in Holland (4.8). This house had to be taken off the site without leaving a trace after the lot's 5-year lease commenced, and be transported to a different site with minimal effort. Thus, it shared similar transportation issues with our project. The architects stated that they met their own requirement to develop a prototype that would equally minimize costs, weight, amount and length of building elements. The amount and length of building elements was of primary consideration to us, since it constitutes the genesis of a coordinated kit-of-parts. The foundation of the Almere house required a minimum number of contact points to the site as the lot ultimately had to be converted back to its original state. Therefore, a space frame system was employed, effectively separating the house from the ground, and enabling the transfer of all building loads onto four leveled concrete pavers (4.9). The minimal attachment to the ground matched our requirement for an optimized connection to the host building underneath: the fewer connections required, the less thermal bridging would occur, the less labor would be necessary, and the lower the costs would be. The lightweight, structurally efficient space frame was attractive to us since minimizing weight on an existing roof would increase the feasibility of our proposal. Therefore, the case study's transfer system had many qualities we were looking for.

The students started their research on prefabricated space frame assemblies, and soon understood their higher costs due to sophisticated joinery and thus greater effort in processing. The budget was insufficient to cover this system for the house as well as the surrounding garden, which continued the same structural arrangement. Thus we had to change our plan, and our way of thinking, which led to studying transfer systems for mechanical installations, such as water towers on building

4.8

4.7 Detail: Dunnage system transition node.
4.8, 4.9 Almere House, Benthem Crouwel Architects.
4.10 Schematic design: the transfer system to the host building sets up the overall geometry for the modular assembly.

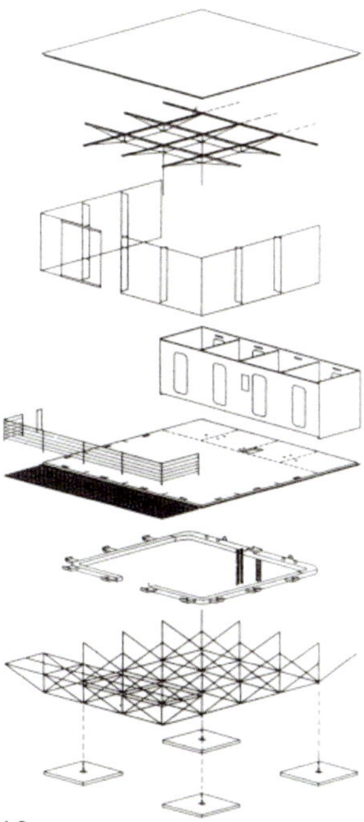
4.9

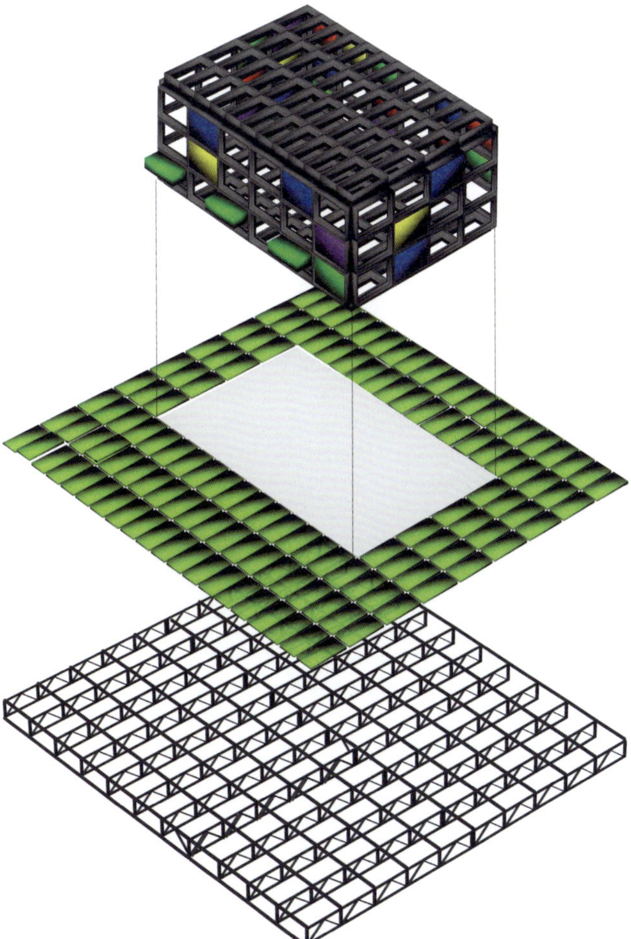
4.10

4.11

4.12

roofs that were not intended to support such weight in their original design. Most scenarios utilize a roof dunnage, which consists of steel beams spanning parallel load bearing walls at the roof level, essentially creating a bridge that can support bulky installations. We also realized that our preferred space frame system, resulting in an elegant floating effect, would be concealed by the host building's parapet walls. Therefore, it was not worth spending a significant amount of money on this item. For the exhibit in Washington we decided to expose the end of the dunnage beams at the front edge of the ensemble (see 4.14), since we recognized the importance of this system in the design process. We were also concerned about the increase in weight due to relying on the heavier dunnage steel beams.

Concerns emerged regarding the tools and handling it necessitated. The prototype perimeter dimensions were about 40 x 60 feet; therefore each dunnage beam was 40 feet long, resulting in beams weighing about 800 to 1000 pounds in total. Investigations made, in effort to reduce these 'dead loads' yielded changing the material to aluminum beams (+:1/3 weight of steel; -: very high embodied energy), or using castellated/cellular (4.12) beams (+: space for plumbing installations; -: higher structural depth). However, neither option was in the end affordable for our application. Thus, we had to live with common carbon steel beams, which were about one foot deep. We spliced (4.13) them in half, sections of roughly 20 feet in length, which could be handled by two strong workers with common tools and installation devices (i.e. roustabouts). To protect these beams against corrosion, they were first coated with several layers of heavy-duty paint, a more economical solution than galvanizing the beams.

The choice of material was directly related to the nature of the project. As a 'one-off' project, taking advantage of the cost reductions offered through industrial mass production was not possible. This is particularly true for the dunnage beams, which are relatively easy to produce as lighter elements (i.e. castellated beams). The choice of material would thus be different if the design were developed as a mass-produced kit.

The decision to switch from a space frame to a simple dunnage system lowered costs, while effectively reducing the amount of labor. However, it increased the dead loads. The concept of using smaller modular pieces, and a transport system that was easy to assemble and disassemble through a mass produced process, had to be adjusted. The common methods of assembly and production associated with the heavy steel dunnage, common to the industry in New York, were prevalent in the design process.

The structural considerations of the Roofpod focused primarily on material efficiency. For instance, since the beams for our prototype were not attached to a roof parapet, we had to resolve bracing the dunnage structure so that the entire assembly

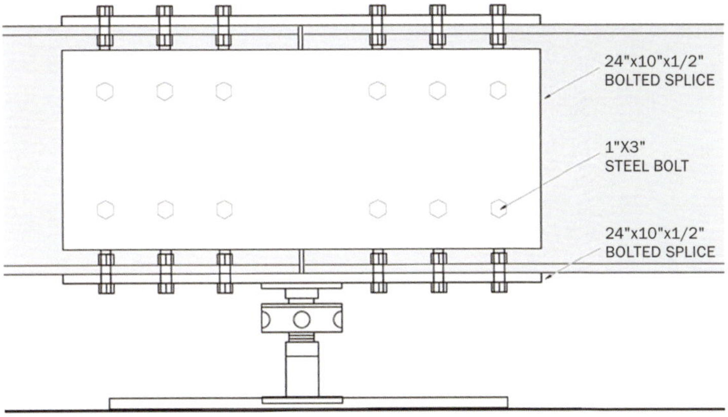

4.13

4.14

4.11 Dunnage beams: The spacing sets up the rhythm for the following construction.
4.12 Cellular beam production with CNC plasma table; the cut halves will be welded back together with full-circle web cut-outs. Light-weighting prefab strategy.
4.13 Splice detail dunnage beams, schematic design.
4.14 Emphasizing the dunnage system for the assembly in Washington, since no parapet attachment was possible in this particular application.

would stay orthogonal. Ultimately, we achieved this through the use of cable cross bracing, which also required additional footings to stabilize the system. In order to accomplish this efficiently, the dunnage beams running below the pavilion center were eliminated, as they didn't carry any of the loads. Unfortunately the pay-off was more footings, which increased the difficulty of leveling the dunnage system and were a significant disadvantage in moving the assembly to Washington.

For such a generalized building system, a more regular configuration with fewer footings, or permanent roof stanchions, was required. Furthermore, the tension cables used to brace the structure could essentially be eliminated. The regularity of the system (4.10) serves the concept of a repetitive kit-of-parts. This is particularly true of the roof garden planting system. The roof-scaping breaks down into units, which can be dropped directly into the dunnage system, thus providing another example of a subordinate system, in this case an armature, that serves as an ordering device.

The visitor ramp, providing access to the Roofpod, was not part of the existing system and was designed exclusively for the Solar Decathlon exhibit in Washington. The ramp was built as an annex to the system. The stringers were fabricated out of glue-laminated beams, and the remaining surfaces and trim were made from cypress. The ramp's assembly sequence relative to the dunnage was of particular interest, and had to be planned ahead. After the dunnage was in place, all the building elements that followed had to be placed from the inside of the lot to the outside in order to maintain access to fasteners, and to be capable of assembly without obstructing subsequent elements. Therefore, after the dunnage was placed and adjusted, the pavilion was positioned on top of it, followed by the surrounding landscape elements, and finally the ramp.

The installation sequence had to be understood well in advance, for if it was not accurately followed, then the system would be rendered obsolete as its parts would not accurately join. The details developed were also based on this particular sequence.

The dunnage functioned initially as the dimensional 'grid' and accuracy in setting up its beams and connecting the subsequent subsystems was therefore vital for the success of the assembly.

For application on a generic New York roof, many pieces, such as, for instance, the access ramp or crossbracing-related details, wouldn't have been necessary. For the kit-of-parts approach, the transfer system should be kept as regular as possible, so that prefabricated manufacturing could be utilized.

The dunnage for the prototype thus became much more complicated than originally intended. However, the main aesthetic function of setting up a 'rhythm of modularization' (4.11) stayed essentially intact.

4.15

4.16

4.17

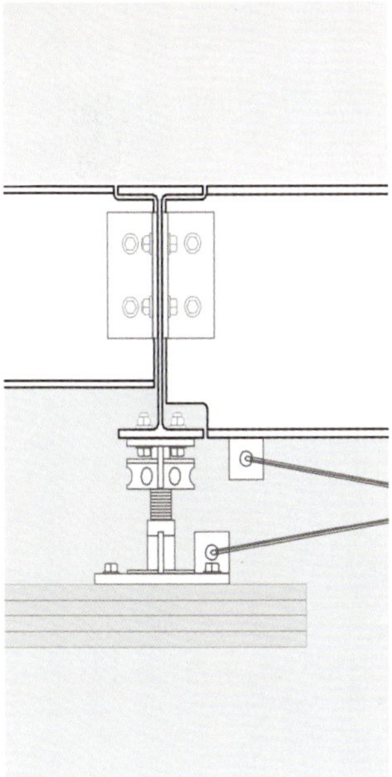

4.18

4.15 Steel dunnage assembly, with splicing.
4.16 The main frame below the house is set up; double footings for extra weight below the Roofpod. The crossbracing will be tensioned later.
4.17 Detail of adjustable footing and shim plate. The plate is necessary for load transition and to roof slope adjustment. The second steel layer is for the ramp support.
4.18 Dunnage detail for the Roofpod construction on campus; typical detail.

In order to set up our urban building system for mass production, the fire-rating of the assembly had to comply to multiple applications and offer the necessary functional flexibility. For reasons that were also related to budget concerns, we decided to configure an 'armature' made of steel, to comply with these requirements in an economic fashion. A mounting system consisting of non-combustible structural members, assisted the space-configuring non-structural 'building blocks' to form the envelope. All the steel was fireproofed by using intumescent paint to comply with the required fire rating.

Although in the end invisible, the steel 'armature' sets up the entire geometry for the house and is related to the appearance of the modular system proportions and the subordinate detailing as well. The design process has to go 'back and forward'. Conclusions about the desired appearance have to inform the structure; structural and 'constructability' criteria have to inform appearance and detailing decisions too. Only with this method can both beauty and practicality be found. Accurate dimensioning for fabricating this crucial subsystem is key.

The steel frame configuration is heavily dependent on rigging the house modules by crane, as this was necessary for an expedient assembly in Washington DC. Therefore, different load cases had to be considered. The suspension from the crane is very different to the typical assembly on the dunnage system. Stability and practical connections, to allow for a safe and speedy assembly, were the main concerns in the design phase. Steel construction results in overall lightweighting and also superior connections, due to strength and diverse welding/bolting possibilities.

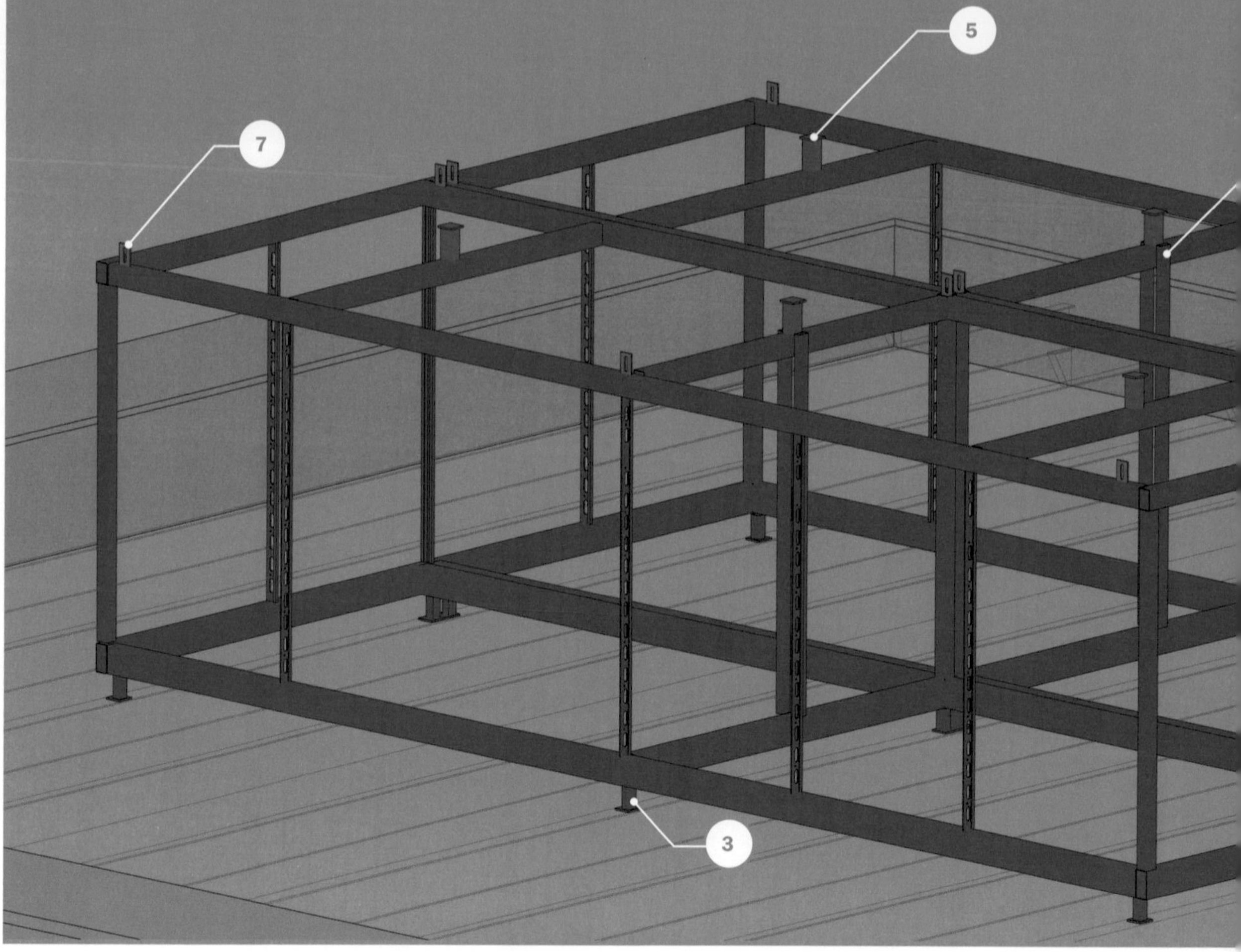

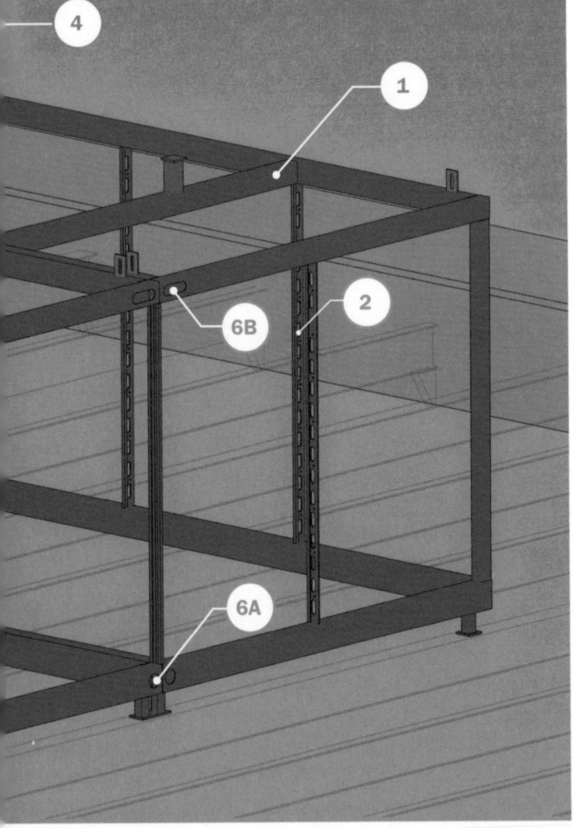

STEEL FRAME: 'ARMATURE'

3

1. MAIN STEEL FRAME, HSS TUBES, 4" X 4", 4" X 6".

2. 'STEEL FINS': MOUNTING SYSTEM FOR 'BUILDING BLOCKS'; LATERAL STIFFENING

3. 'MINI-COLUMNS', MINIMIZED ATTACHMENT TO DUNNAGE

4. SPLIT COLUMNS TO ACCOMMODATE CORE-INTEGRATED SLIDING DOORS

5. STANCHIONS, ATTACHMENT FOR TRELLIS SYSTEM

6. MODULE SPLICE;
A) LOWER
B) UPPER

7. 'LIFTING TAB', ATTACHMENT FOR RIGGING OPERATION

3

SUBSYSTEM 3: STEEL FRAME – STRUCTURAL 'ARMATURE'

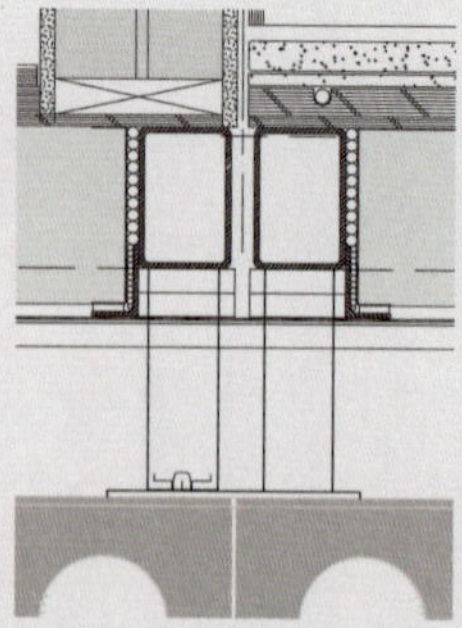

4.19

In the early schematic design phase, we set a goal of building a modular structure wherein its building blocks would configure space and structure at the same time, much like Legos. These blocks were to be used universally, as not only the vertical wall system, but also the floors and ceilings. A similar system, the general panel system (GPS) had been designed by Konrad Wachsmann in the 1940s, after several years of work.

Although the stakes were high to develop this complicated strategy, we tried to make it work. The interconnection of 'building blocks' was of the essence. Several differentiations were discovered: For instance, spanning between support walls meant that only the horizontal blocks had to deal with tensile forces, while vertical blocks could just be stacked. The perceived solution for these horizontal tension blocks was to thread them together with additional tension rods (see 3.13) by providing system holes at the bottom of these blocks (similar in concept to transporting many books in between your hands by applying pressure). But two concerns became apparent:

1) Even though post-tensioning elements to a structural configuration is not totally uncommon, most often these are research projects, developed in controlled testing facilities, with extensive safety precautions. Neither did we have these facilities, nor did we think that such a complicated system had sufficient market potential for the building kit we wanted to develop.

2) The prefabrication concept was governing our structural rationale. Consequently, the structural elements were chopped into smaller pieces, then ultimately reassembled into a continuous structural element again. It appeared that this would not necessarily constitute the most efficient structural utilization of material. Also, the labor and precision necessary to produce a multitude of perfect pieces was of concern.

However exciting this post-tension strategy was, many issues began to surface. First, in order to construct post-tensioned connections, sophisticated equipment is required. Furthermore, the act of post tensioning can be dangerous, potentially causing injury if a tendon snaps. For this reason, the building department would only allow the use of such a system if it was extensively tested and approved by standards and testing agencies. It goes without saying that this would be a far too expensive and time-consuming undertaking.

Also, the energy efficiency of a system that is configured out of many frames is inherently compromised, since the multiplying and joining of structural members produces a redundancy of thermally weak areas. Structural materials are usually not the best insulators. One could argue that wood is an exception to this rule, but only if placed at a depth in the structure so that the material creates sufficient thermal resistance. However, the deeper the structural members become, the more dead weight they accrue. A wood system in this instance would be relatively heavy, not fire-proof, and not recognized by the building code as part of an appropriate construction type. So it would not qualify as a construction system applicable to most of New York's roofs.

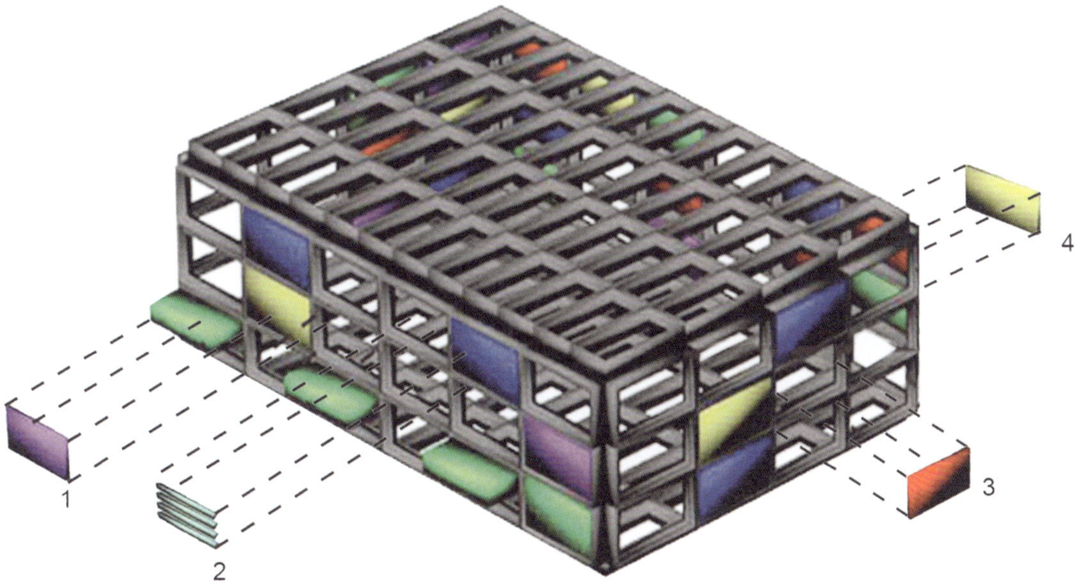

4.20

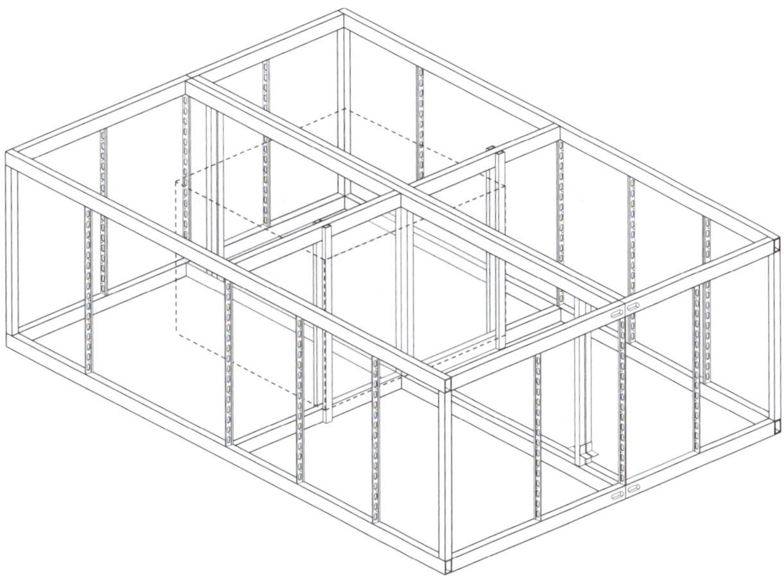
4.21

4.19 Detail: Mini-columns, location finder.
4.20 Conceptual sketch of the initial building block idea during schematic design (with customizable inserts 1,2,3,4).
4.21 Steel 'armature' concept; the result of multiple iterations (code, material efficiency, assembly, transport, etc.).

Finally, dismantling and reassembly of the system for transport to and from the Solar Decathlon exhibit in Washington had to be considered. Point blank, the more pieces required for reassembly, the more danger and complication involved in putting the pieces back together, and the more time-consuming and impractical the system would appear.

It was clear that there was an excessive accumulation of negative factors from a variety of fields, issues such as code, structural efficiency and weight, energy efficiency, and practicality. Thus, it was necessary to rethink the system so that it remained buildable while addressing the issues above. The most important issue was that of code compliance, as it would only make sense to develop a prototype of the Roofpod if it was compatible with the city. And if actual building sites were to materialize, the more possible applications there were, the greater its impact would be!

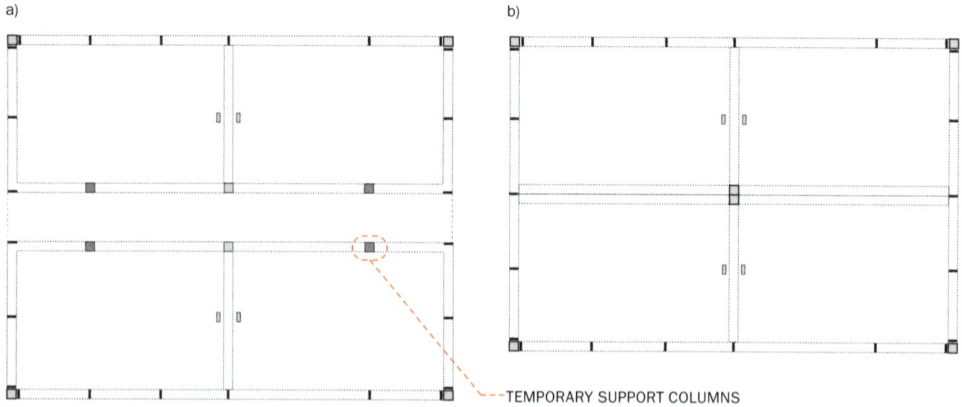

4.22

Since wood was not capable of meeting the project needs, steel was chosen for its affordable, non-combustible properties, and because it is a common material in urban construction. For many reasons, structure and space-configuring elements had to be functionally separated and geometrically coordinated with the use of steel. Much like our previous idea to separate the tension rods from the rest of the structure, a structural armature was created to satisfy code, transportability and assembly requirements, thus giving more freedom and leverage to the main function of the building block: to generate space and thermal performance for an attractive and comfortable interior space.

This being said, the steel armature had to serve and support the block structure's geometry. And for several reasons it was preferable that the armature would use the minimum amount of steel possible. First, it unfortunately conducts heat much more than wood, which is an unfavorable condition. Second, steel corrodes in the presence of condensation, so any amount had to be fully protected by insulation. Third, steel is more expensive than wood.

In order to minimize the steel members, they had to be utilized in the most optimal locations. However, the geometrical system permitted inherent structural members at just the perimeter junctures: corner transitions between floor and wall, wall and wall, and wall and roof, configuring the exterior frame of the building. But these locations were too far away from each other to transmit the existing loads, which concentrate in the center of a structure.

The potential to resolve our need for structural support closer to the center was found in the service core, in the middle of the interior space. This element was created to organize the programmatic functions of the interior, and is accessible from all sides. Suddenly, the service core became a critical element for the structural system. As the design process progressed, more of these overlaps and connections of seemingly unrelated and acutely analyzed building elements began to emerge.

The core columns were one of these evolutions and revelations wherein a design problem can only be solved if considered both holistically and comprehensively. Questions emerged as to whether or not these columns are part of the building's nature, and whether they conceptually belong to the design. Or, to pose the question in a more generative way: how can the design process be steered in a direction to ensure that these columns make sense in a holistic way, not just structurally? The intention was always to keep these columns invisible and repress the steel armature in an effort to emphasize the building block as a modular system. In that respect, the core columns were no exception to the entire steel building frame. Furthermore, the roof was structurally dependent on the center of the building to efficiently transfer the loads. From a functional standpoint, while the core was a vital design element, as it efficiently organized the interior, taking structural advantage of it, the integrated columns do not hinder this functionality. In fact, one aspect of the column's configuration reinforces the functional and flexible qualities of the design.

4.23

4.22: Structural armature, based on transportation sizes: **a)** separate modules, **b)** connected modules; Two modules (W/L/H) of 12'x34'x14' were the right size to be fit on low-boy trailers, for their separate transport to Washington. **4.23** Site installation of steel frame; measuring diagonals to assure orthogonality. **4.24** (next page): Modular system of steel frame and fins, relative to the geometry of the building block design.

The center columns were split into two, with a four-inch space in between, allowing for pocket doors that essentially separate the interior space into two equal halves. This flexibly permits each of the two individual spaces to be climatized differently (see chapter 4.9). The sliding doors mark the middle of the space, in accordance with the structural columns' location. Thus, synergy between the core and structural columns marks another important design development: a buildable, generic solution had been found. Admittedly, this design solution was not as daunting as the original and more ingenious post-tensioned frame system, but several factors in combination generated the obvious and most feasible solution. These factors were geometry, structure, logistics, code and economics.

The steel armature system entails two major elements. The main element is the perimeter frame, consisting of center and perimeter beams and columns, and fabricated from hollow steel sections (HSS) of 4" width and typically 6" depths. The secondary elements are vertical steel fins. These fins are integrated within the vertical joint of two building blocks, and function twofold. First, they resist lateral loads, i.e. wind forces induced on the façades. Additionally, they systematically connect the building blocks with the steel frame, and therefore amalgamate the two systems into one. These interconnections are of the essence, as they are designed to ensure a practical and continuous assembly process between all elements. The steel frame header to the fin, the steel frame sill to the fin, the building block to the fin and the building corner to the building block had therefore to be designed in a way that considered techniques of assembly (4.24). An additional critical aspect to the fin design was that it required a stiff edge to resist bending. This manifested as a hollow square tube of 1x1" joined to the interior side of the fin, necessitating a notch in the building block's jambs to receive it. This groove proved to be a practical element for stacking and aligning the blocks, but posed a challenge in sliding the blocks in between the fins once they were in place. This influenced the sequence of inserting the blocks so that they were placed from one side of the armature to the other, adding the fins one bay at a time. Therefore, the fins had to be removable and their fasteners had to be accessible to both tools and hands.

A special fin had to be designed for the transportation joint, following that the building had to be split in half in order to be transported to Washington DC. The steel fin at this joint was a customized solution to accommodate this separation of the bays. It had to be doubled-up to hold the building blocks safely on each side's end. Several alternatives were developed, but the criteria that eventually led to the final solution addressed the following: structural requirements, preferred assembly sequence, potential tolerance resolution, reassembly after transporting the house, and the proposed fabrication methods.

Another challenge was fine-tuning the thermal efficiency of the assembly. The 6.5-inch deep cavities of the opaque building blocks were completely filled with insulation, greatly exceeding typical thermal standards (see chapter 4.5).

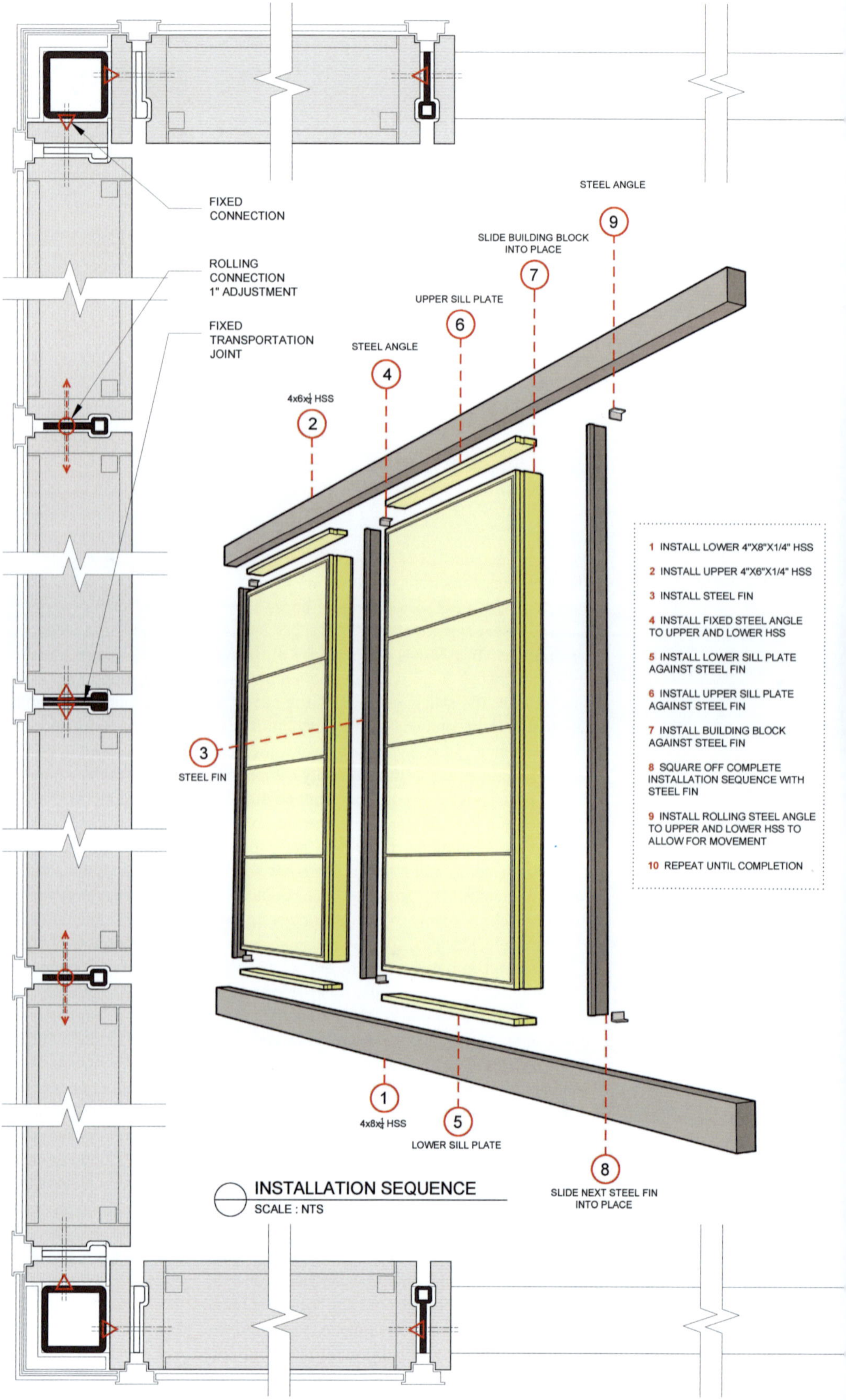
FIXED CONNECTION
ROLLING CONNECTION 1" ADJUSTMENT
FIXED TRANSPORTATION JOINT
STEEL ANGLE
9
SLIDE BUILDING BLOCK INTO PLACE
7
UPPER SILL PLATE
6
STEEL ANGLE
4
4x6x¼ HSS
2
3
STEEL FIN
1
4x8x¼ HSS
5
LOWER SILL PLATE
8
SLIDE NEXT STEEL FIN INTO PLACE
1 INSTALL LOWER 4"X8"X1/4" HSS
2 INSTALL UPPER 4"X6"X1/4" HSS
3 INSTALL STEEL FIN
4 INSTALL FIXED STEEL ANGLE TO UPPER AND LOWER HSS
5 INSTALL LOWER SILL PLATE AGAINST STEEL FIN
6 INSTALL UPPER SILL PLATE AGAINST STEEL FIN
7 INSTALL BUILDING BLOCK AGAINST STEEL FIN
8 SQUARE OFF COMPLETE INSTALLATION SEQUENCE WITH STEEL FIN
9 INSTALL ROLLING STEEL ANGLE TO UPPER AND LOWER HSS TO ALLOW FOR MOVEMENT
10 REPEAT UNTIL COMPLETION
INSTALLATION SEQUENCE
SCALE : NTS

4.24

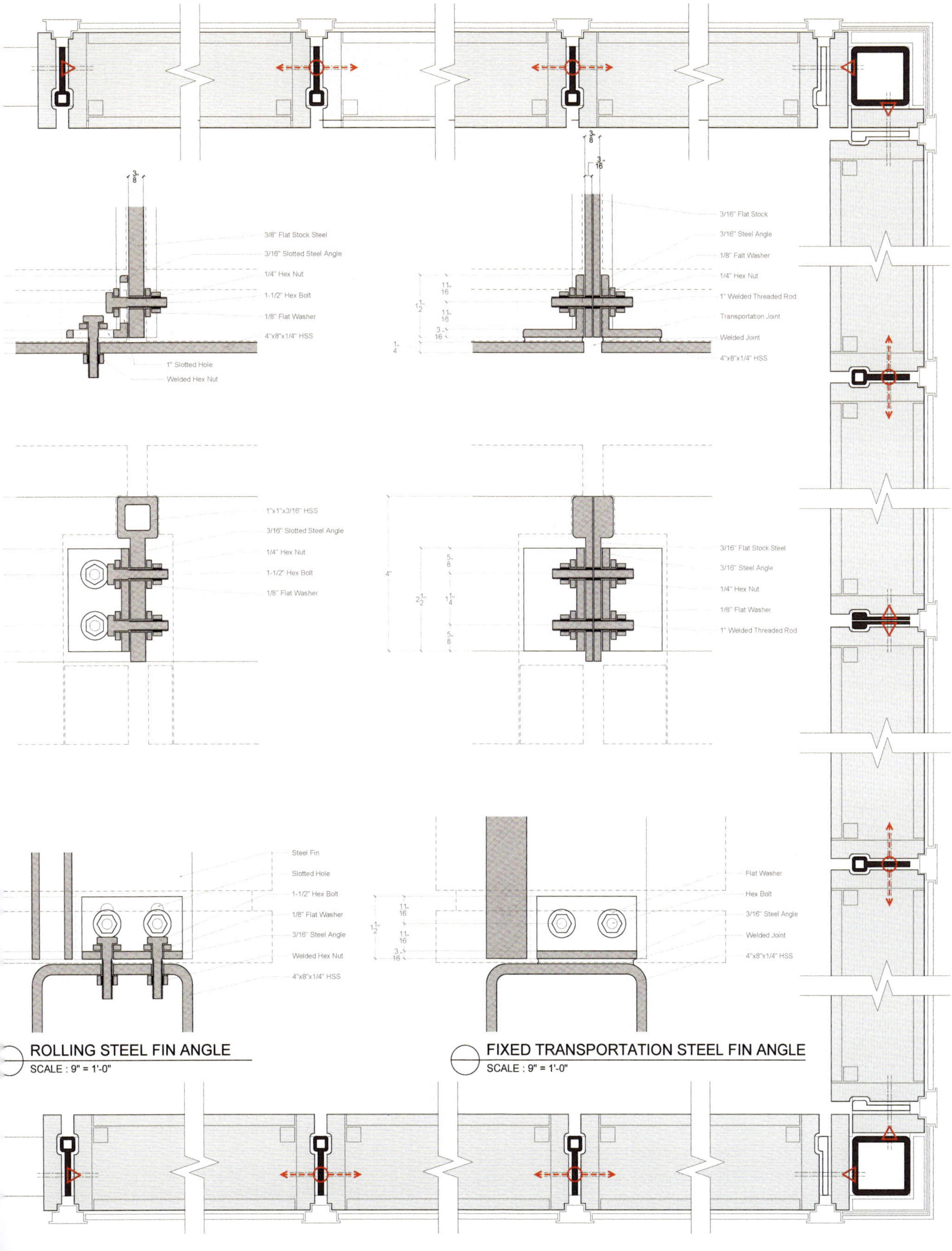

ROLLING STEEL FIN ANGLE
SCALE : 9" = 1'-0"

FIXED TRANSPORTATION STEEL FIN ANGLE
SCALE : 9" = 1'-0"

4.25

4.26

However, the interstitial connecting system of the steel fin, sandwiched between the jambs of the building blocks, was of greater concern since steel is about 250 times as conductive as wood. The position of the fin within the assembly, although indispensable for structural reasons, does carry the risk of thermal bridging and potential occurrence of condensation if it is not placed at the correct depth within the envelope. Two tactics were chosen to solve this issue. First, holes were cut out from the fin areas that were not critical for connections in an effort to reduce the amount of steel used. Admittedly, this generates more effort in the fabrication process. But if the building elements were mass-produced, such as through use of water jet CNC cutting, these pieces would be manufactured quite efficiently. With the reduction of this unnecessary material, we reduced the conductive area to a minimum. Secondly, the fins were positioned relative to the blocks to achieve a 2" space on the exterior side that was to be filled with insulation.

Therefore, the fin itself was always at a temperature that would avoid condensation. That move was essential, since any moisture within the block was likely to get trapped with dire consequences to the longevity of the building. It would cause deterioration of the wooden frames, corrosion of the steel frame, and most likely prompt mold growth. Over time, the building could effectively harm the user and destroy itself. But, it was our ambition to create a building with good enough genes to lead a long life.

The steel frame had to be designed to meet many conditions: structural efficiency and a long life on its final site location, it needed to be lifted and transported on truck trailers, and it needed to endure relatively quick reassembly for the exhibit on the National Mall in Washington. Many construction details had to be specifically designed in order to accomplish this. For instance, the steel frame had to be doubled to effectively make two structural transportation modules that function independently of each other (4.22). Also, the connection between the two halves had to be designed to allow for structural integrity and speedy assembly. In an effort to make sure that the building halves would relate to each other precisely the way they were originally constructed, we designed location finders (4.24) that guided the 40,000-pound modules to their exact positions. These were designed as locator pins at both ends of the transportation joint with a hole on the bottom of the received column, which locked the two parts into a defined connection (see 4.19).

Of course, the sequence of lifting the modules and the detailing along the transportation joint had to be done in such a way that the finishes would not be destroyed during reassembly.

Furthermore, the frame had to be designed so that it could be suspended in mid air with minimal deflection. Any deflection in the frame would cause glass breakage, or movement that would prohibit settling of the major building elements. Also, the attachment points had to be calculated in such a

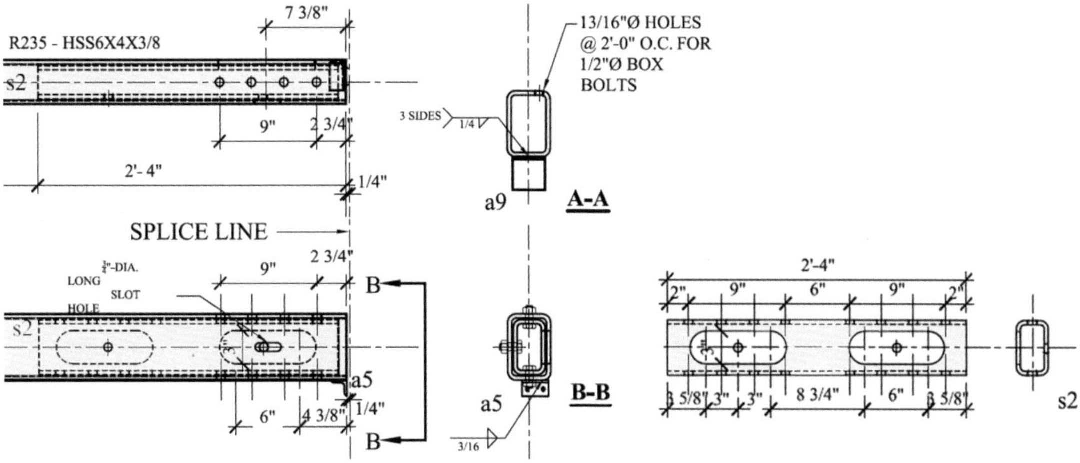

4.27

way that they were sufficient to carry the anticipated loads. The biggest challenge in this rigging operation was the subdivision of the two structurally independent modules.

Instead of four corner columns as initially planned, we were now in need of eight, one at each corner of each module. However, this conflicted with our continuous 'building block' façade. The continuity of these blocks, appearing from the inside as well as the outside, was therefore most important to us. The support column had to move to a different location. We decided to move it 2 feet inwards, into the areas of the living room and bedroom, which worked well in terms of the required accessibility. The holes in the finished floor and the dropped ceiling would be filled in after the removal of the columns (4.22; also 5.15).

The columns, and any evidence of them, could be removed after the two modules were structurally reconnected after transportation. In order to test the system and the construction procedure involved, the house was constructed in two separate sections and connected eight weeks before our deadline, simulating the reassembly in Washington to come (see chapter 4.12). The construction test was undoubtedly justified, as it was discovered that removing the temporary columns was more difficult then expected. The design was subsequently refined.

However, before these temporary columns could be removed, the perimeter beams, which were disconnected during transport, needed to be reconnected as a continuous beam in order to make the column obsolete. The hollow nature of the steel profiles used for these beams was advantageous (4.26) as they allowed for the design of the indispensable reconnection. At the point of the transportation splice line, smaller moveable steel tubes were permanently inserted (4.27). These interior tubes remained moveable and accessible from the exterior structure through access holes so that these pieces could be precisely aligned when bridging the splice, then bolted into place. The nuts for these bolts had to be permanently welded to the outside of the beam, since they were finally permanently concealed. Accessibility to these tubes for the reassembly was identified as an additional detailing criteria.

The most critical details for transporting the house to Washington DC were the lifting tabs used for the rigging. During the process of designing the Roofpod for this rigging operation, its influence on the general assembly strategy became clear.

It made sense to fabricate the house in two large sections, which was conducive to both transport to and assembly at Washington DC, as well as future installations on Manhattan roofs. Essentially, the two prefabricated pieces were designed to fit and be transported to the site on low-boy trailers. Furthermore, using a moveable crane enabled the installation to take place within half a day. After seeking the opinion ofseveral specialists in modular building construction, this was confirmed as the most economic solution for installation. Despite the costs of rigging, this method would decrease construction

4.25 Location finder: Sketch of principle.
4.26 Structural tubes: Principle.
4.27 Structural tubes: Roof edge splice detail (steel shop drawing).

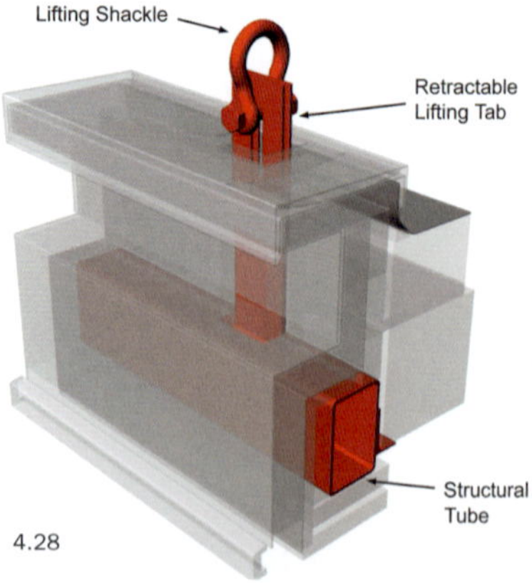

4.28

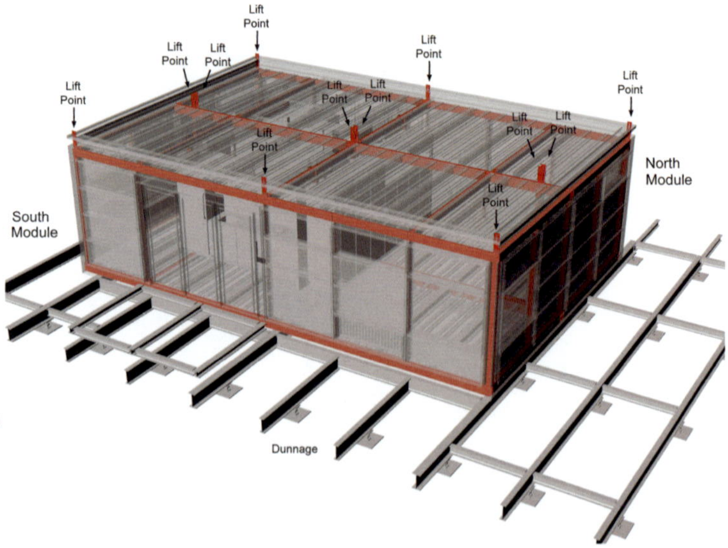

time, and thus also be preferred by the host building property owners. A crane-assisted installation would eliminate the need for use of the internal elevators and subsequent headache to all residents. The lifting tabs (4.28) were therefore a systematic detail, potentially designed for all installation scenarios.

In an effort to avoid unnecessary deflection while lifting the house, the decision was made to have six lifting tabs in total: four on each corner, and two supporting each mid-span beam. Because the lifting tabs were on the perimeter structure, solutions were sought so that they would not be exposed to the elements or require complex waterproofing. Through discussions with a structural engineer, we came up with a solution employing retractable tabs. We discovered that small, thin sheets of steel (3/8" thick, 6" long) were actually strong enough to suspend the Roofpod. Six of these tabs attached at specific points were able to suspend 20 tons of weight: the efficiency of steel was clearly revealed.

Most of the finishes of the house had to be installed in advance, to ensure that everything was prepared for the exhibit in Washington. Therefore, it was crucial that all the finishes would reconnect the same way as they were originally installed once the house was put back together. Each half of the house weighed 20 tons, making it difficult to maneuver the pieces accurately. For this reason, it was necessary to invent a device that enabled the two halves to line up accurately when dropped into their final location with the crane. This investigation started at the points where the Roofpod attached to the dunnage. Due to the need to avoid thermal bridging, each half of the house was placed on six mini-columns made from square steel tubing.

At the two ends of the transportation joint line, where one can control best how the two halves of the building come together, we developed special 'mini-columns' that were interdependent on one another (4.19). One of the mini-columns, the one for the building module set onto the dunnage first, had a plate on it, extending towards the area where the mini-column of the second building module was supposed to sit. On this plate, in the center of the location of the second mini-column has been a small ¾" diameter pin welded to the plate, which at the upper end was slightly tapered. The underside of the mini-column of the second building module has a plate welded inside of the hollow space, at the bottom of that element. In the center of this plate is a ¾" diameter hole, receiving the pin of the first building module: through this device it was assured that the two building halves always meet each other accurately.

The last element related to the steel frame's structural function is the attachment of the concluding trellis system. Despite its light appearance, the trellis system is exposed to high wind loads

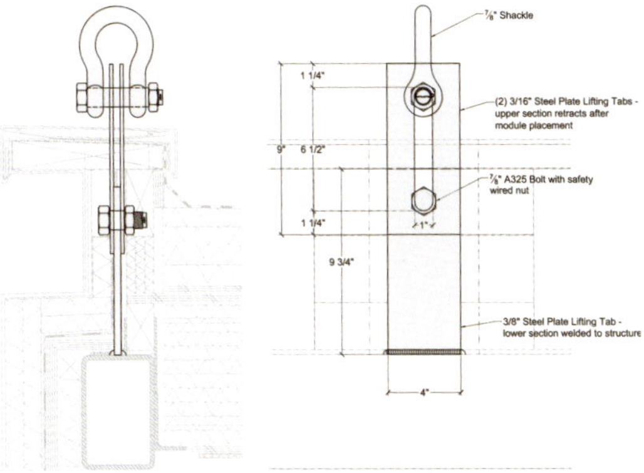

4.29

4.30

4.31

4.28 Module lifting strategy: Six lifting tabs per module, concealed in the building structure.
4.29 Detail: Retractable lifting tab and shackle attachment.
4.30 Test rig: The lifting tabs in action.
4.31 Location finder: the extended 'mini-column' plate of the first module receives the second modules column.

and has to be sufficiently connected to the building structure in order to withstand storms and also not cause uncomfortable vibrations due to wind. Two additional beams were inserted in the ceiling structure in an effort to receive the loads uniformly from the space frame's six bearing points. Stanchions that penetrate the roof, additionally insulated to minimize thermal loss, received these points.

All the details discussed took us at least half a year to work out on paper. Ironically, in the end they were all invisible, clad with insulation and break metal. However, without these intensive details, the entire building assembly would not have worked, neither for transportation nor for the rigging operation. Ultimately, the steel armature made the entire set-up time efficient, as the 'building block' assembly (the 'LEGO' strategy) would have needed to be broken down into its individual units. This was unthinkable given the 6 day time restriction imposed on the building assembly in Washington DC.

Furthermore, feasibility is the context of any construction project: time is money! It is vital to come up with a time-efficient and thus affordable system. This is particularly true in the design of a prototype that needs to be mass-produced.

The steel cage strategy proved to be the most time-efficient and code compliant approach. This system affords the potential of many different assembly strategies. The building blocks could be either inserted into a set steel frame, or as an alternative strategy, the completed prefabricated building modules could be rigged up to the roof from the street level, as they fit comfortably on a trailer. The latter strategy would be appropriate for projects with time constraints. Since design and construction criteria change from project to project, both approaches can be practical.

Often in modern architecture, steel is used as an 'invisible helper'. However, there is also an aesthetic idiosyncrasy to employing steel, in that its strength allows for a cross section that can be easily concealed. Thus, steel is often used in a way that makes masses and volumes visually float, without any reference to how it is done. The semantics of steel embrace its invisibility and foster a fascination for its weightlessness and horizontality. The Roofpod design also shares one of these poetic moments in its trellis system. Here, the connecting stanchions are invisible from the observer's vantage point. It appears weightless, cantilevering to the edge of the building and forming the floating cornice of it.

Once the steel frame is complete, it must be fireproofed, so that it can relate to the host buildings' fire-ratings.

Then, this structural armature is ready to receive its interdependent building blocks, enclosing its newly configured space.

4.32

4.33

4.34

4.32 - 4.35 After completion of the steel frame and dunnage, all elements must be painted, especially the exposed dunnage to provide protection against corrosion. Both dunnage and steel frame also have to be fire-proof: intumescent paint protects against structural deterioration in case of a fire.
4.36 Steel frame detail on dunnage, for the Roofpod construction on campus; typical detail.

4.35

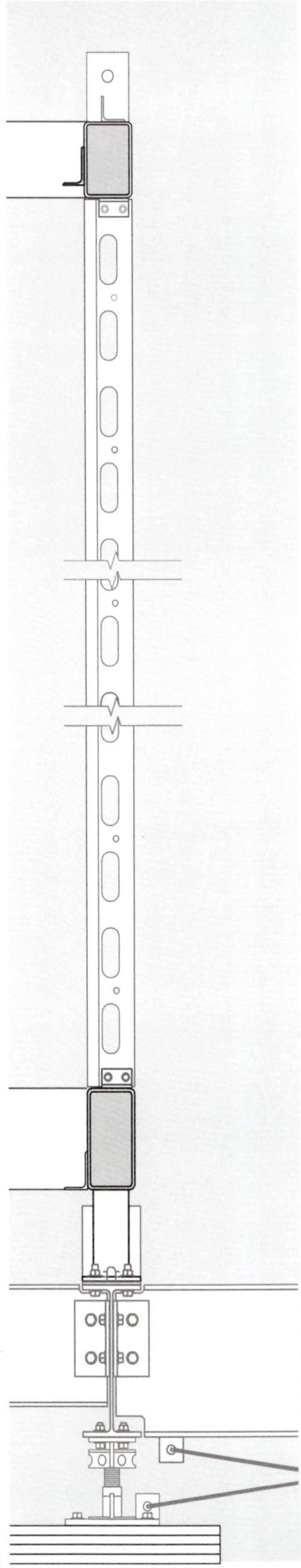

4.36

Prefabricated insulated floor panels with structural load capacity are readily available. They are 'composites', unifying by lamination structural material with material of high insulation value. In our case, non-combustible light-gauge steel channels were embedded within polystyrene insulation so that the steel could not conduct heat through the panel. Since these elements were in the end to be concealed, time-savings and performance data were decisive in taking advantage of a compatible industrial product.

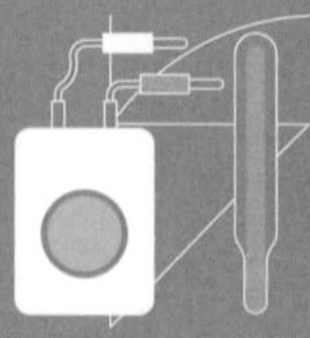

Energy efficiency becomes key for a house reliant just on solar power: Preserving a comfortable temperature inside the livable space is therefore the ultimate goal. The superior thermal resistance of elements configuring the envelope is the determining factor. Expanded polystyrene, used in the floor/roof panels, has a high thermal resistance, contributing significantly to a superior envelope with an R-30 rating (r-value). When calculating the entire wall and roof assembly, we exceeded the energy code benchmarks by far.

When considering insulation for a slim construction budget, often compromises are made. However, additional insulation will pay back over time, since the operational costs to heat and cool the building will be less. The initial cost of insulation is thus well-invested: Not only is money saved in the long run, the environmental impact can also be reduced if the right materials are chosen. The monetary damages associated with global warming can only roughly be estimated; and, increasingly scarce oil will get more expensive over time.

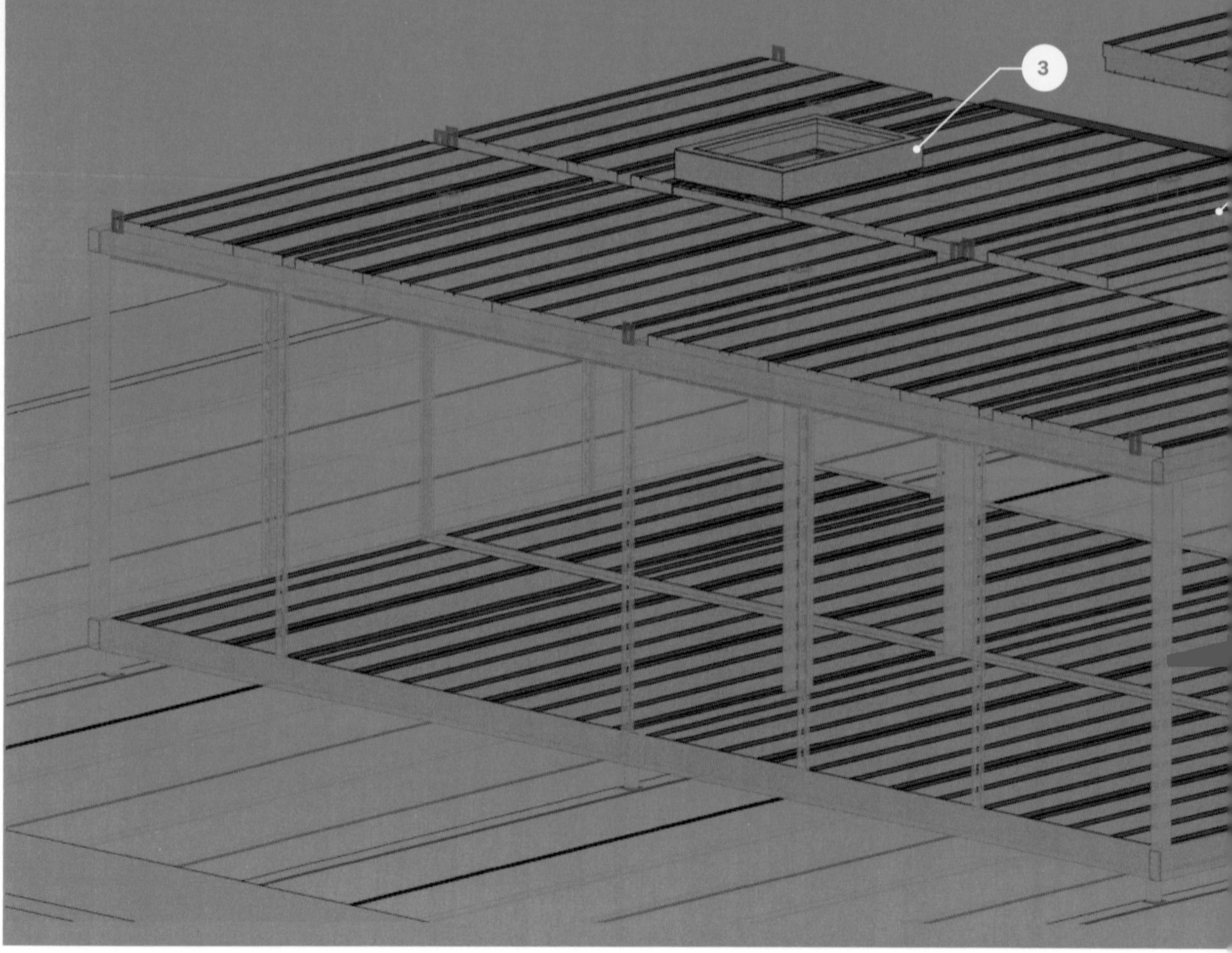

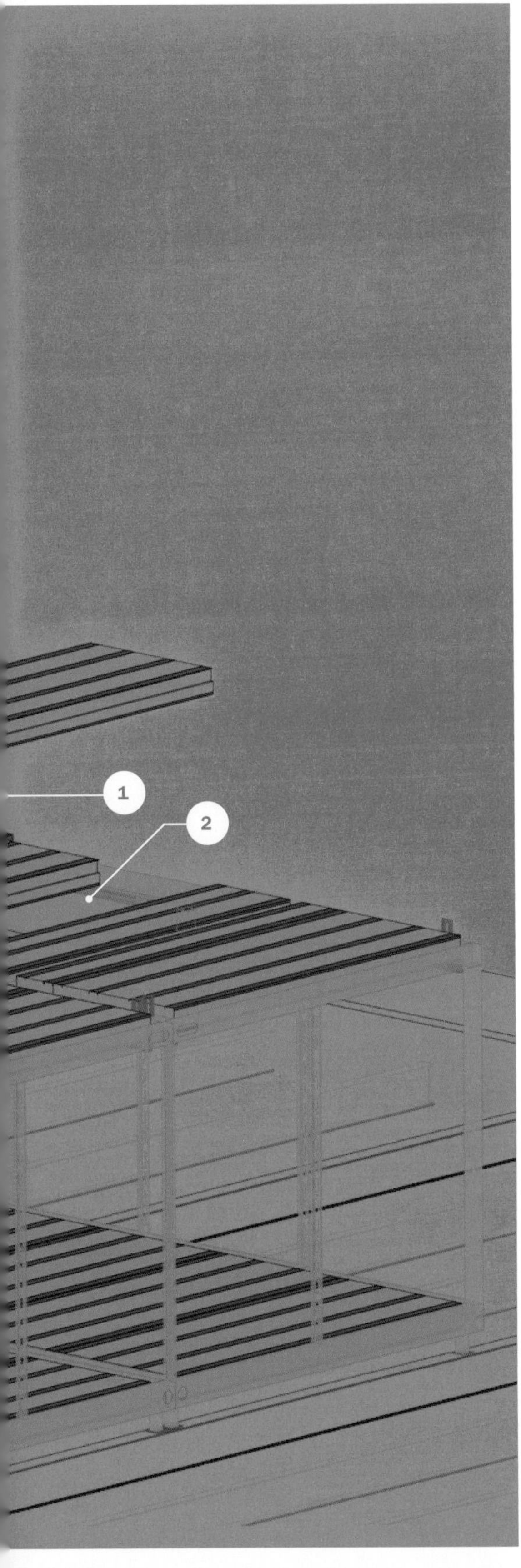

STRUCTURAL INSULATED PANELS

4

1. EXPANDED POLYSTYRENE PANEL W/ DISCONTINUOUS STRUCTURAL LIGHT GAUGE STEEL MEMBERS, 4'X10', TYP.; DEPTH: 7.5"

2. ATTACHMENT LEDGE, WELDED TO STEEL FRAME; PROPRIETARY SHEET METAL SCREW – PANEL CONNECTION

3. SPECIAL PANEL, W/ INTEGRATED FRAMING FOR SKYLIGHT OPENING

4

SYSTEM 4: FLOOR+ROOF –STRUCTURAL INSULATED PANELS

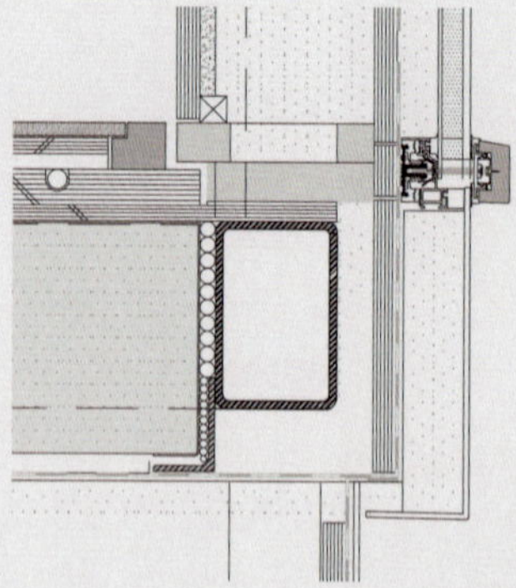

4.37

The initial plan was to build an envelope system where floor, wall and roof units were all the same. Additionally, the initial intention was to mass-produce these units ourselves in an effort to develop an energy-efficient prefabricated modular envelope system. However, it was soon recognized that floor and roof units had very different structural requirements, since spanning from one support to another would make these elements act as beams. Most likely, they would be concealed at the end of construction, as the floor panels will be covered by the radiant floor heating system and finished flooring, and the roof area will be concealed with electrical conduits and plumbing pipes (monitoring and control system, lighting, smoke detectors, sprinkler system, etc.). It would thus be in need of a suspended ceiling to distribute and cover these flexible and retrofittable installations in the future. Since the appearance of the insulated floor and roof elements were thus not as important, the effort to prototype these units would only make sense if similar units weren't available on the market.

Prefabricated, insulated floor panels with structural capacity have been developed since the 60s, when new lamination methods and materials allowed for development and production of such composites. Typically, they are composed as sandwich panels, with a core of expanded polystyrene and plywood-like oriented strand board (OSB) panels on each side, which are laminated to the foam core and act like flanges in a beam. These units are called 'structural insulated panels', or SIPs. They are superior to conventional framing, since there are no framing members acting as thermal bridges. However, since the plywood flanges act as the main structural member, there was concern about its structural performance in the event of a fire. In our research we discovered that similar panels have been developed that use cold-formed C-channels molded at the top and bottom of the unit for structure (4.36). These structural channels do not touch each other since the neutral axis in a beam is neither in compression nor in tension: it is structurally inactive. Therefore, the structural members can be both structurally and thus thermally separated. If the channels are shifted slightly to form a shiplap at the intermediate edge of each panel, they are effectively unified into one continuous plane. This C-Channel system allows for the strength of a structural floor, but with the insulation values seen in SIP construction. While they are slightly more costly than plywood SIPs, the system is more durable and compliant with Type II construction of New York City's building codes. This compliancy was most important for our goal of adapting the Roofpod to the urban roofscape of New York, and since we were focused on prototyping our design for a greater application, this code consideration ruled the decision process.

Furthermore, the energy-efficiency of SIPs is far better than what we could have constructed with our own means of production. We do not have the expertise nor the tools necessary in the industrial process to produce the proprietary channels, mold and laminate them with the foam, etc. The simple version of stressed skin panels that we would have been capable of producing out of common dimensional lumber would be inferior in terms of fire

4.38

4.39

ISOMETRIC INSTALLATION STRATEGY

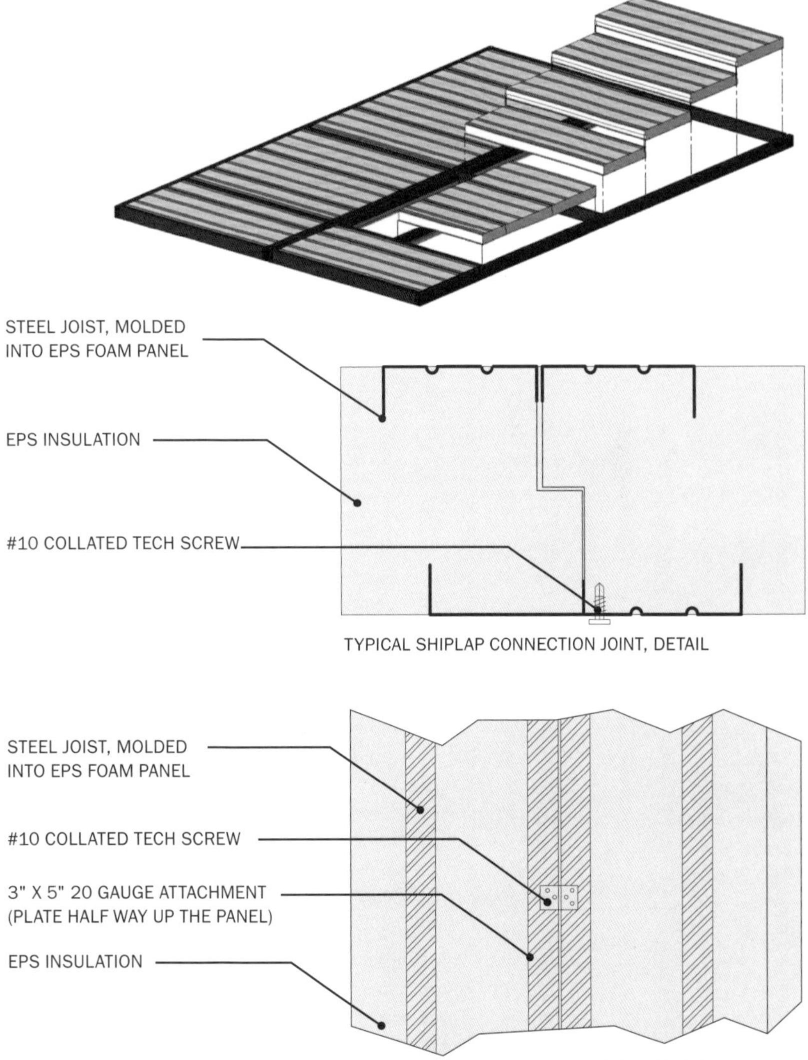

4.40

4.37 Detail of steel angle attachment and sheathing, tying together frame and floor plane.
4.38 Attachment of SIP panels to steel frame; proprietary self-tapping screws were used, see detail.
4.39 Panel principle: The steel channels molded into the polystyrene can withstand applicable forces, while being thermally separated.
4.40 Floor panel installation, preparation for proprietary connections; student drawing.

4.41

resistance, thermal performance and light-weighting. They would also require more work, which entailed additional labor that we became more and more cautious of given the overall scope of work. It was more important to invest our labor into meaningful, visible features that contribute to an overall positive design appearance.

The ability to take advantage of mass production and standardization is an important skill for producing modern architecture. Often, off-the-shelf products save time and money and are capable of supporting the architectural idea. But conversely, they can make the aspiration of creating an idea by means of drawing obsolete. Research on the availabilities of products at first replaces the drawn investigation. It is therefore a fine line to walk: if the standardized product becomes the dominant factor for the appearance of the design, there is no originality left in the approach. The prototypical and individualistic nature of the building would be at stake. If used in context with the conceptual goals and within the overall materialization strategy, the prefabricated items can support the conceptual idea very well. The research into appropriate products, as well as the evaluation skills to understand how these standard products fit into the overall idea, lies with us. These skills are necessary to practice contemporary architecture.

The research and evaluation skills required for industrialized items must also include superior communication skills to allow for better collaboration with manufacturers. In early Modernism, manufacturers were willing to collaborate with architects to customize their products but these alliances are now far less common. Architectural firms that heavily engage in material research jointly with manufacturers, often develop extraordinary opportunities for architectural expression. The potential for collaboration between architects and industry has to be redeveloped and improved, so that the parties involved can experiment with economic, technical and aesthetic considerations. Architectural education should prepare for these collaborations.

In the case of our floor and roof panels, the decision was relatively easy, since it was clear that the prefabricated elements performed in a superior manner while being invisible. Therefore, they had an impact on the overall appearance of the Roofpod, while still improving its performance. The panels also afforded us general experience in industrial materialization strategies. The idea of separating the structural elements into tension and compression zones, and the use of this separation for improving thermal performance will shape the students' design insights for future projects. To see an idea concretely materialized and tested instantly conveys its practicality. Despite the test certificates, the weight these panels could actually support was at first questionable, since the majority of material is just expanded polystyrene. A 4' x 10' foot panel weighed merely 70 pounds and was easy to carry. However, the ease of installation and the structural capacity were astonishing: it took us only a day to finish a floor plane.

When the plywood sheathing on the top surface completed this assembly segment, we were surprised at how much strength these panels actually develop. This taught us much about prototype development, which is a process that focuses on repeated applications. Throughout the design process, problems among similar projects are likely to have standardized solutions. The ability to learn and take advantage of the emerging know-how in industry is critical. In the case of the Roofpod, understanding high-tech lightweighting and advanced energy-efficiency proves the point that standardized products and techniques must be part of modern material thinking.

The panels were perfect for our application, since the steel frame span was less than the maximum available length of these panels. We executed precise shop drawings (4.42) to coordinate the fabrication with the manufacturer. No mistakes were allowed, since the results were not easily changeable. We placed the order on time, and received the panels days before our scheduled installation date. After installation of all elements in less than two days, we were aware of how much time and labor we had saved with a lighter and superior thermal envelope for our Roofpod.

4.41 Shiplap connection of SIP panel; sample obtained from manufacturer.

4.42 Panel layout and panel take-off of roof plane. Student drawing. Dimensions must be absolutely accurate, since the structural steel members laminated to the polystyrene cannot be adjusted later on; tolerances must also be anticipated.

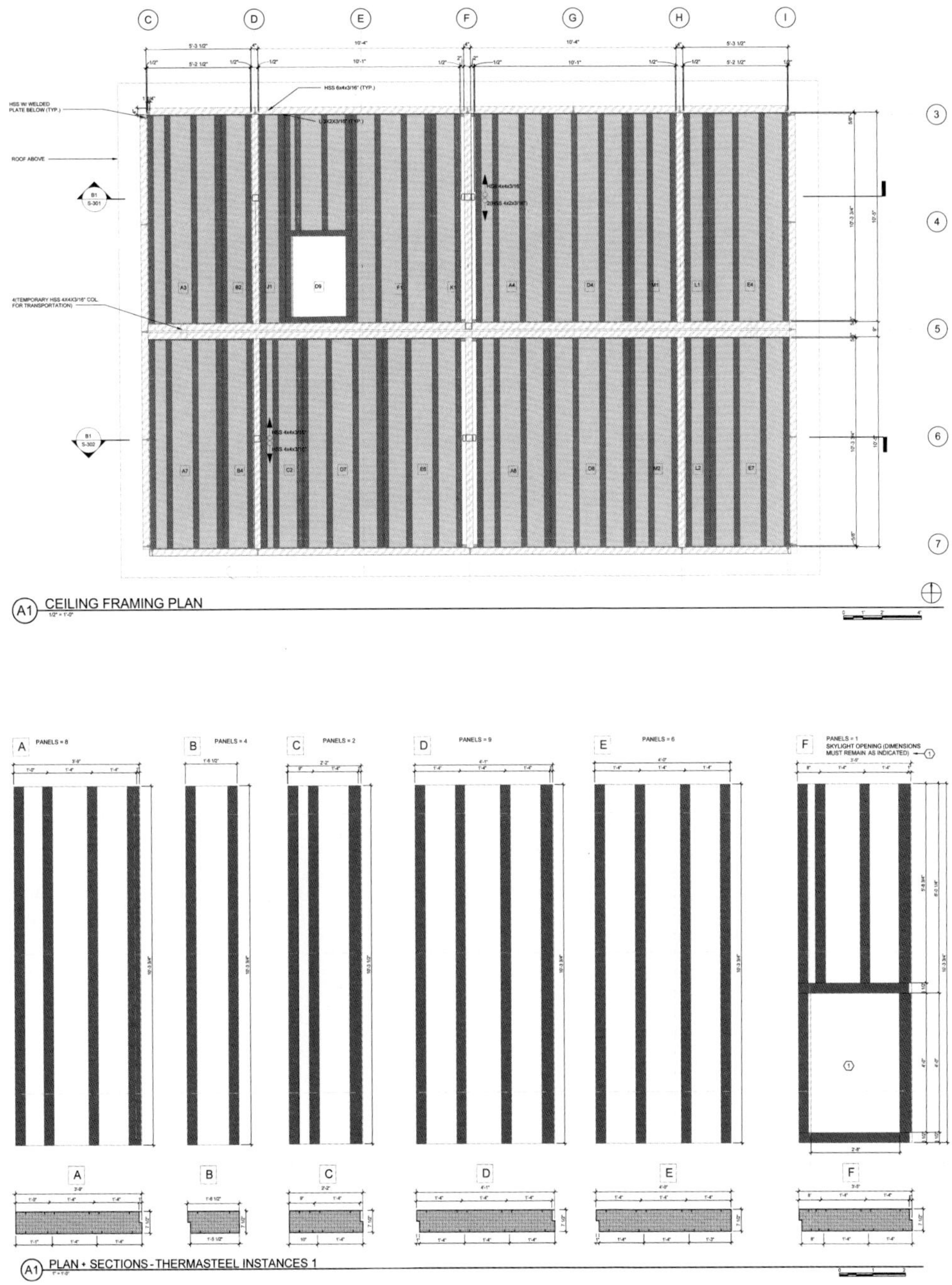

4.42

4.43

4.44

4.45

4.43 Mounting structural insulated panels (SIPs) into the steel frame at the roof level. Even though the panels were very lightweight and easy to handle, they were strong enough to support a person's weight while mounting them.
4.44 - 4.45 Mounting the floor level was slightly easier than the roof level. Panels were fixed to the steel frame after insertion (see 4.38).
4.46 Floor and roof panel insertion, for the Roofpod construction on campus; typical detail.

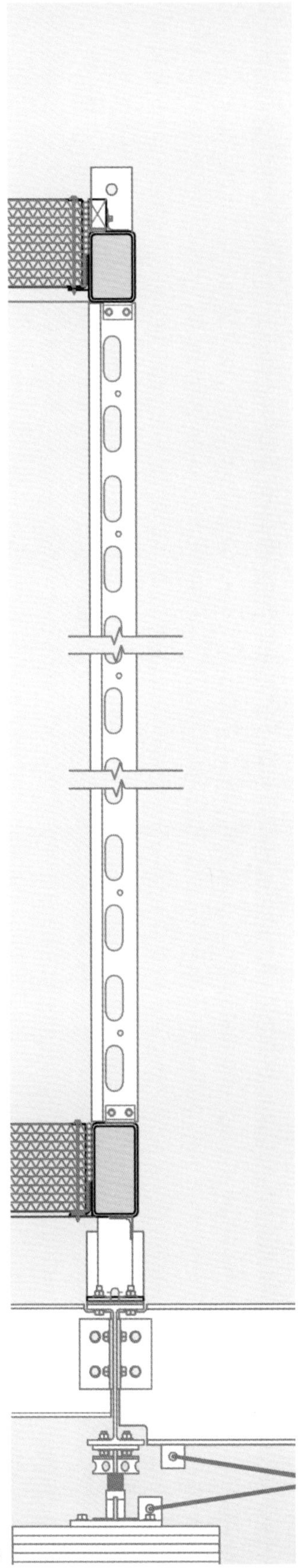

4.46

Our 'building blocks' were not available on the market, and therefore posed a welcome challenge to prototype a building 'subsystem' ourselves. The goal was a modular customizable assembly that responds to climatic requirements as well as personal preferences; elegant in its appearance and superior in its performance. The system evolved from singular blocks to preinstalled units of 8'-0" by 5'-4". This move kept discrepancies smaller and reduced the amount of labor, too. Transportation in an elevator to the roof would still be possible. The method of preinstalled units suggests a limited amount of combinations, still broad enough to allow for a range of solutions for an individualized expression.

The assembly had to be installed in a particular order, due to the nature of the alternating steel fin – building block interrelationship. In order to give the completed assembly enough lateral stability and keep it orthoganal, an exterior layer of sheathing had to cover the block assembly. In order to use the sheathing material, OSB board, most efficiently, sheets of 4'-0" by 10'-0" were acquired so that minimum waste would occur. Since the curtain wall attachment would be located close to the sheathing panels' edges, a solid wood strip connecting the blocks, was introduced, which reiterated the modular pattern again. This layer provided the subsurface for the subsequent waterproofing.

The façade idea of the 'Roofpod' is intended to express 'choices'; to create related architectural impressions based on favorable performance combinations. The exposure to the environment: sun, light, air and also views, was of particular importance. Although the entire ensemble of blocks presents itself in a rather enclosed configuration – advantageous, since the opaque units have superior thermal characteristics – yet the glazed instances happen specifically, where they work best: clerestory windows to redirect light to the ceiling, displacement ventilation through lower and upper vents positioned across the room, and views in locations related to the main circulation axes.

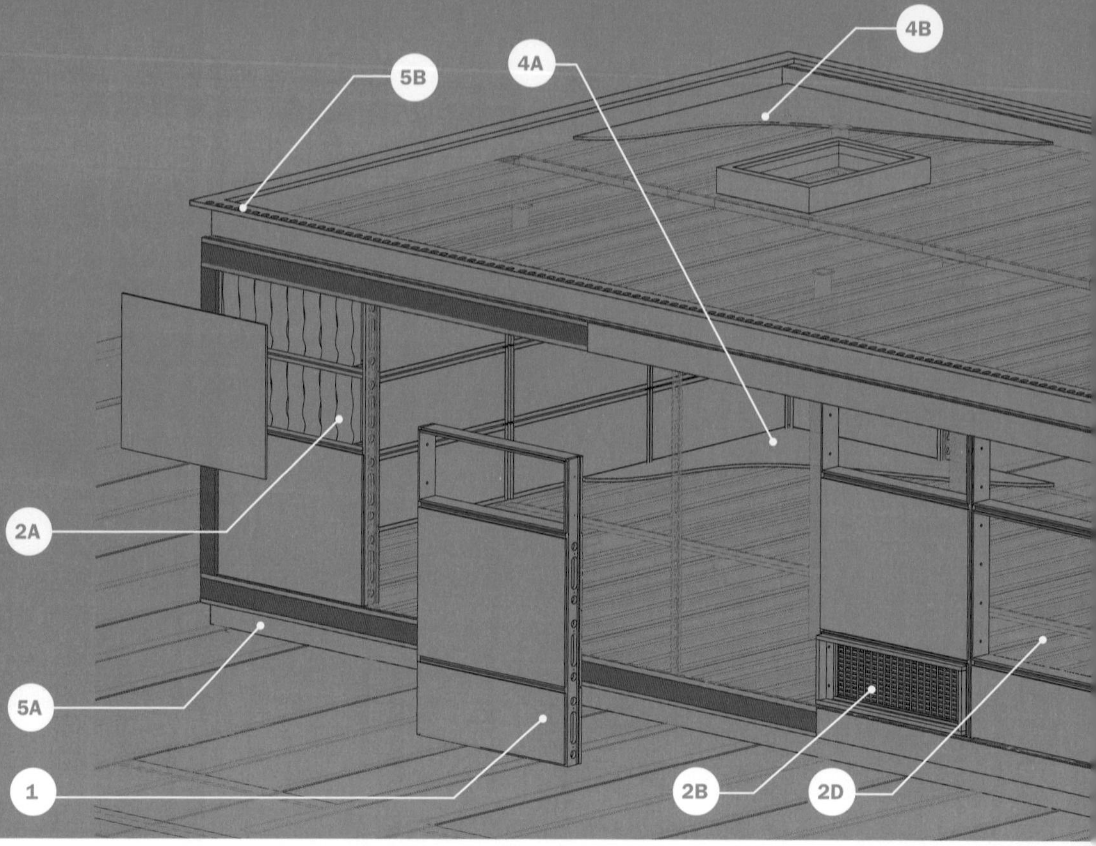

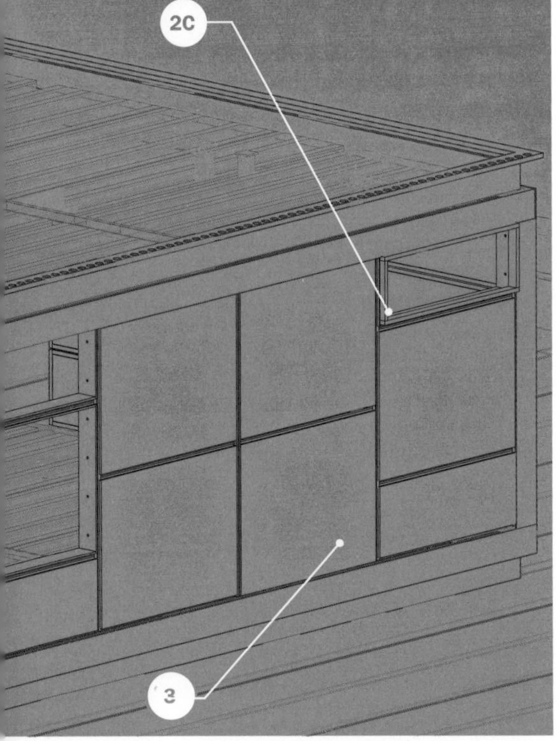

'BUILDING BLOCKS'

5

1. 'BUILDING BLOCK' UNIT
2. INSTALLED 'BUILDING BLOCKS'
   A) FOAM SPRAY-INSULATED BLOCK
   B) LOWER VENT BLOCK
   C) UPPER VENT/CLERESTORY BLOCK
   D) "VIEWING" BLOCK
3. COMPLETED SHEATHING OF 'BUILDING BLOCK' ASSEMBLY
4. HORIZONTAL SHEATHING
   A) FLOOR
   B) ROOF
5. FRAMING FOR BUILDING TRANSITIONS
   A) SKIRT
   B) PARAPET

5

SYSTEM 5: FAÇADE ENVELOPE –THE 'BUILDING BLOCK'

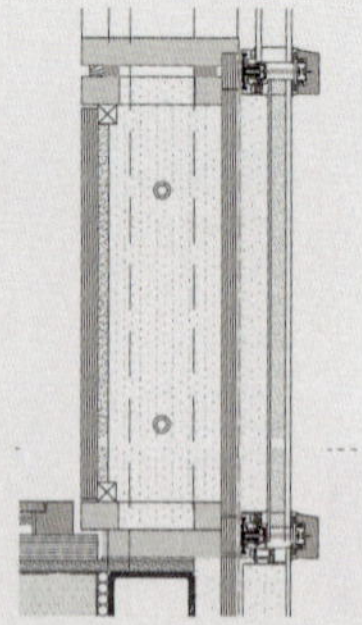

4.47

When designing the façade system, we discovered that, despite being considered modern, curtain walls are not very efficient in terms of thermal resistance. The main issue is that glazing has a considerably inferior thermal resistance compared to opaque materials, even when composed as a multilayered insulated glass unit (IGU). Also, unwanted heat gains in summer are a problem. However, due to the fact that they are easy to install and replace, they are one of the most utilized envelope systems in urban environments such as New York City. At the same time, they are often used in a boring, anonymous manner, as can be seen in uninspired and repetitive buildings.

Therefore, two criteria were defined as main topics in the Roofpod's envelope investigation: performance and expression. Both criteria go hand in hand, since an array of environmental factors can also have a positive effect on the interior space and its related micro-climate. By manipulating façade openings spatially, one can also manipulate the different functions that the openings manage: heat transfer, day-lighting, and air-movement. These openings, acting like filters, are the key to a successful passive design. Passive design is the positive manipulation of environmental impact by means of clever disposition. It is consistently superior to other methods, since little maintenance is required and energy consumption is reduced greatly with no additional effort. The better a building works by passive means of operation, the smarter its architect. Therefore, considering orientation and seasonal solar variations, shading patterns, and prevailing wind directions is absolutely paramount.

To calibrate the impact of these factors, one has to know how much, where and in which combination to use these filters. Naturally, this is very difficult, as these factors are related to the building's location, latitude and climate, and are also somewhat subjective, since this approach is related to a specific design context. In addition, our individual reactions towards glare, heat, cold, draft, views, privacy, and interior-exterior relationships are all related to personal preferences. We therefore decided that the 'building block' system can customize the approach to establishing a relationship to all these factors.

In the example of automobile marketing these personal preferences are emphasized, although the product of a car is a mass-produced item. You can choose between alternatives in accessories, equipment and motor strength: alternatives, which are exchangeable within the system. We envisioned applying the same strategy for our façade.

The façade elements are intended to be combined as a block system which complies with building codes and required lateral loads when inserted into the steel frame. Each block is 5'-4" wide and 2'-0" high. Four of them stack to a height of 8'-0",

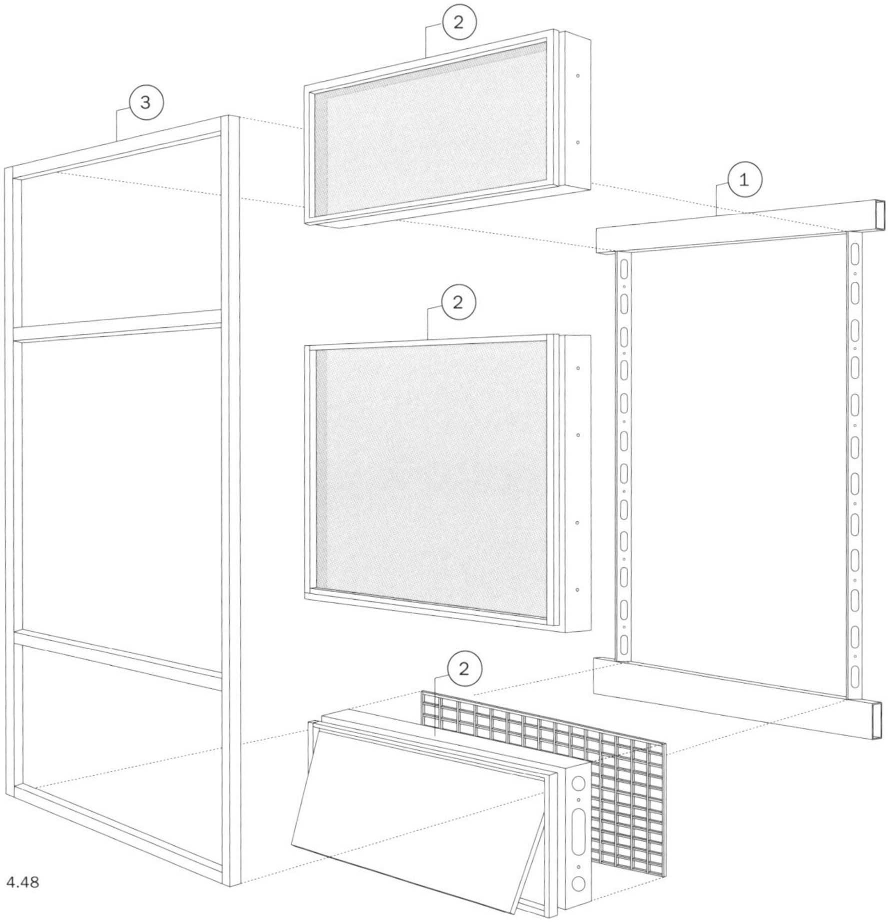

4.48

4.47 Detail of 'Building Block', as part of the exterior wall envelope.
4.48 'Building Block' system: Schematic. Design: (1) Steel back-up structure; (2) Building Block combination; (3) Caps.

which is the standard height for a one story room. As an exception and an expression of the varied options, a double block with twice the height was introduced to create a more generous opening. Initially, these blocks were to be mounted on site, but it was soon recognized that the more pieces you have to combine, the higher the risk of divergence within the system (...and also the more labor involved).

Therefore, we decided to pre-assemble four vertical block segments that, when combined, were still light enough to be handled by two people. These super-blocks were then glazed and capped on site, in order to seal the joint lines and make the system waterproof (4.48).

Several mock-ups were built to understand the connections and assembly sequence (see chapter 3.3). Proprietary systems with similar tasks were also analyzed for their approach to the issues of tolerance and adjustments. Particular attention was paid to visually emphasizing the proportional system within the interior elevations. Inversely, common systems do not address this aesthetic problem, since all bays are usually equally filled with glass. However, our system had many insulated openings, which are bays that significantly improve the building's thermal resistance. These bays were predominant in designing the Roofpod prototype, and thus established the aesthetic consistency of this subsystem.

Up to this point in the design narrative, the opaque, insulated block has been the staple of the Roofpod's system. Overall, it is the dominant, least expensive element and probably the most efficient one, deterring heat and cold. Since the other glazed blocks are somewhat more high-tech and more costly, they should be applied in a way that optimizes their efficiency. That means that day-lighting is enhanced if the windows are in the top area of the system, wherein the descending angle of light can be better distributed in the room.

4.49 Operational façade strategy: Diagram; different passive functions and their favorable locations within the envelope.
4.50 'Building Block Catalogue', addressing exchangeable options.

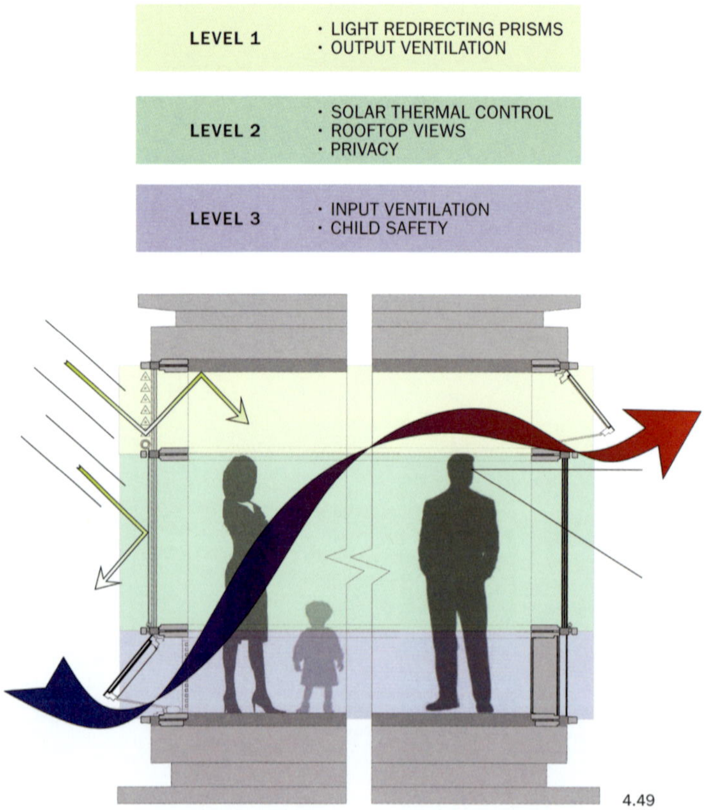

4.49

A unique insulated glass unit was discovered that employs louvers, integrated in the unit's cavity, to redirect the light towards the ceiling in winter, and bounce it back outside in summer (4.50).

A similar product was used for the 'double-blockviewing' unit, which is unique in that its louvers retracted into the interstitial space of the insulated glass unit. These windows were the four-feet-high units, which emphasized views towards the skyline. However, this opening could be detrimental to the overall performance of the system because the openess permits substantial heat gains that balance the benefits of the gorgeous views. The retractable blinds are therefore motorized and operate with the 'Smart House' monitoring and control system (see chapter 4.8), so that an optimized outcome and a minimum amount of unwanted heat gain can be achieved.

The blinds are integrated within the insulated glass unit, which underwent performance negotiation in the design process. The best performing blinds, with respect to their ability to ward off heat gain, are exterior blinds. This performance is related to the 'greenhouse principle', by which glass is transparent in the short wave (light) spectrum, but not in the long wave (heat) spectrum. This means that any light striking a surface, becomes captured within the glass barrier, since part of the energy is absorbed and eventually released as long wave radiation.

Exterior blinds reflect the light before it can pass this glass barrier, therefore the energy does not get trapped inside. Thus, exterior shading systems are much more efficient, warding off 80-90% of the energy impact compared to interior systems which perform at around 50%[58]. Because wind loads on top of buildings are relatively high, many

4.50

Block 1: Clerestory, Light Redirecting, Fixed

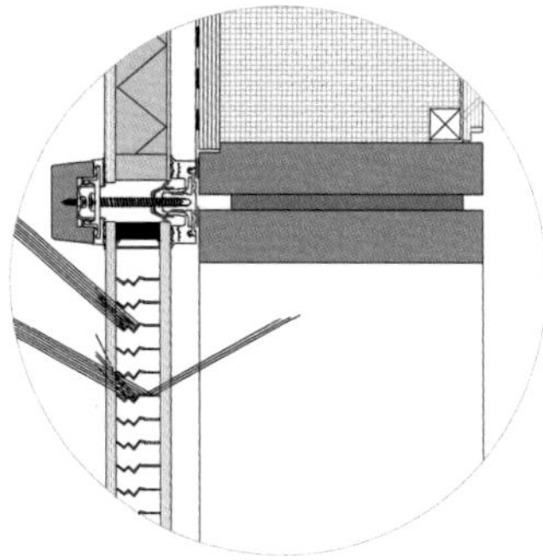

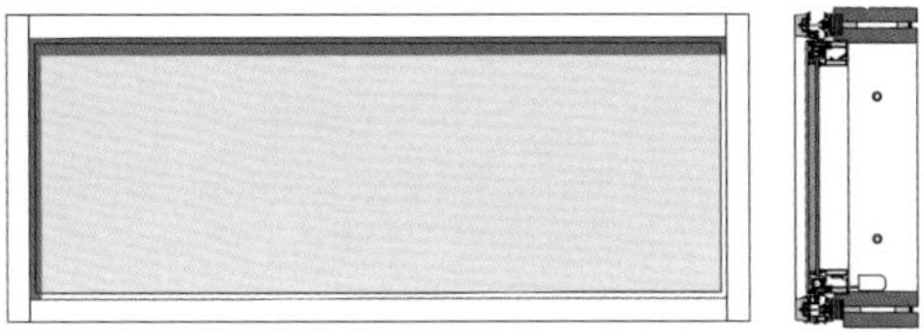

Block 2: Upper Vent, SHGC Reducing Interstitial Micro-Shades

Block 3: Lower Vent, Vacuum-Insulated Spandrel Glazing, Inner Grill

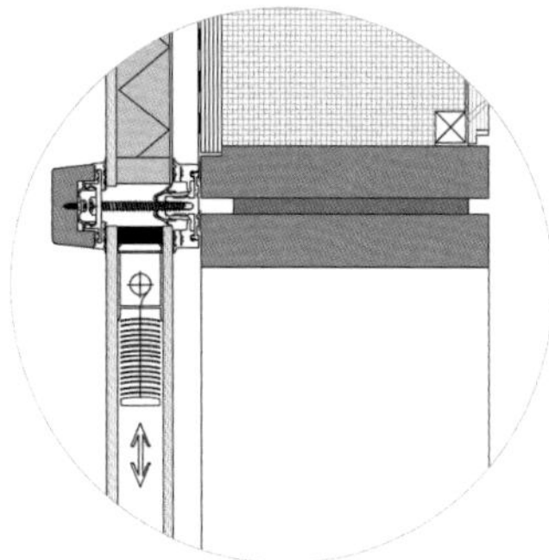

Block 4: View Window, Interstitial Blinds, operable

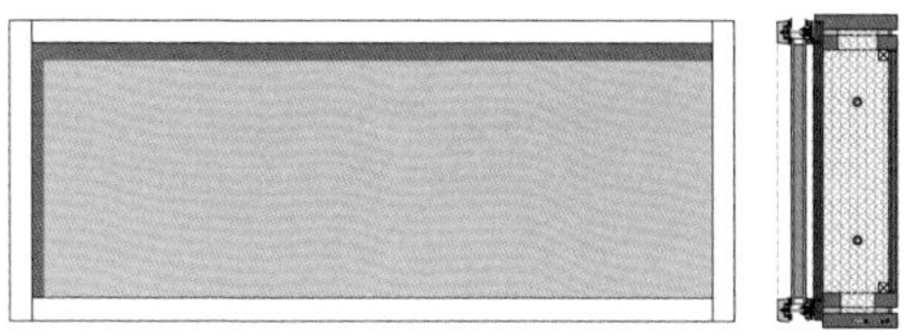

Block 5: Insulated Unit, Vacuum-Insulated, Spandrel Glazing, R-55

manufacturers of exterior blinds withdraw any warranty since the risk of damage is high. There are also heavy-duty products for high wind loads, most of them from Europe, but these are bulky, and unsuitable for a small structure. The best aesthetic solution would be to place the louvers within the insulated glass unit, in the cavity between the glass. The advantages are a slimmer, integrated system, which is well protected and doesn't require cleaning, since it is sealed. The efficiency of shading in this strategy is superior compared to interior shading systems.

Regarding the concept of natural ventilation, in order to provide natural cooling and air movement, we took advantage of the fact that warmer air rises since it is lighter than cooler air. In order to implement this positive convection process in the interior space, two different types of building blocks are necessary: a lower 'input' and an upper 'output' vent, that ideally are across from each other, and as far apart as possible. The Roofpod's vents are all motorized by chain motors, and connected to the monitoring and control system. This ensures that natural ventilation is only introduced when necessary. For example, if there is too much moisture in the ambient air carrying latent heat, mechanical cooling will become more difficult because there is more matter present to be cooled to the right temperature. In order to operate these motors, all wiring has to be run between the computer system and the windows and accommodated within the building block system. The interior area between the blocks was identified as the best location to run the wire, which also visually demarcated the blocks with a ½" reveal. If wiring has to be replaced or added on to, one can access the cable run by just removing

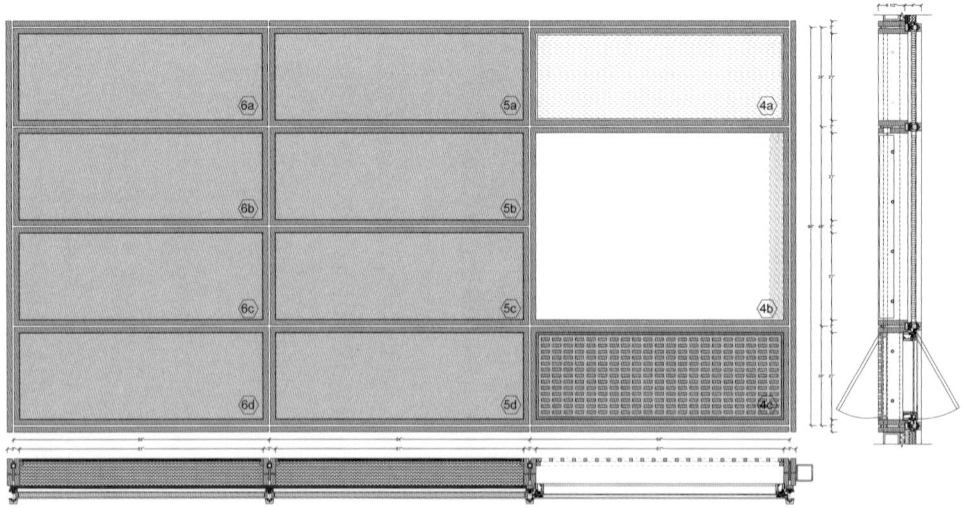

4.51

the reveal cover strip instead of having to take the entire block structure apart.

The main challenge for this upper window was to restrict heat and maintain daylight distribution. The glazed upper vents had the same heat gain exposure as the viewing windows previously mentioned. However, operable blinds or light-redirecting devices cannot be installed in these vents. The operable units cannot accommodate the thickness of glass necessary, nor can they house the wiring required, which would also be easily damaged over time. Therefore, the upper vent presents a compromise: the desire to have a transparent opening in an effort to maximize day-lighting had to be coupled with the heat gain associated with unprotected glazing, which poses costly cooling loads. In this particular block type, ventilation materialized as a third function. While ventilation and day-lighting go well together, because both require an opening of some kind, avoiding heat gain will always be an issue when glass is involved. The same conflict of function can be avoided for the lower vent, where daylighting is not necessary, because its proximity to the floor doesn't present efficient light distribution throughout the room. Therefore, the same vacuum-insulated lacquered spandrel glass, used for the opaque super-insulated block unit, was specified instead of transparent glass. Instead of a thermal resistance of R-2, we were able to achieve a factor of R-20, which is ten times the resistance to heat loss than that of a conventional insulated glass unit (IGU).

Since these vents were motorized and connected to the 'Smart House' control system, it was not necessary to manually operate the window. However, we realized, that its location right above the floor level generated another issue; that these openings were not child-proof. If toddlers were ever interested in 'what's on the other side', they could hurt themselves, since the exterior floor level was almost a foot lower than the interior finished floor (4.49). A favorable visual continuity was also seen in the overall configuration of blocks, when the interior surface was flush with the block frame (4.51). Because of the daylight strategies, the upper blocks were predominantly glazed openings, therefore not compatible with the desired visual continuity, since openings represent interruptions. It became clear that the lower ventilation blocks could easily reinforce this continuity and be child-proof through a covering with a poplar wood grille, as a window guard, made from the same material as the block panels and frames. We invented a system wherein strips of poplar were notched so that the intertwining of the vertical and horizontal members would form a planar grid. The distance, and resulting space, between these members was calibrated to have the same proportions as the full building block. The variations inherent in the grain of poplar can range from grey to yellow to green and purple.

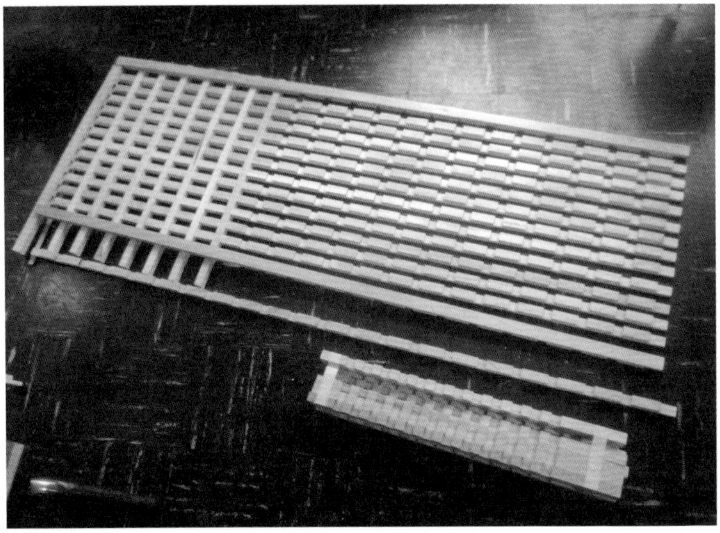
4.52

4.53

4.51 Partial interior elevation with lower vent grille, visually continuing the paneling of the opaque blocks.
4.52 Lower vent grille: The notched members are ready for the grille configuration.
4.53 Prefabricated 'Building Blocks', waiting for their assembly.
4.54 (next page): Building block insertion: Most of the time was spent on adjusting between units, in order to achieve the desired precision.
4.55 (next page): 'Building Block' insertion, for the Roofpod construction on campus; typical detail.

These striations, in combination with the woven texture of the system, contributed to an interesting detail in the overall interior elevations. The paneling itself also provided a similar eye-catching detail in that the poplar plywood acquired for the paneling had similar purple striations on it. The irregularity and repetitiveness of these marks added another layer of sensual ambience.

The paneling was produced with a minimum amount of waste. Using a module size of 5'-4" x 2'-0" and subtracting the frame width and reveals, resulted in a panel size of roughly 5'-0" x 1'-8". So we were able to cut 4 panels out of a 10'-0" by 4'-0" sheet, and used only 10 sheets of plywood for the entire interior of the building blocks.

The last and most expensive aspect of the building blocks is not related to ventilation or daylighting. The layering of the block has not been set up to avoid heat gain or loss. Instead, it is capable of generating energy. In the past decade, glass-laminated thin-film photovoltaics, which are amorphous and much thinner than the common crystalline versions, have been produced for façade panels, among many other applications. While these panels, called 'Building Integrated Photovoltaics' (BIPVs), are not the most efficient form of solar power generation, they establish a connection between performance and space; also, technical advances in fabrication methods will produce a less expensive product over time than the typical crystalline version. Thus, in years to come, entire façades could be conceived in such a way to not only ward off the negative environmental impact of solar energy, but also to use it to their advantage by producing power. While this strategy is still expensive, we found it important to express its potential for future improvement when developing our prototype.

During the process of developing this system, it was understood that many criteria had to come together in order to shape a reasonable approach. Different kinds of program were assigned to the blocks depending on their location in the system's wall and their environmental demands. These programs included heat gain and heat loss, ventilation, daylighting and energy generation. Assessing multiple configurations to address this programming verified its systematic approach. It also had to become an attractive assembly, not only in its elements, but in their overall configuration. However, systems can be so overwhelming that people don't like to associate with them. The aesthetic approach must be subtle but decisive. The interior surfaces have to be elegant and bright in color to efficiently reflect daylight. The production costs, and thus the fabrication methods, have to be economically viable. The system has to work holistically, taking into consideration its capacity to be interchangeable, energy-efficient, affordable, attractive, and with the imminent potential to grow and improve.

4.54

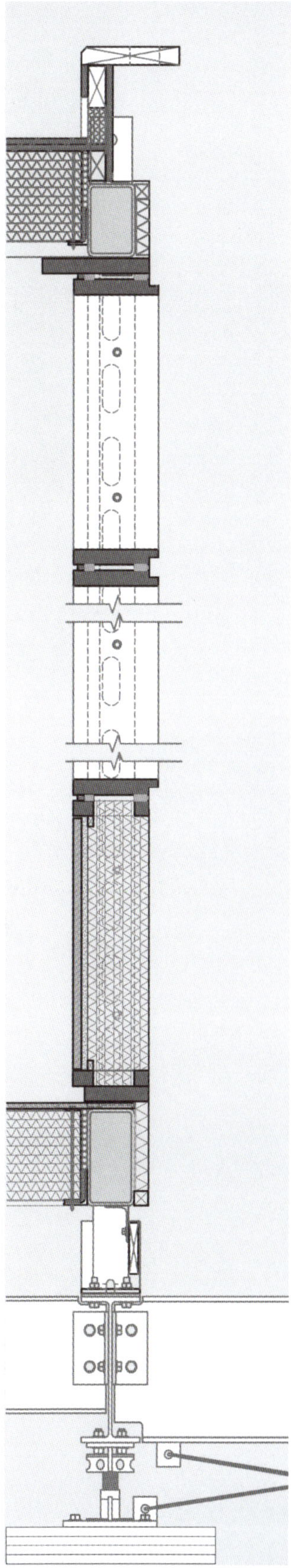

4.55

4.56

4.57

4.58

4.56 'Building Blocks', exterior elevation before sheathing; wiring had to be accomodated to connect to the 'smart house' control system.
4.57 The bolted connection between insulated 'building blocks' was foamed out before sheathing started.
4.58 'Building Block' super-unit, four blocks combined, ready for installation. It was light enough to be carried by two strong people.
4.59 After the blocks were installed, we fixed the steel fins permanently, so that the screws couldn't turn loose.

4.59

4.60

4.60 - 4.61 Interconnecting the 'Building Blocks': The 'steel fin' was situated between blocks, tying the block assembly into the overall structure.
4.62 - 4.63 Installation of sheathing on the exterior wall, to permanently brace the structure.
4.64 Sheathing installation, seen from the inside: The blocks start to configure their repetitive but playful elevation.
4.65 Exterior sheathing of 'Building Blocks', for the Roofpod construction on campus; typical detail.

4.61

4.62

4.63

4.64

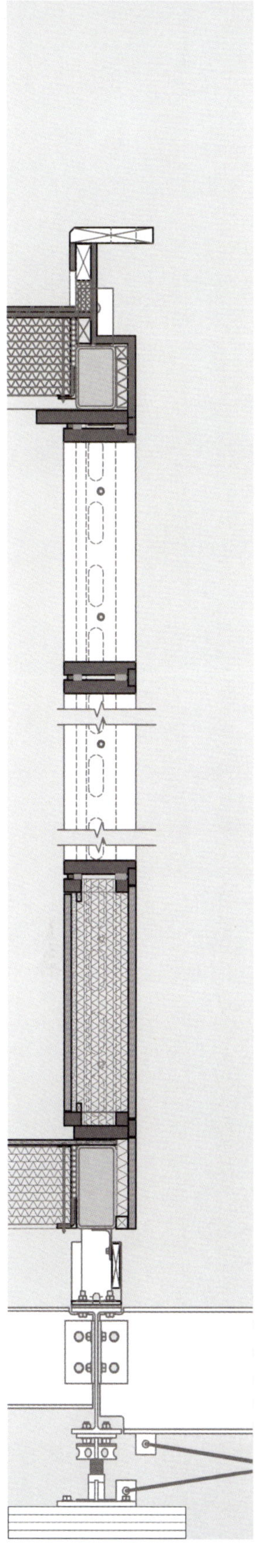

4.65

A building has to open up to its environment, so that its surroundings can be experienced. Especially in summer, when the immediate surroundings become part of one's living space, the physical connection to the exterior becomes important. We looked for design opportunities to connect the house to its roof garden and skyline. Since the entrance to the house – to the north – was already a threshold to the exterior, we created a similar moment on the south side, and made a porch the main access to the garden. We accentuated the opening by integrating a glazed bifold door unit, which could be entirely opened if the weather allowed. This emphasized opening functions as the building's 'eye on the city'.

Sourcing the two bi-fold doors was difficult since they had to fit perfectly into the modular system, within two bays of the building block façade. We learned from the fabricator how important it is to give a certain amount of flexibility to a product line: In this case, fabrication for the doors was set to accommodate different sizes up to a maximum that could not be exceeded due to the structural characteristics of construction and wind loads. Also, the choice of glass could be selected from different options. This was of particular interest, since the biggest glazed area of the building would attract the highest heat gain. The glass we selected responded to that finding. A building block system could in theory be marketed similarly.

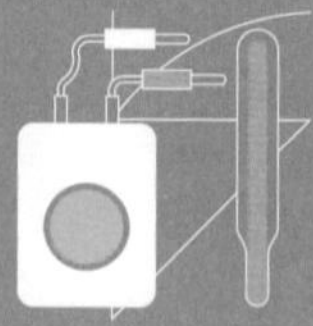

One of the most important performance criteria for an energy efficient building is its airtightness: The most important façade layer performance is the water-proofing membrane, which is not just protecting against water, but also capable of sealing air movement. The most difficult area in terms of keeping this layer consistent is the transition to doors and windows, where it obviously has to be interrupted. Also, any penetrations for electrical and plumbing have to be sealed. To finally check on results, a blower door test can be conducted, pressurizing the building. The air loss is measured in cubic feet per minute at 50 Pascals of pressure (cfm50); a good result should be below 1 ACH50.

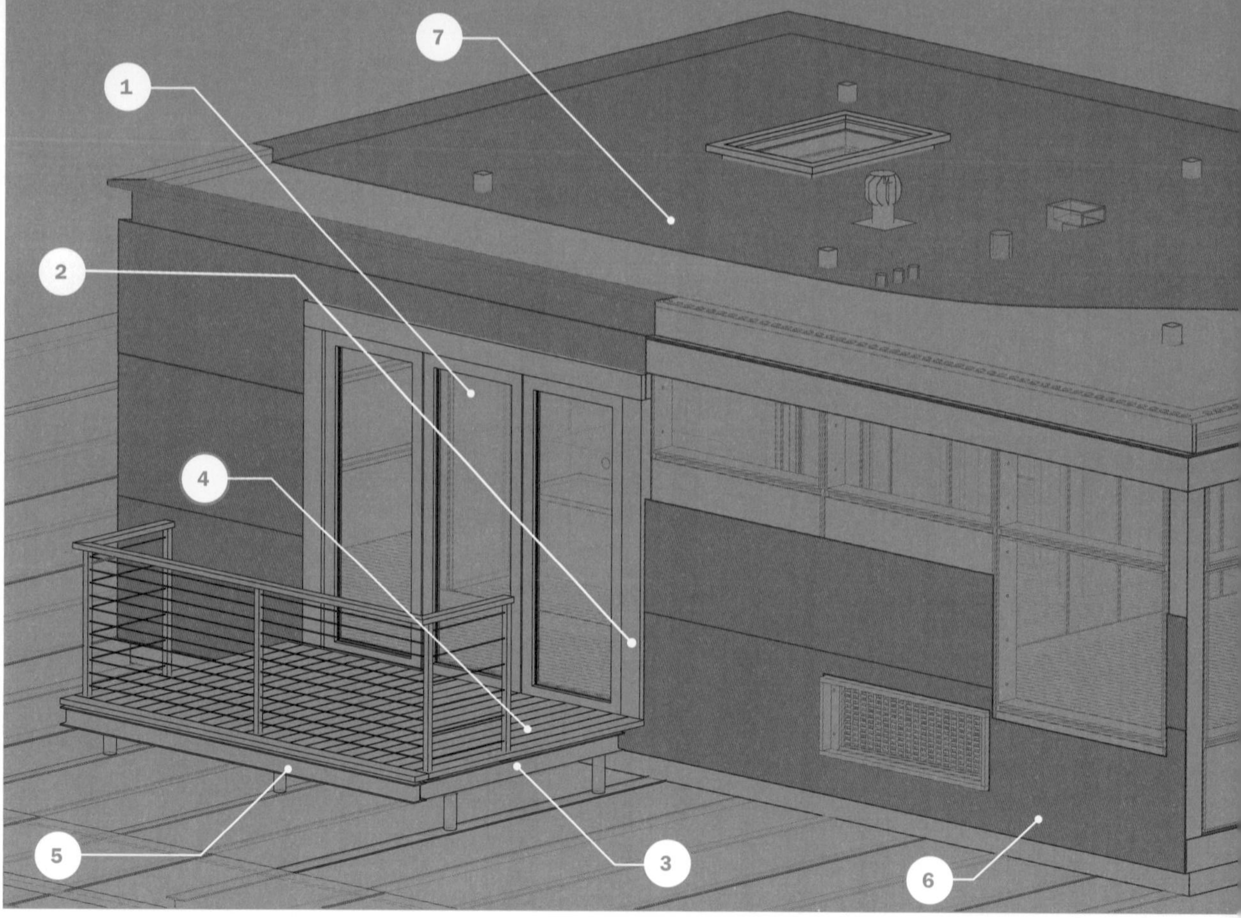

DOORS+PORCH, WATERPROOFING

6

1. BI-FOLDING DOOR UNIT, FULLY GLAZED
2. EXTERIOR TRIM, TO MATCH CURTAIN WALL MULLIONS
3. STEEL FRAME FOR PORCH, SUPPORTED ON DUNNAGE BEAMS
4. WOODEN DECK UNITS, FRAMED; INSERTED INTO STEEL FRAME
5. STEEL FASCIA, FOR PORCH FRONT ELEVATION
6. WATERPROOFING MEMBRANE, SELF-ADHESIVE
7. ROOF MEMBRANE, TPO, 60 MIL; (THERMOPLASTIC POLYOLEFIN)

SYSTEM 6: DOORS AND PORCHES – EXTERIOR SPACE AND WATERPROOFING

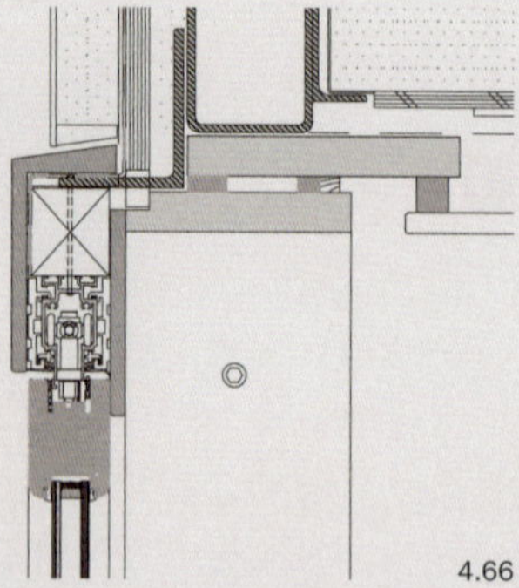

4.66

Our building block envelope system also included operable windows, which relate to passive gain functions and focus on managing heat gains and optimizing ventilation. They also fit perfectly into the dimensional system established for the building blocks.

The requirements the doors had to follow were, however, so plentiful and complex that we could not possibly produce them ourselves, nor could the doors for obvious reasons have been part of the same modular façade strategy. Therefore, prefabricated options had to be sought out once again.

It was understood that these doors were crucial in establishing a spatial relationship between the interior space and the immediate surroundings; the roof garden, which was an important conceptual component. While the windows created a visual relationship towards the garden and the skyline, the doors added an additional dimension in their function as thresholds. By moving from inside to outside, the experiential potential is intensified.

Ideally, the doors specified would be generous and open up to the panoramic skyline. They also had to emphasize the aforementioned threshold moment, by establishing a transitional space (4.68, 4.69). This was important for not only experiencing the roofscape of Manhattan, but also managing the share of visitors to be expected at the Solar Decathlon exhibit in Washington DC. We therefore decided to insert two glazed bifold-door units into our envelope system, one on the north side as the entrance to the Roofpod, and the other on the south side, as a viewing portal to the city. These bifold-doors, whose three leaves could completely fold up, were accessed via a bridge on the north and a porch on the south side. These transitional spaces could be enjoyed in summer in particular, with all doors opened, generating cross-ventilation for the interior. The doors occupied a width of 10'-8", or two bays of the modular façade system, creating a sizable porch for outdoor occupancy.

The doors had to fit precisely into the opening defined by the modular system. However, this posed a challenge, since units have maximum design sizes that relate to the lateral and dead loads produced by the operable leaves, and ultimately to permit smooth long-term operation. The preferred maximum door size of the manufacturer's model was slightly smaller than the opening it had to fill. However, since the door openings were an exception to the modular system, we experimented in our design with a thickened trim surrounding this opening (4.67), and found that it actually improved the façade composition. And, since the header track guiding the bi-folding door mechanism had to be integrated, the horizontal part of this trim was ultimately necessary for technical reasons. Additionally, this trim was advantageous as it acted as the bedstop to the waterproofing membrane, covering the sheathing of the building block elements. The increased width of the trim element made applying the waterproofing layer easier as it is usually most difficult around window and door openings. This layer is important for warding off moisture that ends up behind the façade, and has even more significance in preventing air movement between the interior and exterior.

This uncontrollable air movement is counterproductive for producing an energy efficient building, because along with this air loss, conditioned air of the climatized interior space disappears. Therefore, heating or cooling loads unnecessarily increase. Since this was so significant for the building's performance, we were instructed by professionals in installing the membrane with overlapping joints (top overlapping bottom) so that moisture would be

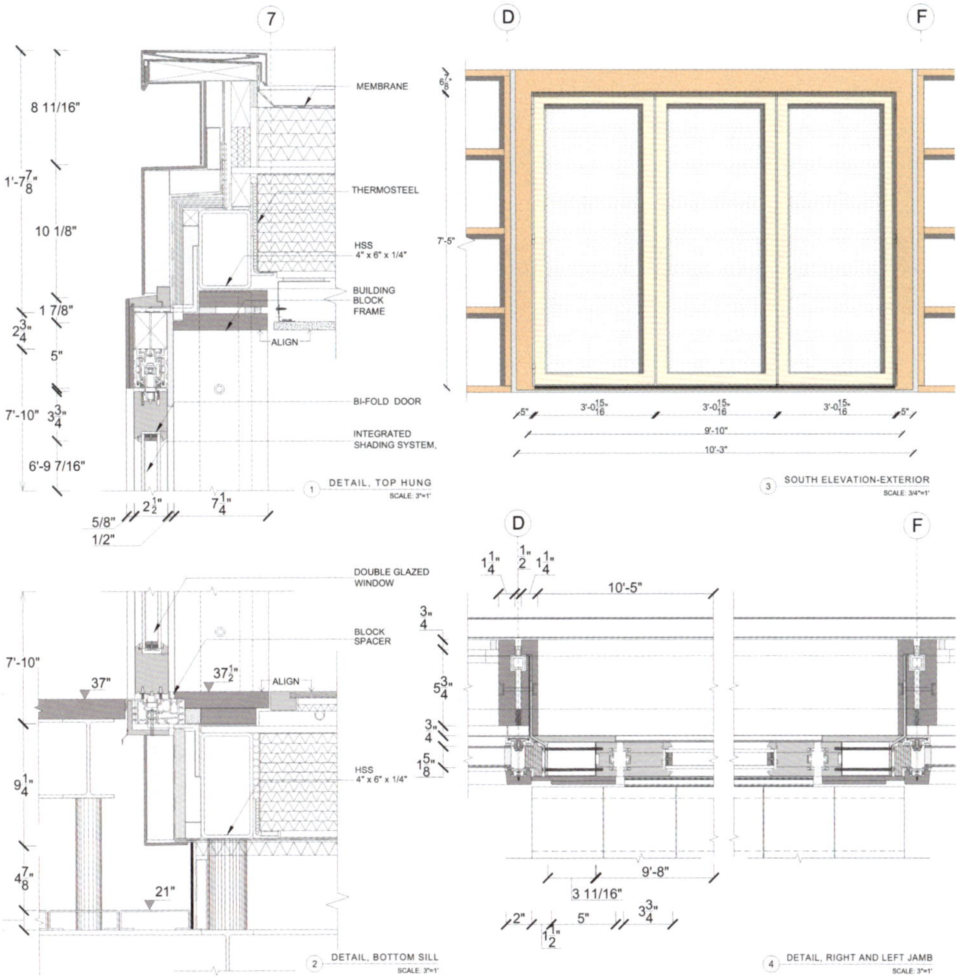

4.67

4.68

4.69

4.66 Detail of bi-fold door track, tying with its emphasized trim into the mullion façade geometry.
4.67 Details for integrating the bi-fold door façade: student drawings.
4.68, 4.69 Transitional spaces: north and south door renderings; student drawings.

4.70

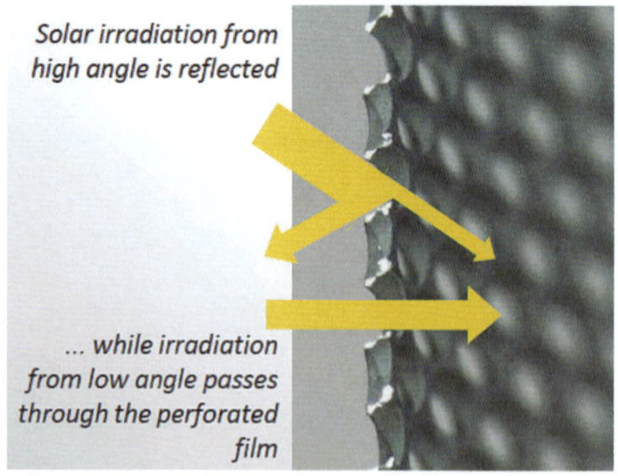

4.71

shed by gravity without penetrating to the inside. The more consistent this layer, the better the performance. And, although not visible while applying it, the results would be revealed in the end, when the house is operating. A procedure that has become common in the quest of energy efficiency, is the 'blower-door-test' (4.70). This is performed with all window and door openings closed, using a device that pressurizes the interior space and measures the amount of time required to return to equal pressure. The air loss is measured in cubic feet per minute at 50 pascals of pressure (cfm50). A good result is below a total air change per hour (ACH50) of 1, meaning that under these pressure conditions the entire air volume would change fully during one hour. When conducting the test for the Roofpod, we had to do several adjustments in order to get close to this requirement. There are always penetrations in an envelope shell. These penetrations, especially for plumbing and electrical, have to be properly sealed, in order to avoid uncontrolled air loss.

Another concern related to the glazed door units was the effect on heat penetration from the increased amount of glass in this area. Although this expansive glazing is favorable for views and relating to ones surroundings, particularly for the south facing doors, unfortunately the solar gains in summer would contribute to higher cooling loads and thus higher energy consumption. While automated operable blinds to shade window glass are quite common, it is more difficult to accomplish the same for doors. The operability of the doors, as well as the long uninterrupted length of glazing, would make this feature more susceptible to failure. As an alternative, a maintenance-free product was discovered that had micro-perforated stainless steel strips laminated between glazing panes (4.71), which efficiently shaded the glass. The perforation was engineered in such a way that the radiation from high angles would predominantly be reflected, while low angles would penetrate through the glass. This allowed for the winter solar heat gains to heat the space passively while avoiding the unwanted gains in summer.

The glazing issue on the north side, although fitting into the identical door unit, was quite different. Here, solar gains were not the concern. Instead, the heat losses of glass had to be addressed, since even an insulated glass unit has inferior thermal resistance. Fortunately, there has been extensive research efforts to better insulate glass. One of the high-tech materials developed is super-dried silicone. It is about 90% light-transmitting, while also preventing the convection of air[60]. It has therefore superior insulation qualities; about five times that of common glass units. However, it is not transparent, but rather has a milky translucent appearance. For this reason, it was used in the context of the entrance door to the house. By taking advantage of its translucent material quality, we were able to provide privacy at the most public side of the building (4.73).

During the glass research phase, many interesting product options were found that could have been used in different contextual situations. Many of them were applied into the cavity space of an insulated glass unit, which is usually around 3/4" thick. Therefore, these options could be used frequently for window applications. They could become part of a glazing catalogue for the Roofpod. They do not only manipulate the thermal performance of the glazed areas, but also alter the ambience of the space.

4.72

4.73

4.74

4.70 Blower-door-test, device, installed at the south doors at the very end of construction.
4.71 Principle of south door glazing screen.
4.72 Aerogel research: high-performance insulation material. Within the cavity of an insulated glass unit, this material can improve thermal resistance five-fold.
4.73 North doors: The translucent silicone fill provides superior insulation, as well as privacy and an interesting shadow projection.
4.74 South doors: The glass-integrated stainless steel screens contribute to the building's materiality.

4.75

4.76

Considering both performance and architectural expression in the selection approach would lead to a superior product.

The possibilities for different frame materials and different glass in-fills offered by the manufacturer of these doors taught us about customization for our own building block system. In order to satisfy different contextual requirements and personal preferences, one must be smart in their customizable choices. This is particularly true when combining both technical and aesthetical strengths, as the potential of generating a more attractive product increases. However, these options must be coordinated so that they generate an identifiable product strategy.

The porches had to be designed with the assembly as well as the kit-of-parts strategy in mind. Therefore, the support system of these elements had to follow the geometry set up by the dunnage system. We thus spaced the steel support beams' for the porches at 5'-4" on center, sitting directly on the dunnage beams (4.75). However, it was not clear whether these porches should be visually identified with the garden or with the house. They were sort of a hybrid, transitioning between the two realms. We decided to make the porches visually float above the garden, and make them level with the adjacent door threshold. In addition, cypress was used in this application, as it was also used for the planter trims in the garden and the mullions of the Roofpod's façade. Through this materialization, the porches tied together the composition of house and garden.

Since the porch support beams were made of steel, which also visually ties into the dunnage beams, a strategy was sited to utilize the beams slender dimensions as an elegant and easy support detail. In this detail, the space between the flanges of the I-beam was used as a track for the wooden decking elements, which fit between the 5'-4" center dimensions and required no fastening (4.77). Two students were easily capable of carrying one of the two decking elements for each porch. In the end, a fascia, in the form of another steel beam, closed-off the assembly and produced a continuous horizontal frame. A floating effect was achieved by simply painting the round tubes between the dunnage beams and the porch support beams black, so that the shadow cast from the porch would swallow these elements. By integrating the wood structure within the steel structure, the depth of the porch became very slender and elegant, and also strong enough to sufficiently support the railing.

The students learned that precision in drawings is different than the reality of achieving it in construction. While it is easy to line up identical elements on paper, the same accuracy in construction requires careful consideration of sequence and ensuing methods. For mounting the decking boards (4.76), the appropriate math and subsequent marks on the joists started the process, followed by cutting spacers that would secure even gaps between the joists.

However, to line them up perfectly lengthwise while keeping the edges straight when installing dozens at a time, is not an easy task and very much related to technique. Carpenters showed us how to install the uncut lengths first to then make one final cut once all boards were assembled.

These 'tricks' were helpful in understanding the thinking necessary to go from a drawn to a materialized result. Experiencing these insights leads to better collaboration and common knowledge between construction workers and designers. All factors must be considered, from technique with tools and processes, to appearance with aesthetics and material. The students excelled in their comprehension of this translation, as conversations now revolved around anticipating the processes of construction.

4.75 Working on the porch construction: One of the most important lessons was the comprehension of appropriate techniques and sequences in order to achieve the anticipated appearance.
4.76 The even spacing of the porch decking was easier drawn than done.
4.77 Detail of the south porch: Taking advantage of the wide-flange steel profile as a track for the wood framing. The steel kept dimensions slim: the reward here was the 'floating' porch.
4.78 Southern porch, almost completed; door and porch configure a unit of geometric and material relations.

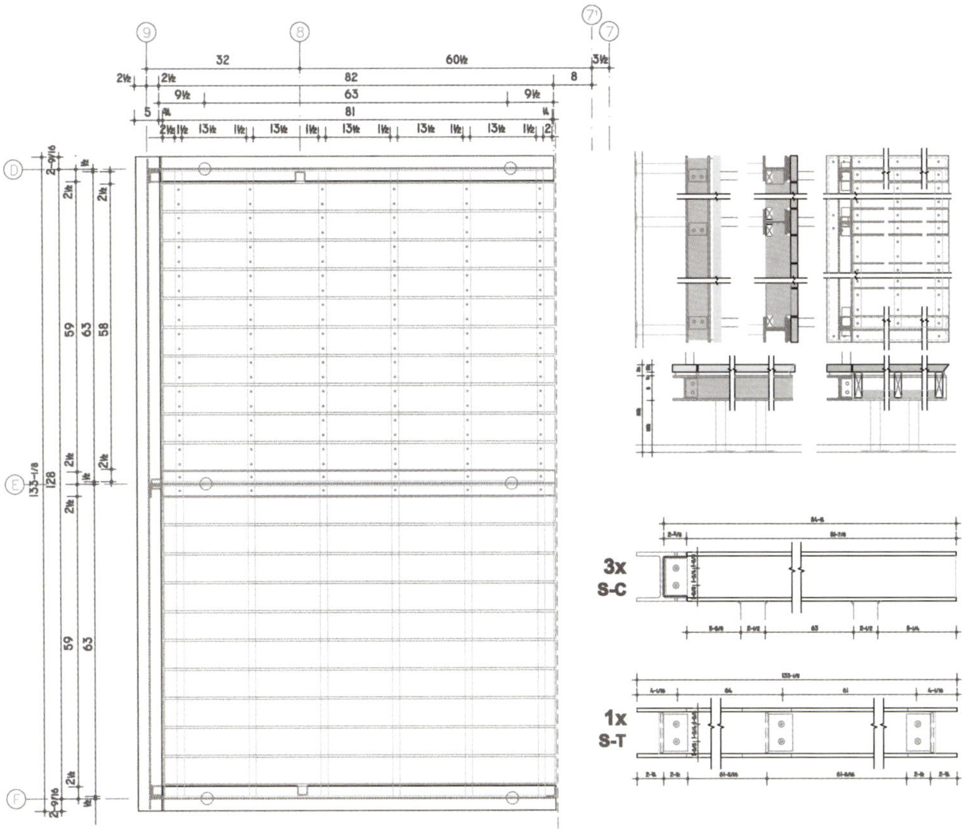

4.77

4.78

4.79

4.80

4.81

4.82

4.83

4.79 Start of door installation: The sill flashing is placed before the doors are set.
4.80 Priming of the subsurfaces for the façade membrane installation.
4.81 Completed façade waterproofing.
4.82, 4.83 North doors, before and after waterproofing.
4.84 Integration of bi-fold door and porch, for the Roofpod construction on campus; typical detail.

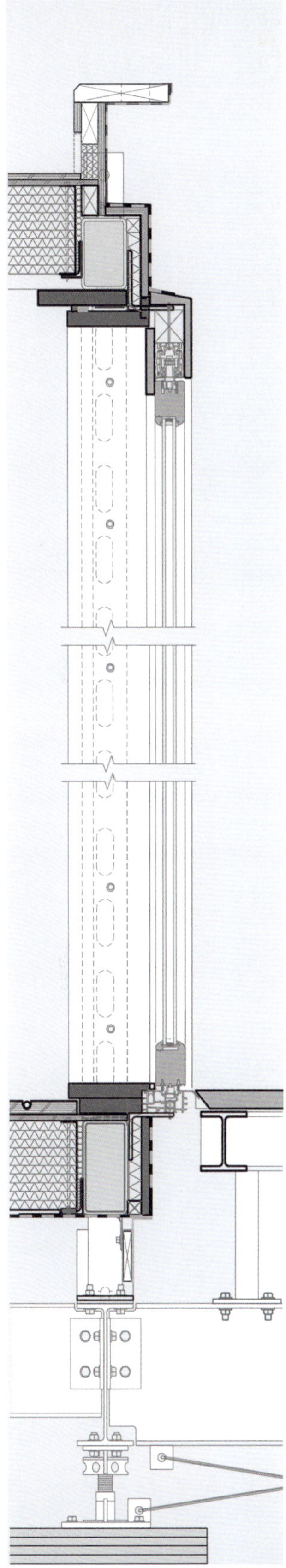

4.84

Glass is very different from other common building materials: Its thermal resistance is only a fraction of that of highly insulated opaque wall assemblies. Due to transparency in the short wave, but not in the long wave spectrum, glass buildings generally heat up considerably, when exposed directly to the sun. In winter, this could promise free 'solar gains', but in summer it adds to the cooling loads. Therefore, some form of regulative shading and 'light management' has to occur for glazed areas. Glass has a very long life cycle and can be 100% recycled. Also, curtain wall assemblies can be easily retrofitted, i.e. for façade integrated Photovoltaics, which become more and more economically feasible over time.

Most of New York's office buildings are made of curtain walls, modern lightweight enclosure systems developed during the past century of urban growth and industrialization. Since the Lever Building's construction in 1950, this envelope system has become ubiquitous in office buildings. Placing this gridded system onto a residential building leads to a juxtaposition between the old monolithic brick façades and the contemporary panelized add-ons on the roofs. These roofscape facades develop their own language out of their 'kit-of-parts' strategy: Although read as a repetitive occurence within the skyline, each one of them develops its own individual appearance in countless customized combinations.

Today's insulated glass panels are highly engineered: they come with triple layers for higher thermal resistance, and with inert gas fillings. They have different coatings to manipulate heat gain and transparency; some of them have clever built-in devices, responsive to seasonal and diurnal fluctuations. We used units with light-redirecting louvers, operable interstitial blinds and perforated screens intended to improve performance while still offering views to the surroundings. We also used opaque enameled vacuum-insulated glass panels with R-20 thermal resistance, which, if used as spandrel panels in all curtain wall applications in New York City could yield substantial savings.

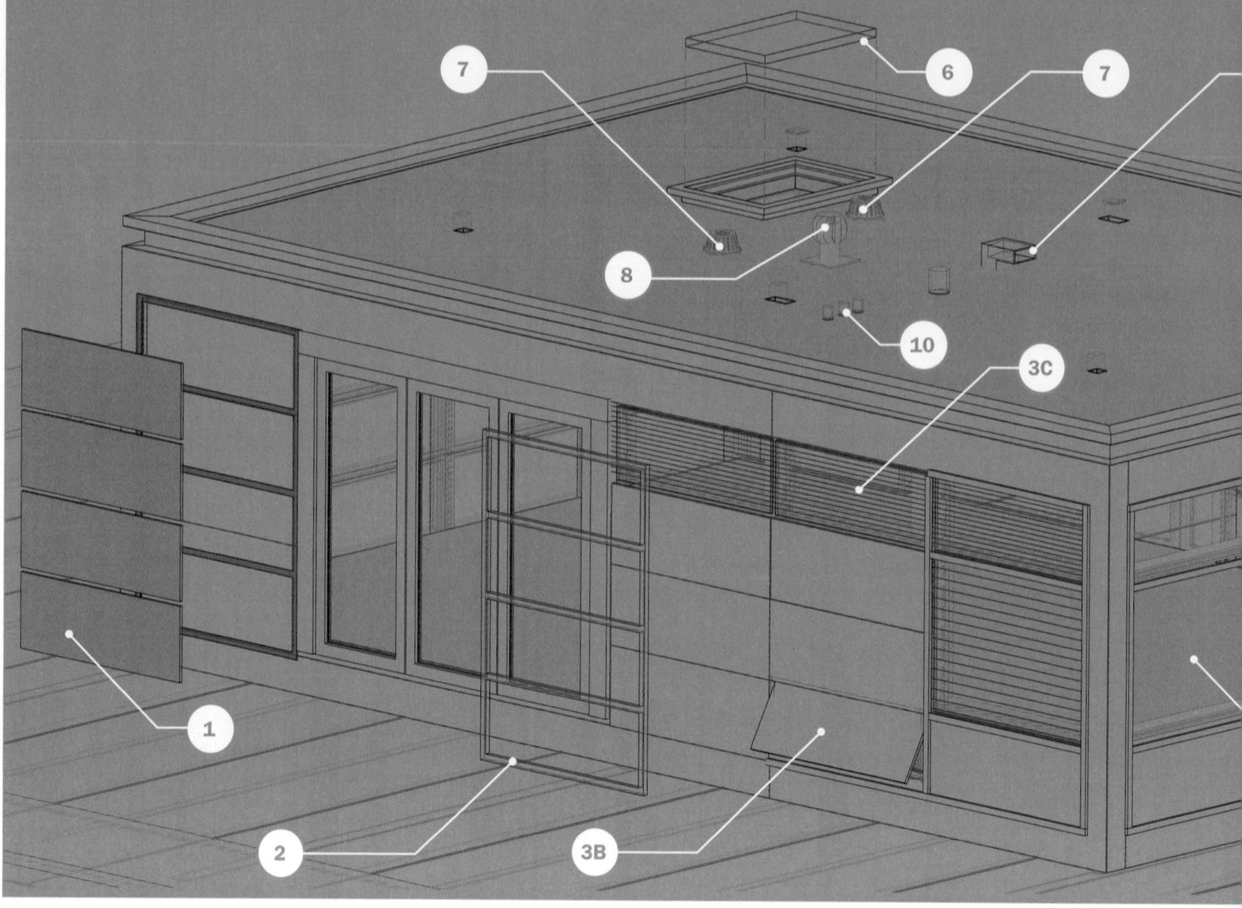

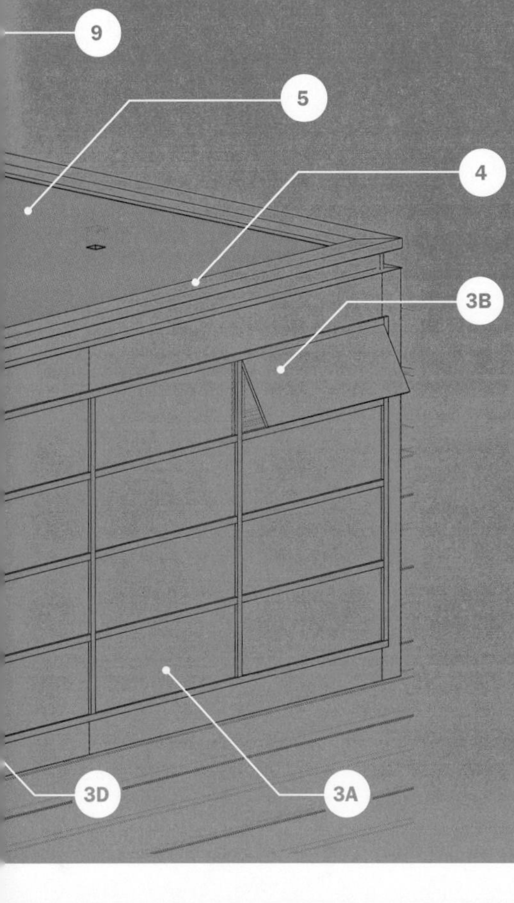

GLAZING +ROOFING

7

1. EXTERIOR WALL, PREPARED FOR GLAZING INSTALLATION
   - WATERPROOFING
   - LEVELED BASE PROFILE CARRIER SYSTEM
2. CURTAIN WALL GLAZING SYSTEM:
   - GASKETED BASE PROFILE
   - GLAZING
   - GASKETED PRESSURE BAR
   - WOODEN CAP
3. INSTALLED GLAZING:
   A) VACUUM-INSULADED FACADE PANELS
   B) OPERABLE WINDOWS: UPPER AND LOWER VENTS
   C) REDIRECTING LIGHT CLERESTORY GLAZING
   D) VIEW WINDOW W/INTEGRATED OPERABLE BLINDS
   E) BIPV GLAZING, POWER GENERATION
   (BUILDING INTEGRATED PHOTOVOLTAIC)
4. BREAK METAL COPING:
   FACADE – ROOF TRANSITION
5. ROOF MEMBRANE
6. OPERABLE SKYLIGHT,
   ADDITIONAL VENTILATION DEVICE
7. ROOF DRAINS
8. NATURAL VENTILATION OF MECHANICAL ROOM
9. AIR INTAKE FOR COOLING SYSTEM
10. ROOF PENETRATIONS
    FOR SOLAR POWER GENERATION

7

SYSTEM 7: GLAZING+ROOFING –WATERPROOFING AND ENVIRONMENTAL MANAGEMENT

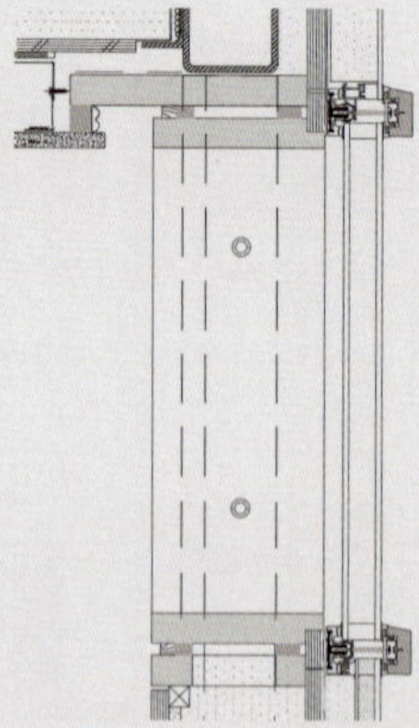

4.85

One of the main fields of material research that the students were fond of was glass. Most likely this is related to a current cultural trend for employing curtain wall glass façades to complement our post-industrial urban context, the corporate idea of expressing transparency, the high recycling potential of this material, or simply to develop a different, more contemporary approach to space, which is more fluent and related to the relationship between interior and exterior. This approach was particularly suitable for the sweeping views expected at many of the Roofpod site locations in Manhattan. However, we were also aware of the performance shortfalls common to most window glass. Among these shortfalls are the inferiority of its thermal resistance (r-value), the potential of 'solar overheating', and the often overlooked increase in cleaning and maintenance.

However, many sensual qualities of glass heightened our interests in the material investigation. These qualities include daylighting, spatial light and shadow effects, solar passive gains in winter, spatial and thermal relationships and certainly views. Ultimately our goal was to improve the overall performance of a curtain-walled glazed building, which, as a typical urban envelope system, was exemplary for retrofitting the existing building stock.

The two thermal concerns listed above were of greatest significance. Thus available products were investigated that enhanced performance through engineering and improving small but significant details. As for the unwanted solar gains, façade shading was addressed first. Exterior shading was preferable, since reflections from solar gains occur in front of the glass barrier, and thus transmit less solar energy into the interior space. Unfortunately, because winds on top of the roofs can be quite significant, the manufacturers of such operable exterior shading systems would not guarantee their product for our Roofpod. However, shading without operability is not satisfactory, since the sun actually contributes to warming interior spaces in colder seasons. And yet shades on the interior minimize the effect of shading because they are located behind the glass barrier and therefore absorb most of the sun's rays. In this case, when these rays hit a surface, including the shade itself, the short-wave radiation of the penetrating light becomes long-wave radiation. Since glass does not permit the absorption of long waves as it is for short waves, the heat becomes trapped!

Shading on the inside is too inefficient for a cutting-edge house, because the heat gain occurs in advance of the shading layer becoming effective. The solution was to use glazing that conceals movable aluminum 'mini-shades' within the cavity of the insulated glass unit (4.86). Although not as effective as the exterior shades would have been, there were many advantages to this solution. For one, the blinds couldn't possibly get tangled up in the wind and would not require cleaning. And as they were perfectly concealed within the glass, blinds were accompanied by a maintenance-free

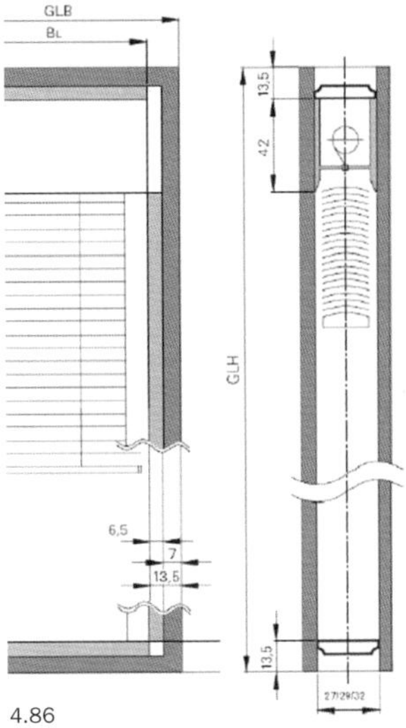

4.86

4.87

4.85 Detail of curtain wall to upper break metal transition.
4.86 Detail of the operable integrated blinds, chosen for dynamic shading.
4.87 Mounting the curtain wall layer, covering the 'building block' joints: Accuracy is key; for a repetitive structure the finishing work has to comply to the overall modular geometry.

motor capable of connecting to our monitoring and control system. Having control over the degree of shading in the house and exposing it to sun when necessary was the device's most significant function. Therefore, selecting these devices would be based exclusively on their operability and optimal shading of the glass. Yet in lieu of this superior approach, for budgetary and technical reasons, not all the windows could function this way. We decided to use these shades only where views could be affected, between 2 feet and 8 feet in height, directly relating to the user either sitting or standing in the room. For the other windows, a more static strategy was sought to protect the interior from unwanted solar heat gain, while still letting the advantageous sunlight enter. We became familiar with other options of glass unit integrated interstitial devices, which considered the sun's movements in order to optimize heat gain in combination with daylighting. These units function in a way that does not require any mechanical adjustment, which means they always bear the risk of mechanical failure. These units may not be as responsive, but are significantly more cost-effective.

4.88

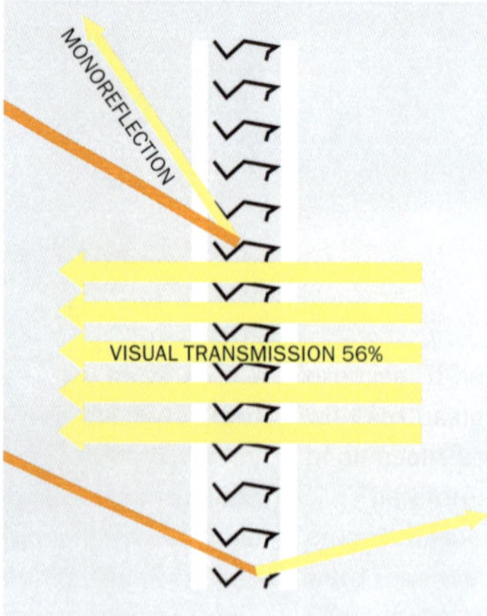

4.89

One of those fixed ideas used for the doors (see chapter 4.6) employs the perforation used in the glass-integrated stainless steel sheets. Here, the perforation is angled in such a way to consider the changing position of the sun during the various seasons. During the summers in NYC (at around 60–70° altitude), the slanted perforations do not let sunrays penetrate, while in winter (at around 20–30° altitude) the angle of these penetrations is lined up parallel to the sun's rays, allowing more heat and light into the interior. However, these penetrations have to be carefully adjusted to site locations, because different latitudes translate to different sun angles.

We discovered another product that employs a similar strategy, yet with better daylighting management coordination. This product uses uniquely shaped fixed blinds (4.88) mounted in the cavity of an insulated glass unit (4.89). While in summer these blinds reflect most of the radiation back to the outside, in winter they let the light enter and simultaneously redirect it towards the ceiling. At the same time, the spacing between the singular blinds allows for maximum visual transmission. We used this glazing type for our clerestory windows, between 6 and 8 foot in height, which was most suitable for taking advantage of the redirecting function described above. Because the distributed light has to find a reflector plane, a white ceiling serves this function well. This, in turn, brings light deeper into the space, and makes electrical illumination obsolete at times, which ultimately saves energy. All related building elements must be coordinated to develop a product that responds to heat gain, daylighting, power consumption and visual connection to generate superior performance.

It was understood that the movable façade elements, like the integrated shades for the 'view windows', did more than just improve performance. These elements also communicate the way they are perceived by the user, and thus the user's comfort is expressed. Even though this happenstance existed, it was not clear how to respond to it in the design process, since the user's evaluation of comfort is not necessarily predictable nor related to optimized operation. Therefore, it became critical to find the middle ground between the user's well-being and an energy-efficient operation. The mechanization system, which manipulates solar impact, would need to be flexible. This could only be achieved through a smart system that would optimally monitor and control the shading and to regulate this operation according to the user's requirements. This concept was crucial for ventilation, because people are particularly sensitive to drafts, fresh air and abrupt changes of temperature, which are all directly related to ventilation strategies. The design began again by identifying locations along the height of the façade module zones, to integrated ventilation devices for optimal efficiency, and according to the existing 2 foot module. The principle of displacement ventilation states that interior used air becomes warmer and rises while exterior colder air replaces it. With this principle in mind, it became clear that the 'intake' and 'outtake' had to be separated from each other. Therefore, the vents were placed either

4.88 Close-up of engineered blinds for light-redirecting clerestories.
4.89 Diagram of interstitial blinds to redirerct light differently during summer and winter.
4.90 Different façade in-fills, according to diverse performance criteria: daylighting, heat gain, ventilation, insulation, views, etc. The location is determined by both environmental specifics as well as personal preferences.

4.90

in the lowest area of the façade module as an 'intake' (between zero and two feet), or at the highest possible location, as an 'outtake' (between six and eight feet). By distributing the vents strategically on all four sides of the façade, uniform ventilation throughout the interior space was ensured.

Low-voltage wiring was integrated between the building blocks so that the ventilation openings could be operated by small chain motors, called actuators. These motors (4.93) push the awning window open when the control system indicates that this is necessary. However, we integrated a function that allowed the user to overwrite these commands if preferable, therefore allowing for flexibility and subjective comfort. If more ventilation was required, then the folding doors could also be opened manually.

The upper vents also contribute to daylighting, since they are located in the upper portion of the room. While the lower vents contribute negligible daylight quality because light entering at that height do not disperse. Instead, an opaque panel for the

4.91

4.92

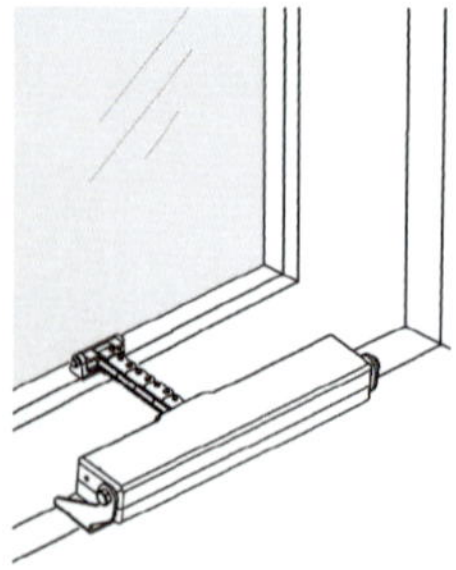
4.93

lower vents was used to prevent heat gain and provide better insulation. On the interior, a handcrafted wood grille visually continues the poplar paneling of the building blocks (4.91).

For a long time, we struggled to decide whether the façade should be clad with material that would not just ward off solar loads, but also use this energy to generate power. Several products were studied, including laminated glass panels with sandwiched thin-film photovoltaics (also called BIPV: building integrated photovoltaic) that contribute to power generation. Applying this material meant that the building's expression would communicate a high-tech approach. Although this direction was attractive, there were two major obstacles: this technology was beyond our budget, and the Solar Decathlon organization had placed a rule stating high budget projects would be penalized. However, we felt strongly that emerging technology should be expressed in artifacts that try to improve energy efficiency. We therefore decided to display one of these panels (4.94); these could be further used in locations necessitating not only transparency, but also opaque panels throughout the envelope, where no active function was required. If applied throughout the façade, this would have added about 2kW to the solar power generation of our house.

Double-glazed options were specified for all glazing. Although triple-glazing yields higher efficiency, the advantages were negligible and the costs were much higher for the Roofpod, rendering it beyond our budget. In its place, high-efficiency double-glazing with stainless steel intercept spacers were used, which successfully minimizes flanking. This strategy, coupled with an inert argon gas fill, achieves a greater thermal resistance. Also, a low-e coating was applied on all transparent glass units to effectively optimize performance within the budget. Low-e stands for Low-emissivity: A coating is applied to the glass that manipulates its emissivity relative to environmental loads.

Another issue regarding glass buildings surfaced: it is estimated that about 90,000 birds are killed in New York each year by flying into the 'mirrors' of our glass buildings[61]. There have been experiments focused on helping the birds regain their sense of orientation. For example, one of our glass product sponsors invented a glass coating, which can only be seen by birds, which are more sensitive to ultraviolet light. With this visual noise strategy, the Roofpod is one of the first buildings in the United States to use this particular product (4.96).

The transparent windows all received optimized glazing, often related to the height in the modular 'building block' grid. Therefore, it became clear that much of performance-related decisions happen in section.

For the neutral options, the question remained: what to do with the opaque panels? Typically, opaque panels manifested in curtain wall technology are called 'spandrel panels' in areas below desk height or in front of suspended ceilings. While often in curtain walls the opaque panels are not significantly better insulated than transparent panels, they do yield huge potential savings. As shown in the introduction's envelope chapter (see chapter 1.3), curtain walls are one of the preferred contemporary façade systems in urban environments,

4.91 Ventilation: lower vent; motorized air intake.
4.92 Actuator/vent/control system wiring coordination during construction.
4.93 Actuator: working principle.
4.94 Building blocks with integrated photovoltaic glass: façade-related energy generation and light distribution.

4.94

4.95

4.96

4.95: Bird-proof glass w/ ultraviolet coating: What we see...
4.96: ...and what the birds see...; pattern visible only to birds' eyes.
4.97: Façade alternatives, student drawings. The placement of windows is related to their most efficient location for daylighting and views. Also, excessive heat gains should be avoided to save energy from operating air-conditioners. Therefore, the extent of openings is kept conservative.

however their thermal resistance is far inferior. Research into better methods of insulating glass produced familiarity with vacuum insulation, which is integrated again between the glass layers. Vacuum insulation is a sandwich-panel of an evacuated porous core material, enclosed in a gas barrier membrane. In the case of retrofitting all of New York's curtain wall façades, the envelope's thermal losses could be cut in half, which is technically possible given that the glass thickness would not change. This investment would result in significant savings in CO_2 emissions for the environment. These spandrel panels would usually be considered window areas in energy code compliance checks, even though they do not contribute to daylighting or ventilation. And thus they would fail in their performance benchmark if counted towards wall envelope requirements. Only through policy changes can retrofit obligations be pushed forward.

With our opaque wall assembly, almost 7" deep insulated 'building blocks' and the additional vacuum-insulation panel cover, the total thermal efficiency ammounted to an insulation value of about R-45. This is twice the value of energy code requirements in New York, and many times the current average of curtain wall insulation. Our strategy had been a hybrid of conventional interstitial framing insulation and high-tech space saving façade-panel insulation. In future approaches to prototyping this assembly, the challenge would be to reduce the overall wall thickness in favor of higher efficiency with lighter weight and more available square footage, which are all favorable to the real estate industry. But for our first attempt, considering the conservative amount of window openings stipulated, we were able to achieve a superior façade envelope. Also, the thickness of the façade, which was visible at all window opening sashes, was just right to give a sense of protection and spatial barrier.

With all these façade decisions, we grew closer to the desired appearance of our building. By blending the transparent windows with the enameled glass of the opaque elements, we achieved a more consistent look. The shiny glass throughout the envelope gave the building its contemporary expression. Conversely, the opaque panels dominated, as they were essential to the building's appearance. The prototype for Washington had a conservative amount of window openings, as the energy modeling indicated that this was necessary to avoid an unhealthy increase of cooling loads.

In retrospect, additional openings would have been favorable for making the Roofpod more attractive as an architectural skyline viewing device. However, in an effort to avoid heat gain, fewer transparent surfaces were used. Although we first tried to emphasize the vacuum insulation, event though the manufacturer was against this, it was ultimately concealed. We also tried to express the bird-proof pattern, since for us, again, the connection between performance and expression was desirable. However, the budget did not allow for realizing this idea.

The final version was enclosed within white lacquered glass, so that the foil is not exposed to detrimental ultraviolet light. The white color for the Roofpod conveyed a classic modern look, while efficiently reflecting solar radiation. Also, lighter colors don't necessitate the same cleaning maintenance as darker colors, as can easily be observed in the case of black limousines.

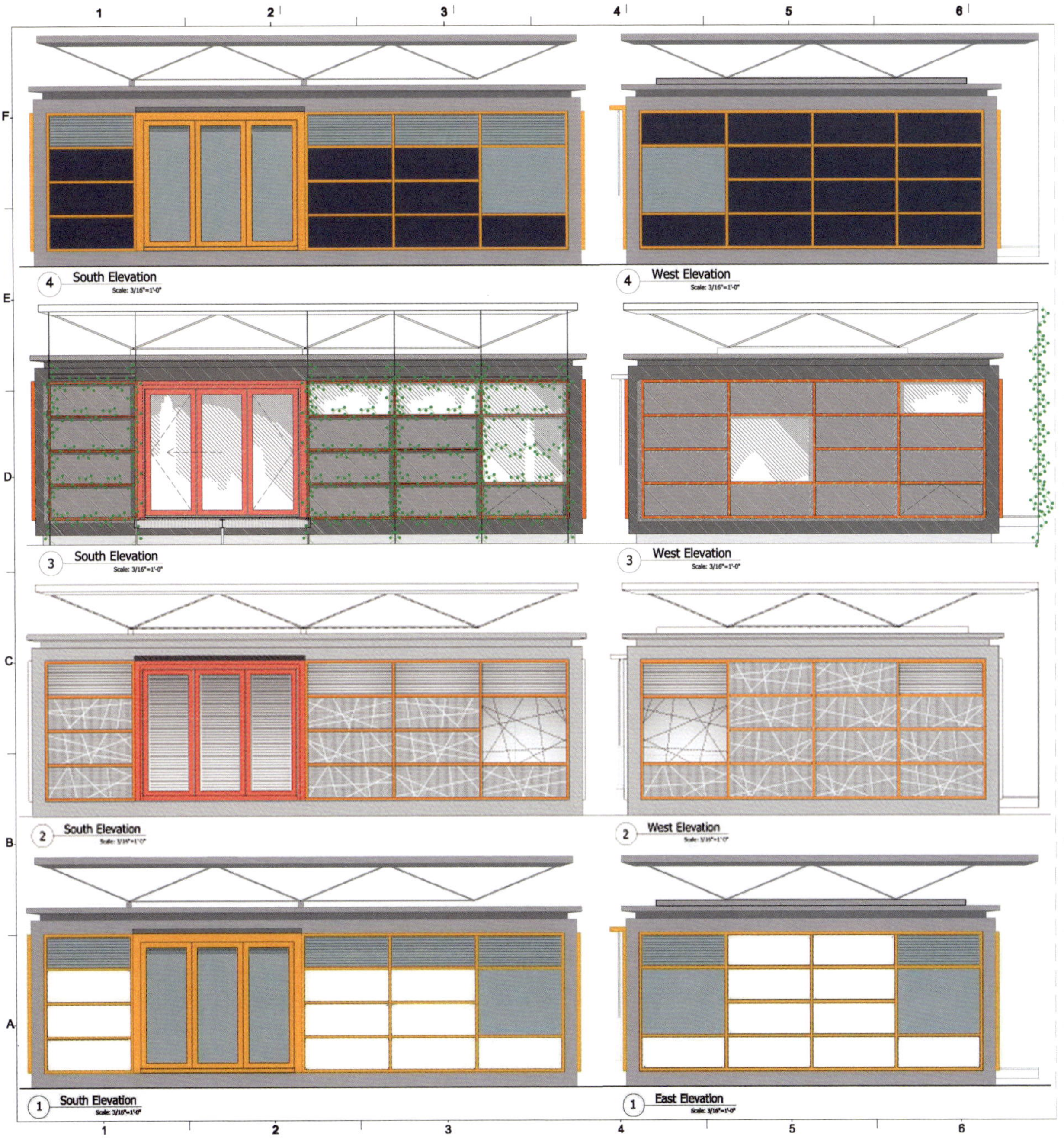
4 South Elevation
Scale: 3/16"=1'-0"
4 West Elevation
Scale: 3/16"=1'-0"
3 South Elevation
Scale: 3/16"=1'-0"
3 West Elevation
Scale: 3/16"=1'-0"
2 South Elevation
2 West Elevation
1 South Elevation
Scale: 3/16"=1'-0"
1 East Elevation
Scale: 3/16"=1'-0"

4.97

A significant design task involved specifying and locating the right glazing to fill in the façade's block structure, and to determine how these glass panels would physically attach to the building blocks. In our research strategy, we looked for products that improve the thermal performance with subtle changes. Several curtain wall manufacturers have attempted to optimize this attachment detail by improving gaskets and other micro-strategies. For example, the attachment of the 'cold' pressure bar to the 'warm' interior carrier molding generally produces a conflict, because this necessary structural connection consistently interrupts the insulation, and therefore creates a thermal bridge. Even though it is only one screw every foot, the thermal loss from these screws accumulates quickly, and causes considerable problems with condensation and ultimately corrosion. To ensure that the screw stays warm and protected a piece of insulation can be inserted into the cavity (4.98). This kind of detail requires significant research by manufacturers, contractors and architects alike to define a practical and improved solution.

We found a committed fabricator to sponsor our project. In the proceeding months we worked on different methods to attach the curtain wall carriers to the blocks. In the beginning we favored an approach that would reflect the nature of the building blocks by using a 'unitized' system, which relates the frame to the glass, therefore doubling two profiles at the joint. We soon realized that a stick system, with only one mullion per joint, would make a more beautiful and better performing assembly. Thus it was understood that the building block system functions as a modular system, while the curtain wall, or anything beyond the waterproofing layer, functions as a unifying 'cap' that seals the joints between the blocks.

Whether this amalgamation implied that each block would arrive at the construction site glazed was discussed at length. However, glazed building blocks are heavy and fragile and too difficult to transport. It also implied that the waterproofing layer would be pre-installed, as it is covered in glass, even though its airtightness would be compromised with so many unnecessary membrane transitions. Therefore, the waterproof membrane must be installed continuously and prior to the glazing. If the waterproofing layer concealed the joint, then any visible layer would have to develop its own appearance independent from the building blocks. So, we developed the 'cap' as a mullion that would draw the building block joints on the exterior (4.101). We found a product that allowed us to install a custom wood mullion as the final curtain wall grid. We liked this approach because it allowed us to shape a significant design part of this system individually, while also fine-tuning the materialization of the exterior using the same wood for the molding, planter trim and other garden elements. This strategy helped to develop a consistent and unified appearance.

A proprietary pressure bar, holding the glass on its exterior surface, allowed for attaching the customized molding. It became a great detail design item, as many students showed interest in developing a grid cap that looked interesting and worked for all geometric and waterproof purposes. The development of the final detail took longer than a semester, with countless mock-ups and drip-nose test to keep the water off the glass (4.100). The interest was mainly based on the recognition that small design items – especially when used in repetition – often characterize and define a product morphologically.

While developing this mullion detail, the roof membrane was being applied on the construction site. This membrane had to be contiguous with the façade's waterproofing layer in order to keep the building airtight. However, configuring the roof is more complicated, for many reasons. First, a slope must be established to direct the water to the roof drains. In addition all roof drains must be fully installed, so that the rainwater doesn't run into the building, which we unfortunately experienced once during a rainstorm. Ultimately, the vast majority of planning coordination was focused on continuing the roof membrane into the façade membrane, which lay under the curtain wall system.

During the design process, we came up with an element that visually completes the top of the building much like a modern interpretation of a cornice (4.102). This parapet detail was important for several reasons. Not only did it visually finish the repetitive modular structure, it was also necessary for installing additional insulation on the roof, which would have exceeded the modular geometry. This parapet, as well as all exterior corners of the building covering the insulated steel frame, had to be clad. In order to keep the cladding thin and allow for maximum insulation we specified break metal, which seemed to be the logical choice for the continuation of the parapet's coping. This metal, anodized aluminum, also complements the glass surfaces of the building blocks, providing a harmonic transition at the building mass's corners. The cornice had to address the challenges of accessing the lifting points, which was achieved by integrating access covers within the coping sheet material. The horizontal and vertical corners of the building dictated equal dimensions to afford a continuous

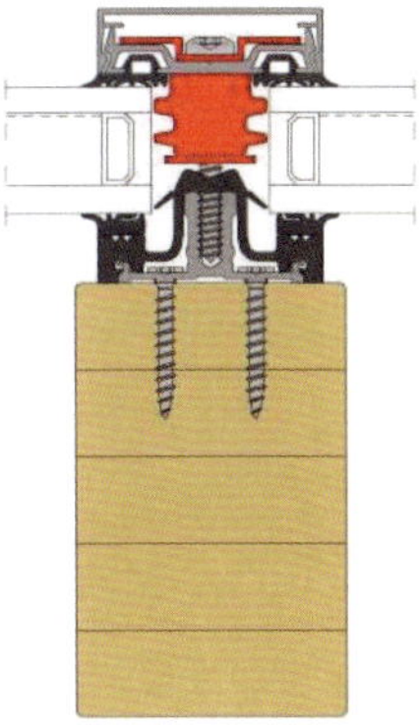

4.98

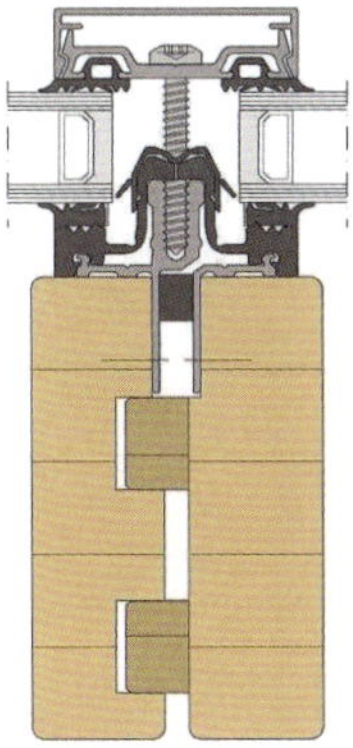

4.99

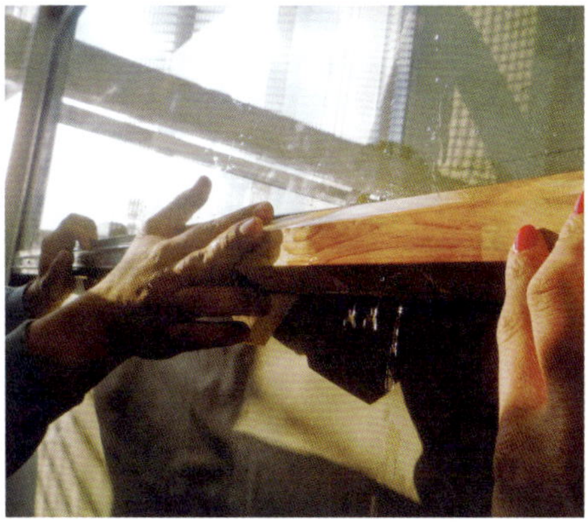

4.100

4.98 Curtain wall: Energy efficiency by means of small detail. The red insert insulates the screws connecting the carrier system with the exterior pressure bar.
4.99 Curtain wall: Split carrier profile. This profile is usually used for movement joints; in our application, it was perfect for the transportation joint transition.
4.100 Mullion mock-up: Drip edge testing in action.
4.101 Custom mullion: Design process drawing.

4.101

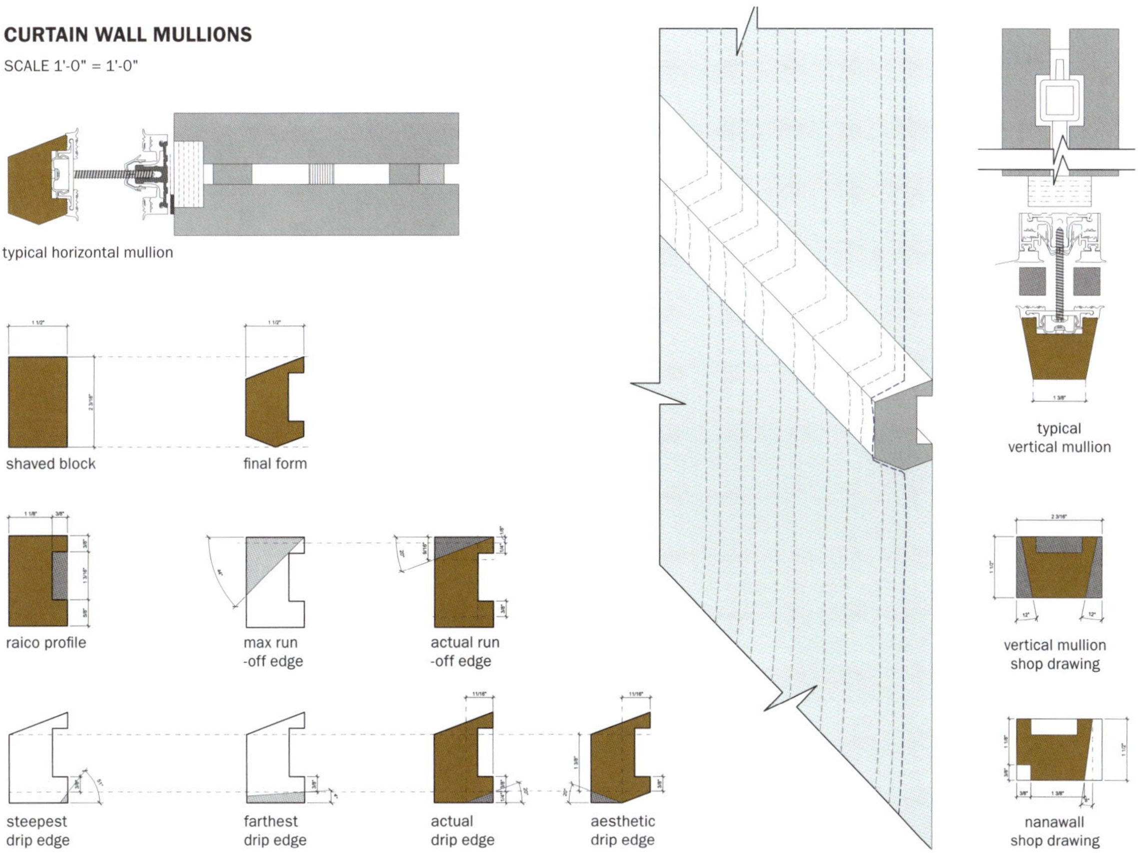

4.102

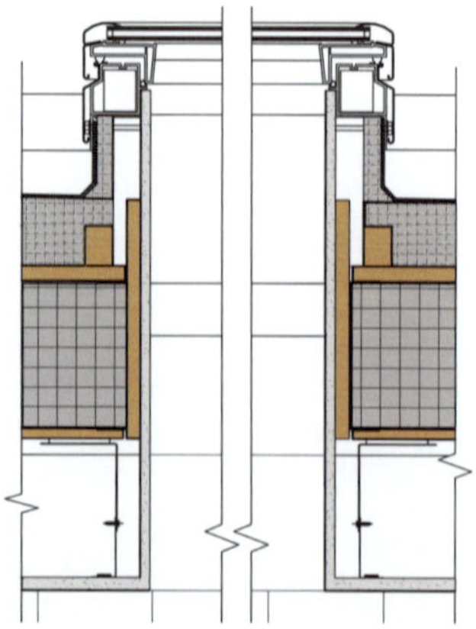

4.103

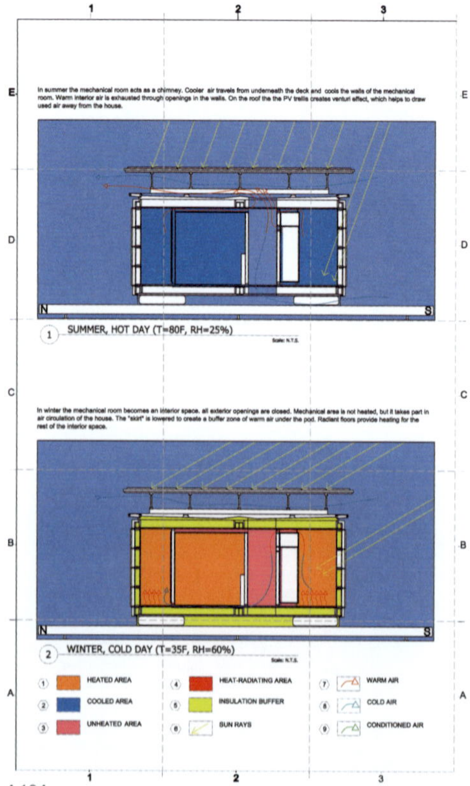

4.104

4.102 Roofpod cornice: Additional element w/ multiple functions.
4.103 Detail of prefabricated skylight, adjusted to specific conditions.
4.104 Passive design studies related to different seasons: The skylight is identified as a vital element for natural ventilation.
4.105 Configuration of the skylight during construction: A wooden curb had to be built on top of the structural insulated panels, to achieve minimum height for the membrane's dam at this penetration.

4.105

appearance to the module frame. It also had to visually tie into the curtain wall system, since the outer mullion's pressure bar held both the break metal and the glass surfaces in place and sealed it against water penetration (4.85).

The roofing itself also required a lot of coordination, as the photovoltaic and solar-thermal collectors required a connection to the mechanical room within the core of the house. In addition, plumbing vent lines, roof drains and a weather station had to be ideally located to connect the roof to the interior without compromising the tightness of the membrane, since these penetrations required sleeves of various diameters to accommodate diverse pipes and conduits. The last and largest penetration was a skylight, which we decided to put above the core-integrated bathroom in order to provide daylight to that small room. This skylight was placed over the shower, so that the solar thermal collectors, which were the source of heat, could be visually experienced while showering. The prefabricated skylight fit nicely over the dimensions of the shower basin. However, much thought went into coordinating all the building elements involved (i.e, from the opening in the structural insulated panels [SIPs], the interior framing of the bathroom, and the shower basin's plumbing). At many different stages of construction, we had to focus on the goal set forward in the construction drawings.

After several studies regarding passive ventilation (4.104), a motorized skylight was chosen. We understood that the skylight was a significant strategy for an efficient passive ventilation, because the roof is usually under negative pressure, which helps to extract air from the building. That made the core element, with its integrated skylight, predestined to function as an interior chimney of the building. Since we had worked on the smart monitoring + control system for the façade vents, it was clear that the core skylight had to be integrated into our system's operation. The skylight became both a spatial feature for the interior bathroom's daylighting, as well as a passive ventilation device based on its practical and low-energy performance.

We realized that innovative façade technology, in particular that of curtain wall knowledge, is predominantly in the hands of specialized manufacturers. Thus, there is not much room for do-it-yourself technology. This is due to the fact that the materials are mostly dependent on industrial processes involving machines and techniques that we do not have access to. Examples of these industrial materials include intricate aluminum extrusions and gaskets, as well as sophisticated specialty glass with the interstitial infill as previously described. Furthermore, operable and motorized devices such as the blinds, façade vents and skylights, are particularly challenging, since their development is typically based on decades of research.

Fortunately, our curtain wall sponsor sent us a tool box that included the specialized tools required for cutting gaskets and extrusions, and instructions on overlap rules, weep hole and flashing strategies. This hands-on work helped us to understand the material approach and sequence of this building system; although we were bound to the methods required by the system, this insight into the fabrication built competency for future projects, and possible opportunities to collaboratively customize such systems with a manufacturer. The architect, who was once more involved in product development, could now be more involved in developing products with industry. It is clear that in today's industrialized world, as architects, we have to take advantage of prefabrication techniques in construction.

However, our typical approach to the problem of custom fabrication is to force something to fit into an unrelated or unnatural reccurring design issue. The problem is related to systems thinking, both in terms of performance and aesthetics.

The ability to evaluate how things fit together holistically, and incorporate them systematically as part of the contextual design, can only be learned if the prefabricated item is understood in detail. However, that is exactly the crux of the problem. The ready-made item has to be rediscovered in detail to specifically identify the design context. The work on the Roofpod's façade was directly related to this kind of systemized and holistic thinking: incorporating prefabrication as a critical part of the design.

4.106

4.107

4.108

4.109

4.110

4.111

4.112

4.113

4.114

4.106 - 4.108 Roofing: Various activities related to roofing and waterproofing. Experts intitially taught students the necessary installation steps; later, the students took charge on their own to complete the installation.
4.109, 4.110 Break metal installation: Again, students were first instructed, then took over the operation.
4.111 - 4.114 Glazing: Various activities related to curtain wall installation. Installing gaskets, laying out the carrier system, placing glass units. The students received an installer's tool box from the manufacturer, and instructions from a local company in the techniques needed.
4.115 Completed façade assembly, for the Roofpod construction on campus; typical detail.

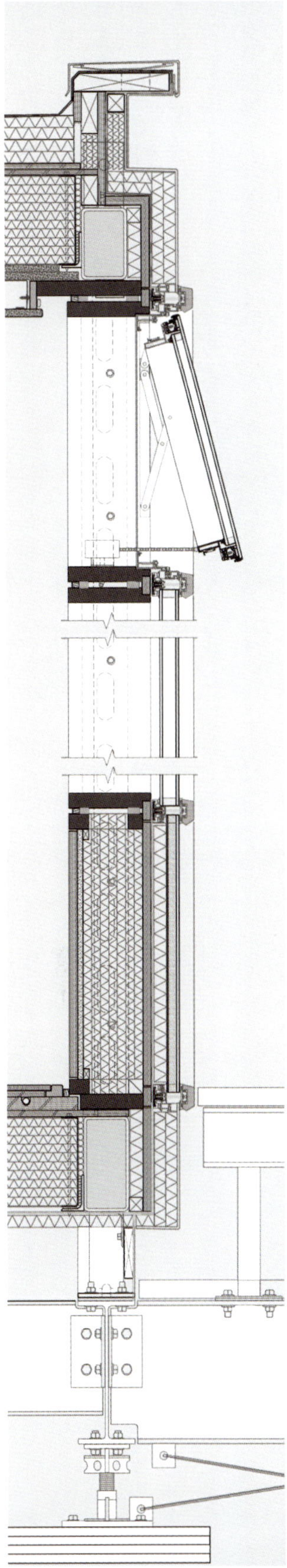

4.115

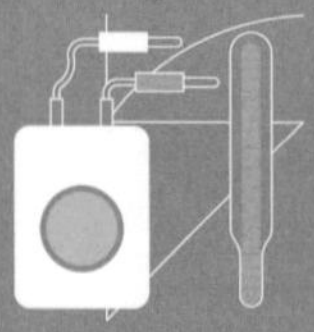

After analyzing a virtual model of the Roofpod regarding its thermal performance, it became clear that the cooling loads would dominate the design of the mechanical systems. While the well-insulated envelope is able to prevent major heat loss and gain, the glazed areas were particularly vulnerable: they would catch unwanted heat due to the greenhouse effect. In a rooftop location, glazed areas are paramount for skyline views. The sizing of the solar-powered cooling system had to compensate for this architectural requirement.

New York's power distribution system is weakened in summer. It usually reaches its peak load at the hottest time of the year, when countless air-conditioning units are used to cool down the city's building stock. However, cooling does not just have to occur with electric power, but can be delivered by using heat as energy source, thus reducing the city's peak load power benchmarks. We integrated an adsorptive cooling system, cooling the space with heat collected on the buidling's roof trellis by evacuated tube collectors.

Any mechanical system has to be planned in consideration of architectural built-ins, so that conflicts of pipe or conduit routing are addressed ahead of time. Our goal was to integrate these elements very space efficiently, like the engine compartment of a car. The mechanical room is accessed from two sides so that main appliances and related valves can easily be reached. Heat is distributed through the radiant floor system, while cool is provided by two fan coil uints, integrated in the cabinetry of the 'core', positioned centrally in the interior space.

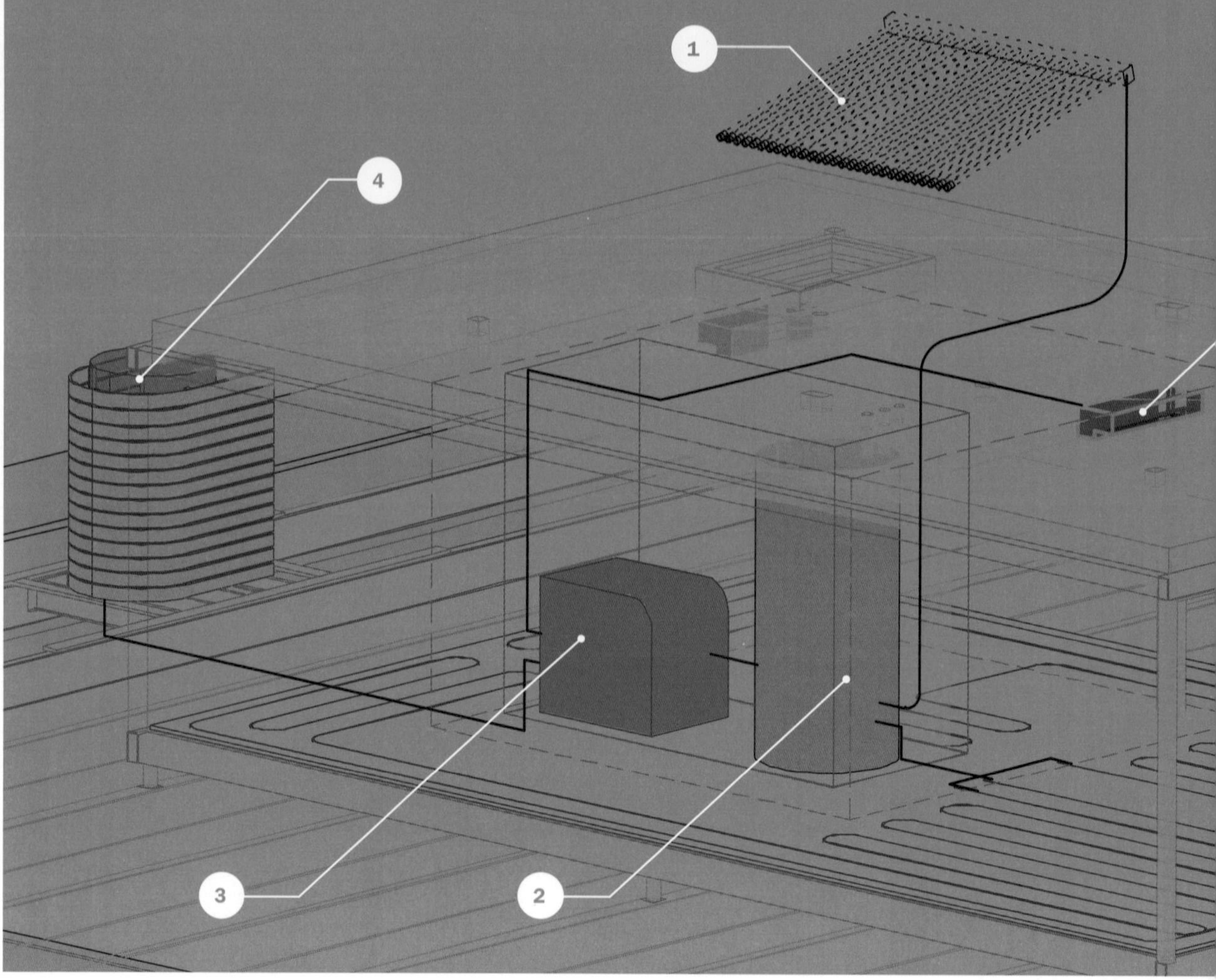

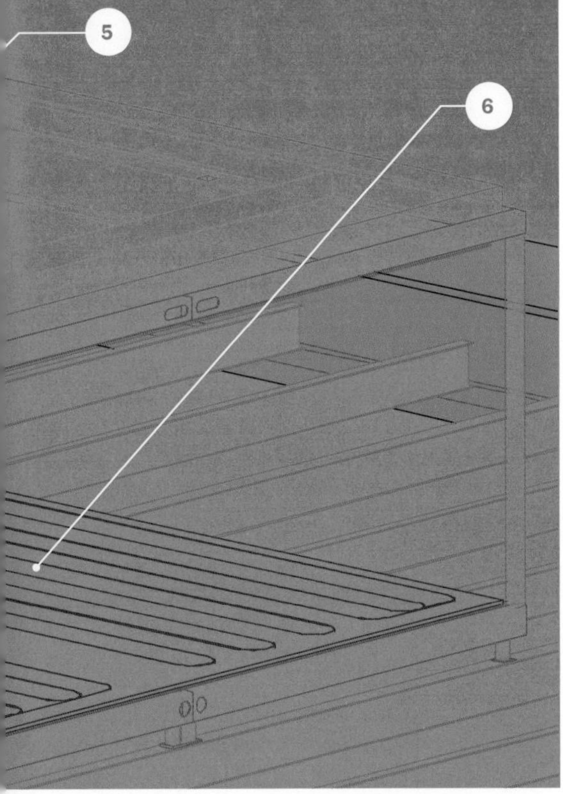

'MEP' HEATING +COOLING SYSTEMS

8

1. SOLAR THERMAL HEAT COLLECTOR: EVACUATED TUBES
2. HEAT STORAGE TANK, 200 GALLON CAPACITY; PARAFFIN-FILLED
3. ADSORPTION CHILLER, 8 KW COOLING CAPACITY, COP: 6
4. COOLING TOWER: HEAT REJECTION
5. FAN-COIL UNITS: COOL DISTRIBUTION; INTEGRATED IN CABINET 'CORE'
6. RADIANT FLOOR: HEAT DISTRIBUTION

8

SYSTEM 8: 'MEP' –CONTROLLED COMFORT: HEAT, COLD, LIGHT, AIR

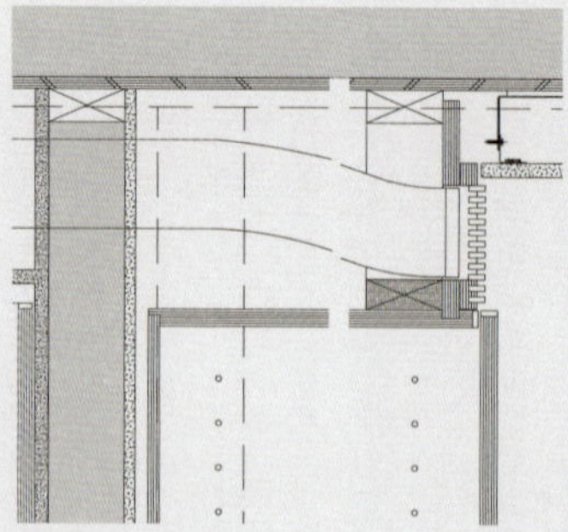

4.116

The decision to organize the service functions of the Roofpod in one concentrated core had significant implications for the mechanical, electrical and plumbing systems (MEP) of our design. In the beginning we considered having the core prefabricated and hoisted into the building as one preinstalled unit, containing the kitchen, bathroom, mechanical room and other storage functions. We soon realized that too many trades and installations were involved in this process. In particular, the MEP distribution, that runs throughout the structure like the veins of a circulation system made the idea of a plug-in core unlikely. Among the many circulation systems that had to be integrated, the following were most crucial: domestic plumbing for kitchen and bathroom, radiant floor heating, fan coil unit cooling, photovoltaic power feed, solar thermal collection to storage tank, sprinkler system, electrical and lighting systems, and the monitoring and control system.

The distribution of all the necessary pipes and conduits required copious coordination, along with vital advice from tradesmen in the involved specialized disciplines.

The modular system of the 'building blocks' restricted the distribution space, since these blocks' interior elevation necessitated full feature blocks for aesthetic purposes. We did not want the blocks to be obstructed by radiators or other equipment, nor did we want to run conduits through these modular units. The conduits would have been detrimental to the thermal resistance. They also would have complicated the 'building block' installation strategy. We were able to integrate radiant floor piping into the shallow plywood subfloor by using a product that came with precut groves for the pipes (4.119). However, for all other distributed systems, we chose one of the following three strategies:

1) Avoid pipe runs beyond the core's perimeter, as the space around its central location can be served directly. This was the case for both the air-conditioning system (4.116), and for the domestic plumbing.

2) Penetrate the roof in the area above the core to make a straight connection between the mechanical room and any roof installations. Since there was also the possibility of dropping the ceiling lower in the bathroom, or using the mechanical room as a distribution space, the flexibility in running pipes and conduits above the core was somewhat greater. The ceiling access panel in the bathroom became the main reconnection area for many of the circulation systems after re-assembly in Washington. The roof penetration approach was used for photovoltaic and solar thermal installations, as well as roof drainage channels.

3) Distribute all necessary pipes and conduits that had to encroach on the main space in a minimally dropped ceiling (4.120). This strategy was employed for the sprinkler system, fire alarm and sensors, other monitoring and control functions and electrical lighting fixtures.

The main technical component of the core was the mechanical room. It was our ambition to keep this room as small as possible in an effort to maintain the house's existing compact footprint of 800 s.f. Our paragon was the engine block of a car, which are designed to minimize space for all necessary mechanisms. The diverse systems are intertwined to take advantage of all remaining interstitial spaces. Often, special tools are needed to

4.116 Detail of core distribution strategy and air-conditioning grille; core-ceiling transition.
4.117 Schematic model for the mechanical room: engine block strategy.
4.118 Schematic section of mechanical room organization.

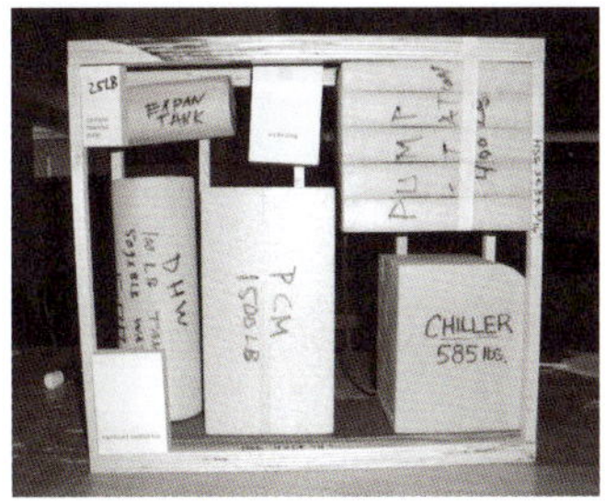

4.117

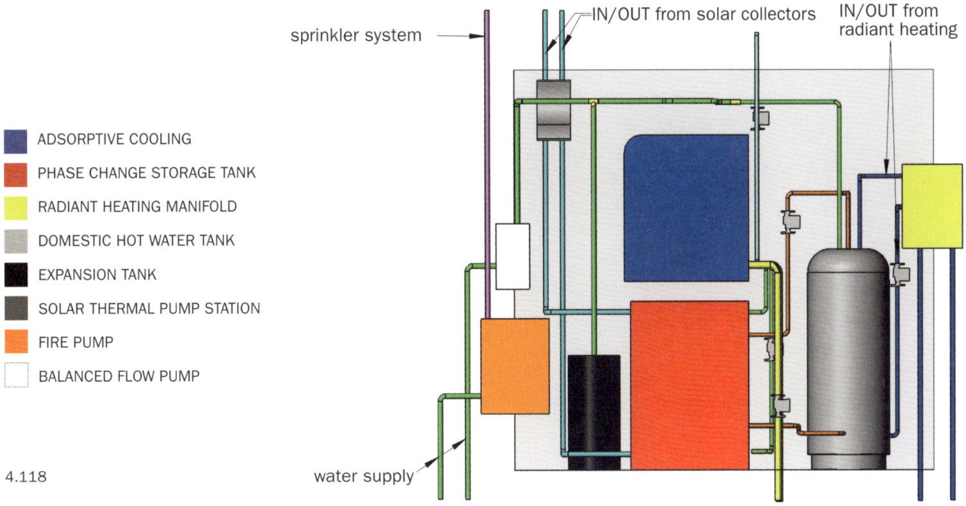

4.118

access certain bolts and fasteners. The sequence to disentangle the system must be known to the mechanics who maintain and repair the car.

We asked ourselves why the use of space for mechanical systems is not more restricted in buildings, especially in an environment such as New York where every square inch of space is usually utilized. However, we also learned that with some of the industrial components specified, there were limits to this concept of minimization, since access panels, operation valves, and pressure gauges cannot be customized as efficiently as they can be in cars, where high and repetitive output allows for customization. In this paragon we saw the potential of improving space management for mechanical installations in buildings.

We managed to organize the entire heating and cooling system, heating manifold, adsorption cooler, storage tank, domestic hot water tank, expansion tanks and several pumps and heaters in a space of 3 ½ feet wide by 10 feet long by 8 feet high (4.118). We first used a physical model and discussed the space optimization with our engineering partners.

We discovered that it was occasionally difficult to reach certain equipment. To improve accessibility, we installed two doors at both ends of the space (4.144), with one facing the living room. Therefore, we were able to present the 'heart' of the house quite efficiently for the exhibit in Washington. We used an original NYC subway door window to present our compact mechanical room to the public, for observation even when the door was closed (4.121, 4.122).

The most unusual system of the house was heating and cooling, which took up most of the free space. Early in the design process we had studied peak power requirements, which are typically during summer months, and the enhanced efficiency of solar thermal collectors compared to that of photovoltaic technology. The efficiency of solar thermal collection can be four-fold that of its electrical alternative, therefore, any system run with heat has a huge advantage. The strategy for winter heat distribution was quite obvious, as it became clear that heat could also drive cooling, which is even more advantageous in summer during peak power conditions.

In the adsorptive cooling scheme (4.124), water evaporates in a vacuum at room temperature and thereby extracts heat from its surroundings.

4.119

4.120

Through this process, cooling takes place in this localized circuit. In a closed circuit, the evaporated water is not released as steam into the surroundings, but re-condensed within the machine. However, for thermodynamic reasons, direct condensation of evaporated water is energetically unfeasible. Instead, the water is first adsorbed by hydroscopic silica-gel, which regulates the evaporation cycle much more efficiently. The condensation of secondary evaporated water from the silica-gel is easily attainable, as opposed to the primary evaporated water from the cooling circuit which is not. The silica-gel can constantly be regenerated as it utilizes the system's warm waste air, which effectively enables the process to continue. We were able to locate a machine based on this principle that was the appropriate size to fit into our mechanical room. We also understood that an additional pump station, much bigger than imagined, was necessary to run this system. The space we dedicated to the mechanical room grew more and more crowded as we learned about the system's actual operating process.

For instance, running such a system requires a lot of heat, which is directly related to the volume of air required for cooling. The Roofpod uses about 6,000 cubic feet of air, which, according to our engineers' calculations, necessitated a huge water storage tank to store this heat, equating to almost half of the allotted space for the mechanical room. However, since we only needed the heat contained in the water but not the water itself, we researched methods of storing this heat more efficiently.

Paraffin was the perfect storage medium because it is a synthetic material, whose melting point can be engineered, meaning more heat can be absorbed by it when it changes phase from solid to liquid. As a result, we were able to reduce the volume of the tank to about 170 gallons, occupying a drum 2 feet in diameter and 7 ½ feet in height (4.126). The paraffin's melting point was set at 170˚F, which was the temperature that the evacuated tube collectors on the roof were able to produce. The PCM tank effectively acted as a big thermal energy 'battery', with the capacity to provide heat, and thus also cold, to the building.

4.121

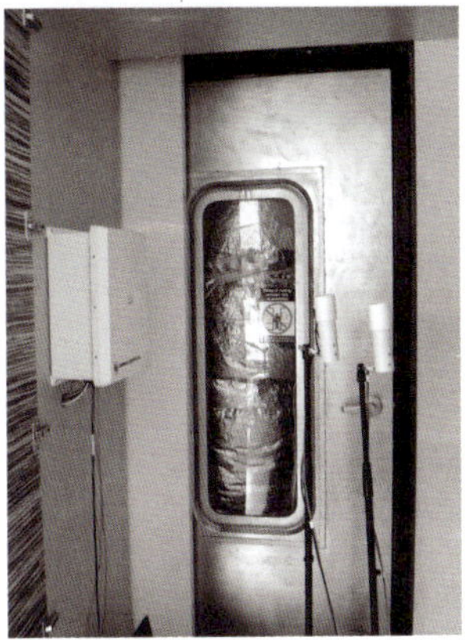

4.122

4.123

4.124

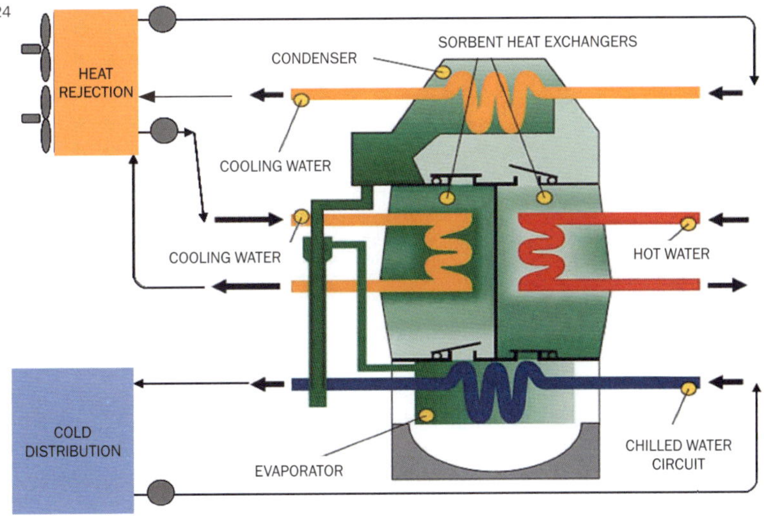

4.119 Site meeting to coordinate radiant floor installation.
4.120 Dropped ceiling distribution: Lighting, sprinkler, monitoring and control wiring in a plenum reduced so as not to obscure the 'building blocks'.
4.121, 4.122 Mechanical room, NYC subway door to expose the system's components.
4.123 Crowded mechanical room space.
4.124 Adsorptive cooling schematic design: strategy for the solar cooling system.

4.125

4.126

The capacity had been calculated so that it could use energy stored earlier in the tank after several days without sunshine.

The students build this 'thermal battery' from scratch after they had calculated the amount of loops that were necessary to transfer the heat to and from the device. A complicated loop system of brass tubing was fabricated according to these calculations. In the course of the competition in Washington, we understood that the design did not perform as intended. Computer simulations cannot guarantee performance. Since the tank was finished late, we had no chance to adjust it to the conclusions formed during testing. The PCM tank would have needed the same mock-up testing phase that other building elements' design development went through.

The PCM tank was also quite heavy when it was finally delivered to the site. It took a considerable time until it could be hoisted into the building and installed in the mechanical room. Since it was the biggest and bulkiest element in that small room, it was installed first. Now all the other pieces necessary to complement the system could be added.

Also, advance planning was necessary to fill the PCM tank with the paraffin. Since the tank went up to the ceiling, in order to maximize its capacity, the tank had to be filled from the roof. In order to access the filler neck at the top of the tank, we had to provide a roof penetration. This penetration was of course part of the roofing scope, which happened much earlier than the installation of the PCM tank. The penetration had to be big enough that one could see the top of the tank when filling it, and it had to be at the right location. Everything went well. After the tank was filled, the hole was used for the natural ventilation roof fan. The cooling system would only work if the heat extracted from the interior space was exhausted to the outside. Thus, we needed a cooling tower to discard the heat to the exterior. Two locations were discussed in order to integrate this cooling tower into the design: either the Roofpod's roof, or the landscape surrounding the house. We decided on the latter, for accessibility and to express the utilities of the house during the exhibit in Washington.

Additionally, the backwards-running power meter fed by solar power was an attractive feature for the competition visitors to view. Therefore, we looked for an architectural element in the landscape that could combine the cooling tower with the meter and power feed necessary for the Roofpod's energy assembly. This landscape element had to express the power of our technical equipment. Not often can backwards-running meters or sophisticated cooling systems be experienced. Thus, we wanted to embrace this logistical necessity as a design and communication opportunity.

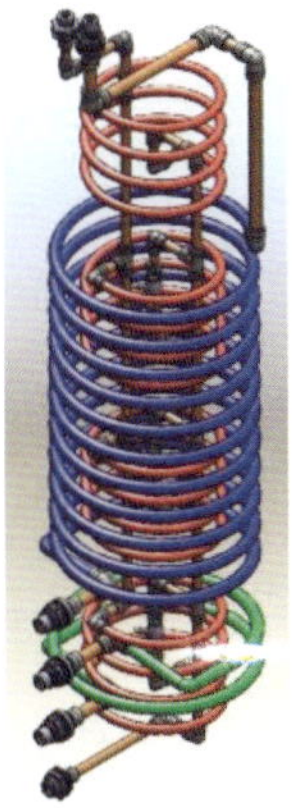

4.127

4.128

4.125 Initial installation of the mechanical room, starting with PCM tank, chiller and pump station; cramped conditions.
4.126 Site delivery of PCM storage tank; custom fabrication. Capacity: ca. 170 gallons. Weight (empty): 100lbs.
4.127 Internal design of PCM storage tank, feed and extraction loop and outtake for domestic hot water.
4.128 PCM storage, design in fabrication.
4.129 Installed PCM storage tank (middle door), insulated to prevent heat loss.

4.129

4.130

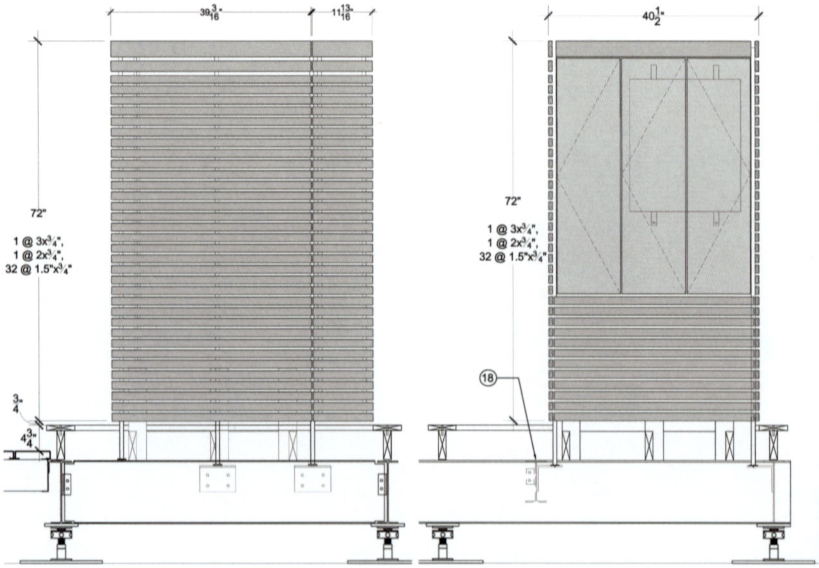

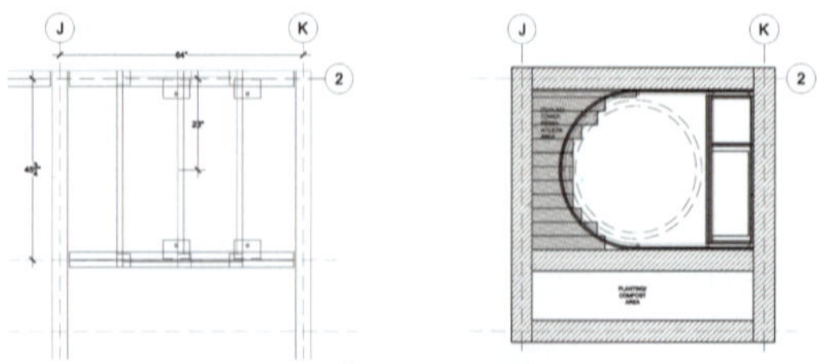

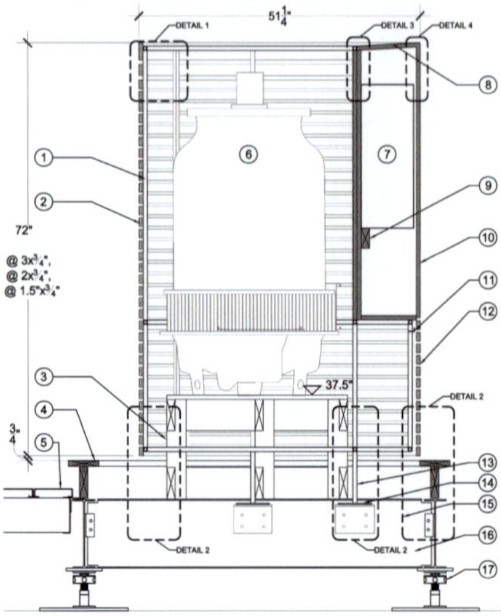

KEYNOTES:

1. ¾" SQUARE TUBE STEEL CAGE
2. 3MM LAMINATED PLYWOOD TO ¾"
3. COOLING TOWER SUPPORT
4. EXTERIOR PLYWOOD CASEWORK
5. SLIP RESISTANT METAL PLANKS
6. COOLING TOWER, 32"Ø
7. ELECTRICAL METER BOX
8. PITCHED CABINET TOP
9. 2x4 CLEAT
10. EXTERIOR PLYWOOD CASEWORK
11. WELDED CLIP FOR ACCESS PANEL
12. ACCESS PANEL FOR PLUMBING CONNECTIONS
13. ¾" SQUARE TUBE FOOTING
14. 1"x6" ¼" STEEL PLATE
15. INTERMEDIATE BEAM
16. STEEL DUNNAGE
17. W12x22 STEEL DUNNAGE
18. L5x5 ANGLE

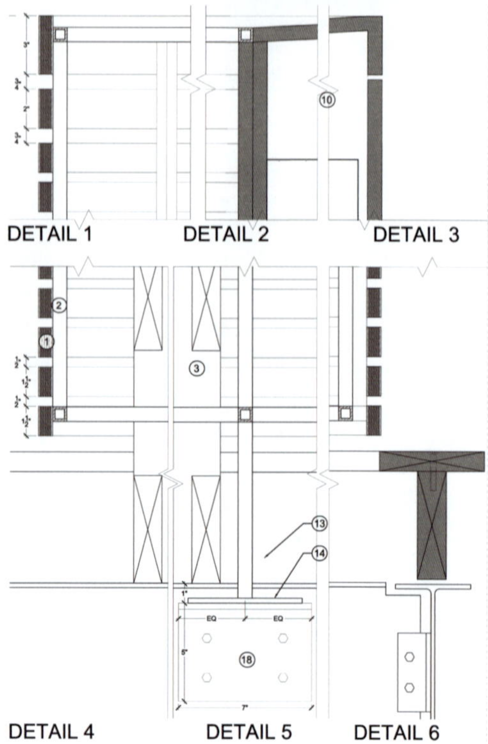

4.131

4.132

4.133

4.134

4.135

4.136

4.130 Cooling tower enclosure: Additional landscape element, incorporating the cooling tower and the backwards-running power meter; student drawing.
4.131-136 Various activities related to cooling tower enclosure construction: steel frame, painting, jig for lamination, installation and final result.

We designed an enclosure that minimized the visual impact while adding an interesting element to the landscape. The round shape of the cooling tower was the impetus for the design, and we added a rectangular box on the end, visible from the visitor's ramp (4.130). This box had to contain the electrical equipment, while the space inside the cooling tower had to be sufficiently ventilated. Thus we sketched an enclosure covered by many slatted ribbons with gaps just big enough to allow for sufficient air circulation, without revealing too much of the industrial equipment.

The first drawings looked interesting, as these ribbons made the landscape element appear light and dynamic. Drafting the curved lines did not require more effort than that of an orthogonal design. However, experiencing its construction allowed the students to understand that the more complicated the geometry, the more difficult it is to build.

For example, the wooden strips had to be laminated to the required diameter. In order to do this effectively, we built a jig in our woodshop class (4.133) to use as the formwork for the lamination process. In our steel shop class, the enclosure's steel support frame was fabricated, and just as much effort was made to produce the accurate curvature. It took weeks to produce all the customized parts. Then, the ribbons had to be primed and painted, and finally be installed carefully.

Ultimately, installing the enclosure on site and its disassembly for transport to Washington were viable concerns, too. Only once this element was finished, did many students fully comprehended the time relationship (3.40) between design and fabrication.

Another significant challenge in terms of developing the MEP system was the smart house concept (see 4.138): It was clear from the beginning that monitoring and evaluating the impact of all energy related concerns was necessary to develop an energy-conscious lifestyle for the occupant of the house. In order to do so, we decided to monitor the following functions:

a) Façade elements: window blinds and ventilation, b) HVAC and dehumidification, c) Greywater collection and irrigation of the planted landscape around the house, d) Electric lights and appliances,

e) Prototyped consumption 'traffic light' indicators, explained later in this chapter.

Many different kind of sensors were needed in order to understand and optimize the real-time performance of the house. Examples are temperature sensors, humidity sensors, motion sensors, luminance sensors, grade meters for production and consumption, a soil moisture sensors and water level sensors for the greywater irrigation around the house, air quality and CO_2 sensors for air quality control, and the typical carbon monoxide and smoke detectors (mandatory for New York City dwellings). A weather station and a carbon footprint monitor, together with its software, complemented the information collected by the sensors listed above. These functions had to be processed so that the user could always be informed about the operational status of the house. Actuators were installed to open and close the ventilation windows, start and stop heating and cooling, operate the shading blinds, in effort to control the performance of the house. In order to make this level of communication feasible, the electrical engineering students designed sophisticated operation circuits with modems and routers. All information was then available via laptops or smart phones. However, the main challenge remained the junction between performance and aesthetics, yielding the question: where should these sensors and actuators be located in order to optimally function, provide accurate data, and simultaneously convey a superior aesthetic and overall concept? After vigorous discussion of viewpoints, it was decided that most of the visible sensors would be placed in one of two predominantly interior locations. The first location would be at the upper area of the cabinet core, which was full of wiring and provided the shortest run to the system; alternatively, they would be located at the ceiling, which was a visually more forgiving solution if feasible bearing in mind that it already contained the sprinkler and lighting system.

As a general rule, sensors were not mounted on the building blocks for the following reasons: the conceptual significance of these elements; our hands-on efforts in finishing the poplar with a striking look; the installation logic of the block assembly suggesting that its air tightness could be compromised by any wiring penetration. It was critical that the envelope should express a repetitive generic system, allowing the core's strength to lie in its ability to adapt to the specifics of use and performance. We soon realized that the space in the mechanical room was not sufficient to incorporate all necessary control system equipment. Therefore, we decided to have all the electrical equipment dedicated to two cabinets, which would be easily accessible. One shallow cabinet would hold the regular fuse box, while a deeper compartment would contain all the low voltage equipment from the energy management system's circuitry.

We used a similar strategy for the electric lighting and the correlated switches used to operate them manually. In this case again, no light switches were allowed on the exterior walls. Instead, they were intelligently integrated within the core. The ceiling mounted light fixtures were easy to place and track lighting was added for some flexibility.

Meanwhile, the core integrated LED task lighting supported activities in kitchen, bathroom, home office and decorative niches.

Outlets complemented these areas of activity around the core. Due to code requirements, we were required to provide more outlets in the flexible space located beyond the core. We decided to use regularly placed floor outlets for this application, which allowed for flexible functions. This harked back to our idea that the Roofpod could not only be used as a residential unit, but also as an office space, conference room, or various other kinds of flexible programs.

The core grew more and more complex throughout its development, especially with the challenges of integrating all the required mechanical systems. The initial principles of how to run any circulation within the house was essential to organize a system that in the end was far more complex than imagined.

The core was initially intended to be the service element in the space. It ultimately became an appliance-like entity, with many integrated pipes and conduits.

Even though we were not able to develop this element as a prefabricated plug-in unit, we now understood the steps necessary to steer it in that direction. The requirements were just too complex, and with the many specialized engineering groups involved, it was not possible to prototype such an idea in the amount of time available. Despite the research and time required, the concept of a prefabricated core had been successfully pursued in architecture before, by architects such as Jean Prouvé (4.139) or Kieran-Timberlake (see also chapter 3.4). While it wasn't possible for us to achieve this concept, these exhaustive efforts would be necessary to develop the house as a more systemized product, if the Roofpod were to go into mass-production.

4.137

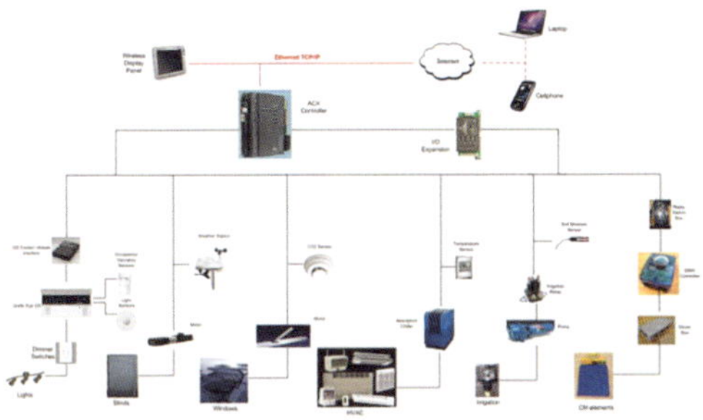

4.138

4.139

4.140

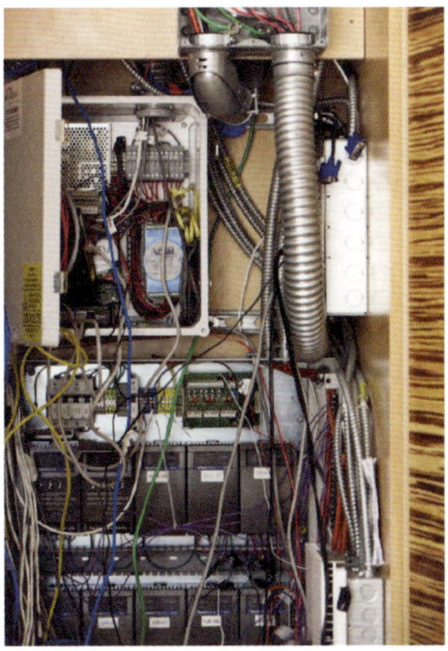

4.137 Presentation of the multifunctional core at the Solar Decathlon public exhibit: many features were related to MEP installations integrated in this unit.
4.138 Monitoring + control system's energy management schematic design.
4.139 Prefabricated core unit: Jean Prouvé: "Maison des jours meilleurs", 1954.
4.140 Monitoring + control system: resulting circuit cabinet compartment in the core; work in progress.

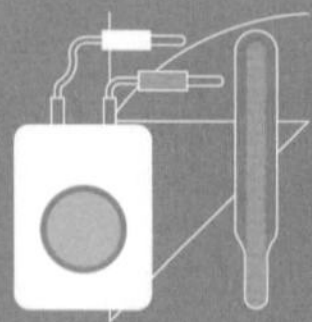

The interior 'core' is the main service element of the 'Roofpod', for programmatic and for comfort functions. It unifies the small spaces, bathroom and mechanical room, cabinets for the kitchen, home office, TV area, monitoring, murphy bed and closet space in a very compact way. Each side of this centrally positioned 'mega-furniture' differentiates the space adjacent to it with its program. The dynamic pattern of the bamboo panelling contrasts with the calm interior finish of the poplar 'building blocks'.

Integrated in the core, at the center of the interior space, are two full-height sliding doors. The doors can completely retract in the core. However, when suitable climatically, these doors can also separate the space in two equal halves. While one half can be cross-ventilated naturally, the other half can be climatized by the mechanical system. In common use patterns, we are not using all available space simultaneously. Thus, climate zone separation contributes to considerable energy and cost savings.

All the materials we specified went through extensive research. For wood materials, we went with fast-growing species that would also support our ambience goals. The dominating interior wall surfaces, the 'building blocks', were fabricated in poplar, which is local and thus low on embodied energy. It is also light in color. Therefore, it reflects more daylight, which saves energy by reducing the need for electrical lighting. The bamboo for the 'core, was made of waste scraps from flooring production.

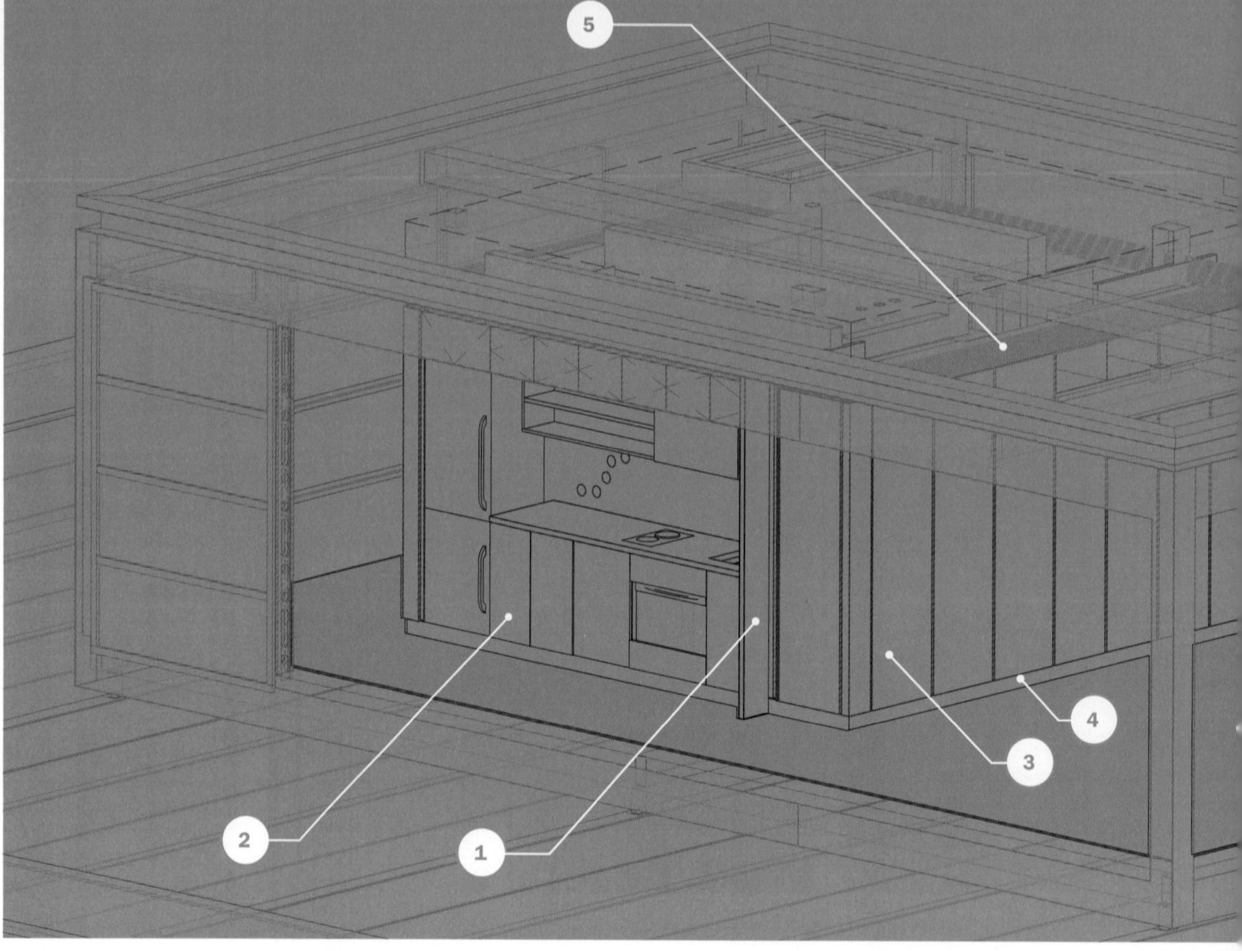

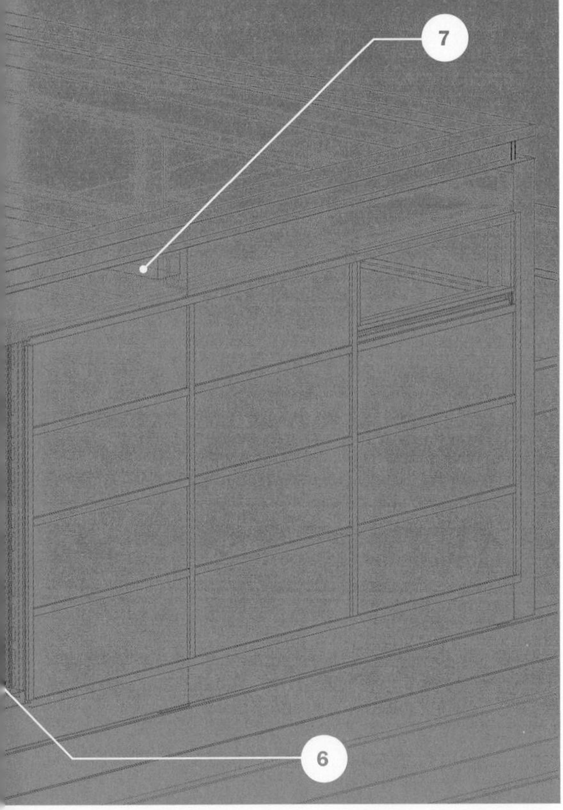

CABINET 'CORE': 'MEGA-FURNITURE'

9

1. SLIDING DOORS: SPACE DIVIDERS
2. KITCHEN CABINETS, COUNTERTOP + APPLIANCE INTEGRATION
3. LIVING ROOM CABINETS, ELECTRONICS INTEGRATION, ACCESS TO MECHANICAL ROOM
4. TOE KICK
5. CORE GRILLE: VENTILATION + TOP EDGE
6. CORK FLOORING, TONGUE AND GROOVE SYSTEM
7. SUSPENDED DRYWALL CEILING, W/COVE LIGHTING PERIMETER

9

SYSTEM 9: CABINET CORE – MULTIPLE FUNCTIONS

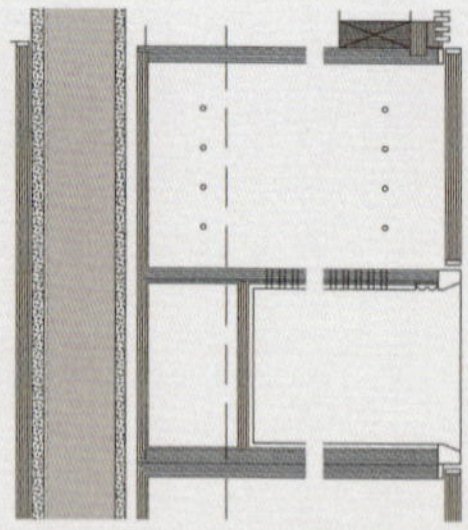

4.141

While the exterior envelope of the building was all about creating a standardized, repetitive system, it was clear that any function related to occupancy that did not fit into this regular system had to be accommodated elsewhere. The core was conceived as this accommodating element, as it spatially and functionally determined the spaces around it. These surrounding spaces are the entrance, bedroom, kitchen and living room, along with their diverse requirements to contain belongings, electronics, clothing etc. While the exterior of the core was conceived as an array of cabinets forming one seamless 'mega cabinet', the core's interior was subdivided into two main spaces, according to the geometry of the transportation joint necessary to bring the house to the exhibit in Washington. The southern half of the core contained the mechanical room, while the northern half included a small bathroom with a laundry niche, equipped with both a clothes washer and dryer. A wet wall formed the separation between these two spaces, containing most of the plumbing and serving both 'transportation modules'. We went through several iterations to configure the core most efficiently, only to realize that there are countless layouts dependent on program, overall size and personal preference. The core is thus an exclusively customizable element specifically adapting the modular structure to fit to the future Roofpod's occupant.

We decided on juxtaposing the core's materiality with that of the building blocks, after we understood how different the core design problem was from the overall modular design strategy. While the building block walls were made of poplar, which is homogeneous and calm in both color and texture, we were looking for a material that had more vibrancy and was still sustainable.

Ultimately, we found a plywood product from Moso bamboo scraps, which is fast growing and thus a relatively 'green' product (in 5 years it grows to 40 feet in height, quite similar to the growth rate of poplar). Even better, the 'zebra-striped' texture of the bamboo product looked very luxurious, reminiscent of Mies van der Rohe's often used Zebrano wood paneling. We decided to use this product as the fascia material of our core, and blend other materials with it as complementing accents. For all materials, the students were required to do research into their origin and fabrication methods (see chapter 2.7). One example of a sustainable and thoughtful element were the interior cabinet boxes, which were built of certified Medium Density Fiberboard, and niched on all four sides to break up the continuous surface and create a more sculptural quality (4.148). These niches were emphasized: They were laminated with linoleum, cork, and a glass-epoxy composite for the kitchen, based on the results of previous material research. They were also used as a presentation area for material samples at the Solar Decathlon exhibit in Washington.

The transition areas between cabinets, as well as between cabinets and niches, had to be detailed by the students. In a woodshop class with a master cabinetmaker, they had the chance to discuss these details and refine them (see chapter 5.4). At the same time, the students began investigating hardware.

In the beginning, they were attracted by videos and animations, wherein hardware was used to fold-out cabinets in very adventurous ways. However, they soon understood that anything movable requires precise attention and a disproportionate increase in planning efforts to achieve success. Although their ambitions were diminished by this insight, the awareness of pivot point requirements for particular applications grew. In the end, we managed to integrate quite sophisticated hardware into the cabinet detailing for such elements as a murphy bed, retractable doors and concealed appliance panels for the refrigerator and dishwasher. In addition, LED lighting fixture strips were integrated into the design to illuminate the presentation niches

4.142

4.143

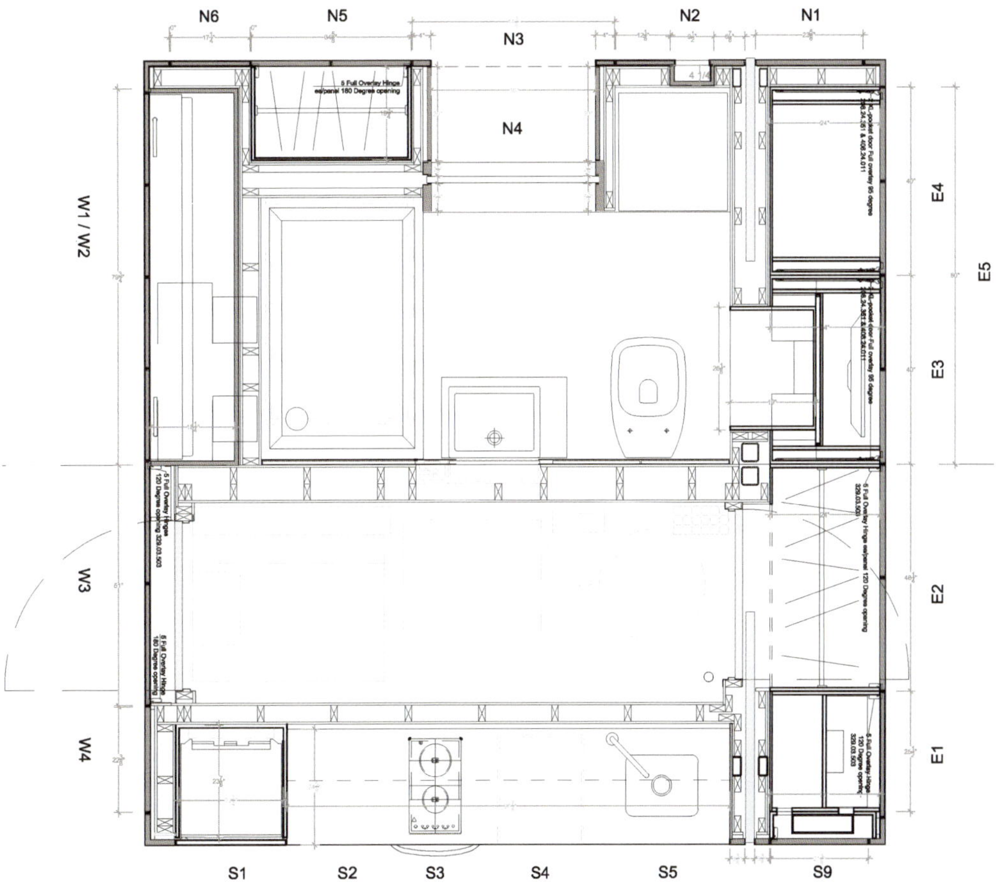

4.144

4.142 Close-up of Moso-bamboo paneling.
4.143 Site coordination: transfer from drawing to building.
4.144 Core cabinets: Shop drawing overview. The transportation joint was crucial for detailing the core as it was necessary to divide the house into two separate modules for the trip to Washington. The south module contained the mechanical room and the kitchen, while the north module enclosed the bathroom (with sink and toilet removed during transport), as well as cabinets and a murphy bed.

4.145

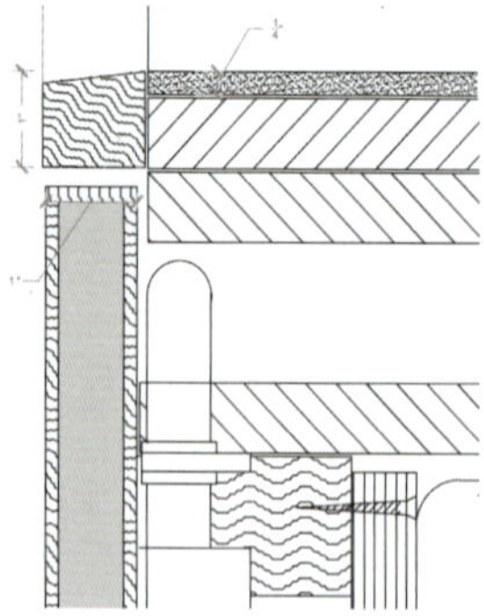

4.146

and the kitchen countertop. The core was originally conceived as one homogeneous cabinet in the conceptual drawings. However, differentiating its wall enclosure, framed drywalls and cabinet boxes, required more thought regarding its realistic material composition. It became clear that the most efficient way to configure the spaces was to approach them from two directions in the drawing. The first approach was from the inside-out, enclosing the interior rooms with framed, sound-insulated drywall. The second approach was from the outside-in, optimizing and controlling storage in the exterior cabinets' perimeter.

These spaces, which included the rooms and the cabinets, demanded careful calibration of the dimensions at hand in order to make them both practical and appear spacious. For instance, a cabinet's depth of 2 feet was influenced by practical considerations, storing items such as clothing, while fixtures and movement influenced the bathroom's dimensions and layout.

The main challenge on the construction site was to coordinate the fabricated cabinets fitting into the framed walls. Rigorous precision was necessary with the initial framing, requiring the students to draw sill plate locations on the subfloor and using the steel support columns and the 'transportation joint' as their starting point. These marks were added to those locating the concealed radiant flooring pipes, in an effort to avoid damage when installing the framing sills. The subfloor thus became a tangible drawing for construction purposes. Installation of the core started with integrating plumbing, mechanical and electrical lines. In fact, the electrical cabinets were the first to be installed, so that all the cable could be run to and from the switchboards before they were concealed (4.145). After framing was complete, all but the finished corners of the cabinets and core paneling were installed. The exact dimension of these remaining pieces were measured on site to accommodate the tolerances picked up during construction after the cabinets and paneling were fully installed. The students experienced first hand the sequences and techniques necessary to achieve precision, much in the same way cabinetmakers approach these issues. The shop drawings were done entirely by the students, and functioned as the communication tool to understand this part of fabrication, the installation sequence and the precision crucial for all dimensioning. Even though shop drawing is seldom taught in school, it is paramount to creating and controlling the desired design. It was noticable how students started to use the drawing as a tool to investigate how hardware would work, and how to configure the cabinet around it (4.146).

The same understanding of precision and control of sequence was also needed for the sliding doors that were integrated into the core. Early on, these doors were a vital contribution to the design, as they provided regulation in both privacy and ventilation (4.149). We decided to use large pocket doors, which enabled them to disappear into the serving core. This function assured spatial continuity while circulating in that space, which was important in discovering the surrounding views while moving around the core. However, these doors had to be considered in both pen and closed position to

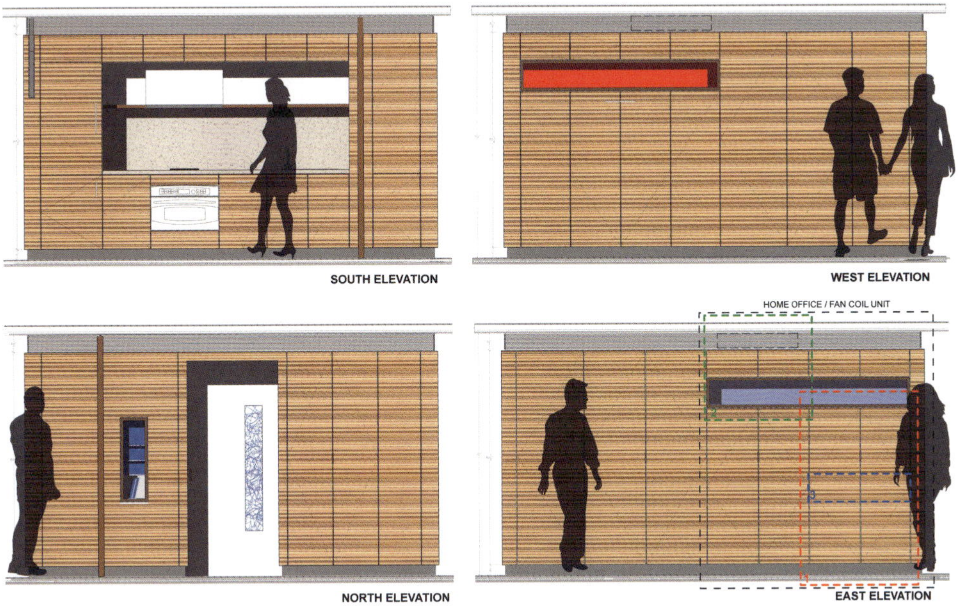

4.147

4.148

4.145 Site coordination: different teams worked on framing, cabinet installation and electrical routing. Electrical work had to be completed before the walls were concealed.
4.146 Detail: core niche to cabinet door transition, student drawing.
4.147 Core elevations; student drawings. Shop drawings linked the process between renderings and final materialization.
4.148 Core elevation living room: final result.

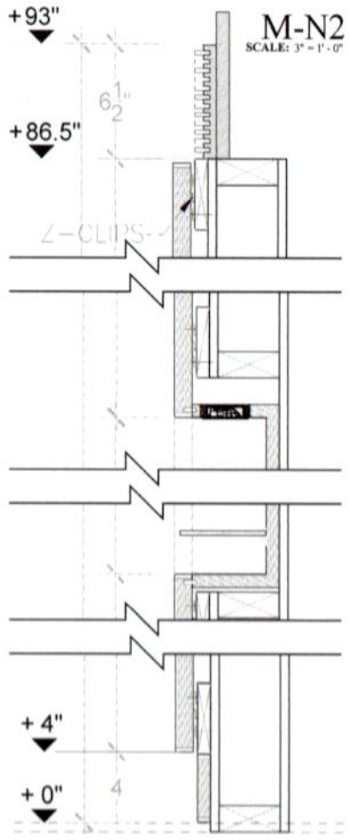

4.149

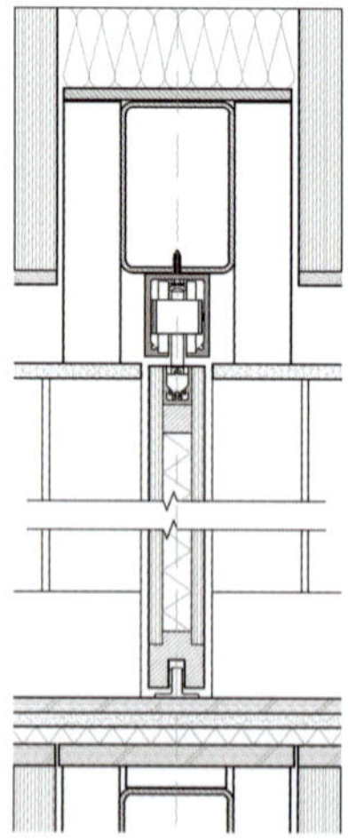
4.150

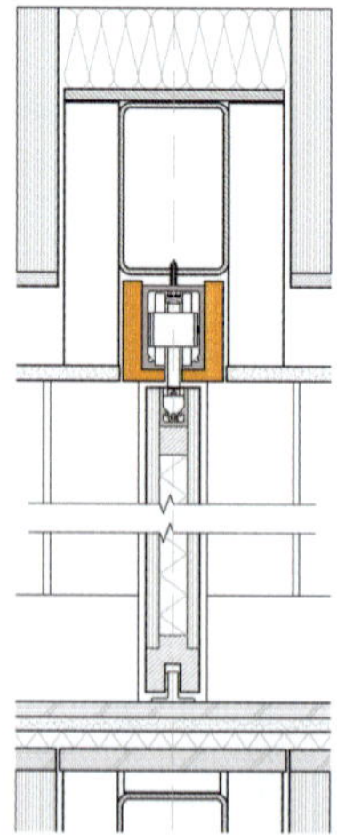
4.151

make them aesthetically appealing when detailed. When closed, they will connect the geometry of the exterior walls with the core, connecting at a particular point on the interior elevations. This was crucial, since the elevations were based on the building block joint, and thus quite challenging in its geometrical rigor. For this reason, and to satisfy the ventilation concept that divided the house into two equal parts, we wanted the door to close at precisely the center joint (4.152). However, because the most important steel columns were located at exactly that line, the sliding pocket doors would be blocked. We anticipated this early on by dividing these columns into two pairs, with enough space between them to configure the pockets (4.144). When the doors were in their pocketed position, they disappeared completely to become flush with the wall, without any evidence of their hardware.

This was not an easy task, since these heavy doors move on tracks suspended from above. When the door is retracted, a gap more than 2 inches wide (the thickness of the door), would show in the ceiling without an effort to conceal it. To avoid this heavy black gap, we considered altering the hardware's application by moving the track up. By moving the track up we minimized the distance to the suspension bolt's thickness (4.150, 4.151). Because this decision would have revealed the aluminum track, we laminated the surface, applying the same poplar wood we used for the building blocks (4.153). In addition, these surfaces shared an important detail, as the sliding door gap was now the same width (½") as that of the regular building block reveal. This kind of customization was possible because of the precise study of previous design conditions and the knowledge of the functionality of available hardware. Some questions asked in this study were: Is the bolt long enough to allow for the connection? Would the carrier be inserted, before the track was installed? In what sequence are the doors hung and the hardware mounted? These questions could only be developed by studying haptically. To arrive at a successful result, one has to fully understand the mechanism at hand and anticipate any possible problems during the drawing phase. The detail at the microscopic level is the vital tool in developing a full understanding of the imminent issues and conditions. Since this tool is very abstract, we made sure that the hardware was ordered ahead of time, so that haptic studies of the installed devices would confirm that the envisioned detail was actually buildable.

The upper portion of the core, where it transitioned to the ceiling, was also customized. In the schematic design phase, we did not continue the core up to the ceiling, since the drawing indicated that this element of 'mega-furniture' was an independent unit and separate from the modular approach of the envelope (see 3.9). However, several reasons made us decide to deviate from this plan. First, it was recognized that dust would accumulate on top of the core, which would be difficult to clean because it was such a restricted space. Furthermore, bringing the core up to the ceiling would reduce the height of the already small bathroom. Finally, most of the mechanical and electrical

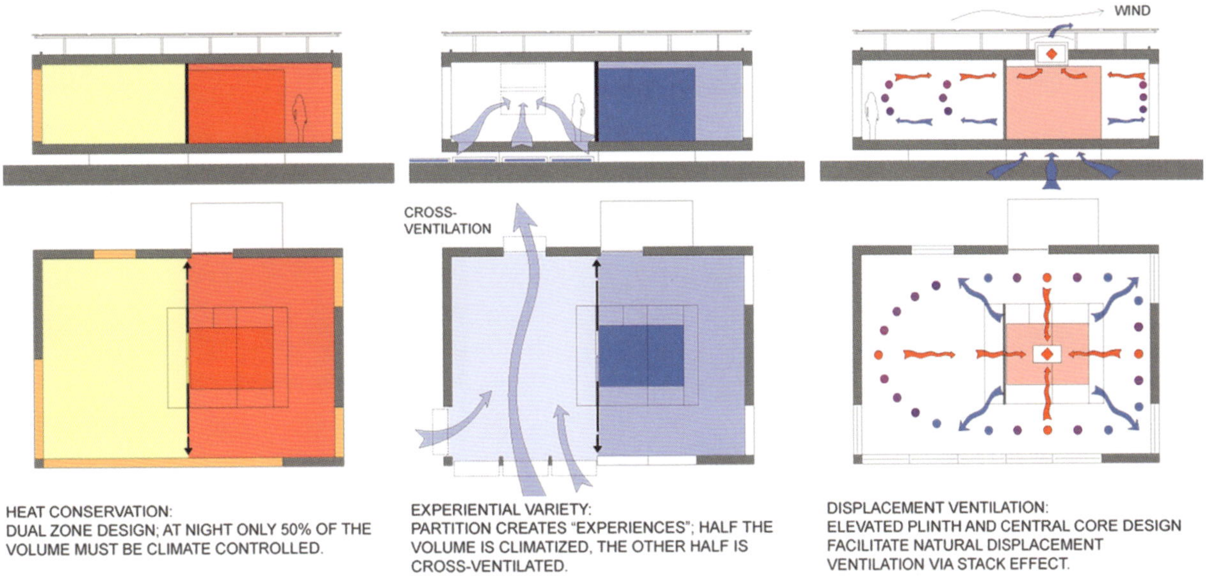

4.152

4.149 Wall paneling with niche and LED lighting. Shop fabrication drawing scale: coordination between wall framing, cabinetry and electrical work.
4.150 Typical detail for sliding doors.
4.151 Custom track detail for the Roofpod sliding doors.
4.152 Climate zone and natural ventilation concept. The sliding doors separate the house into two zones; student drawing, schematic design.
4.153 Sliding doors: From operational strategy to detailing the final result.

4.153

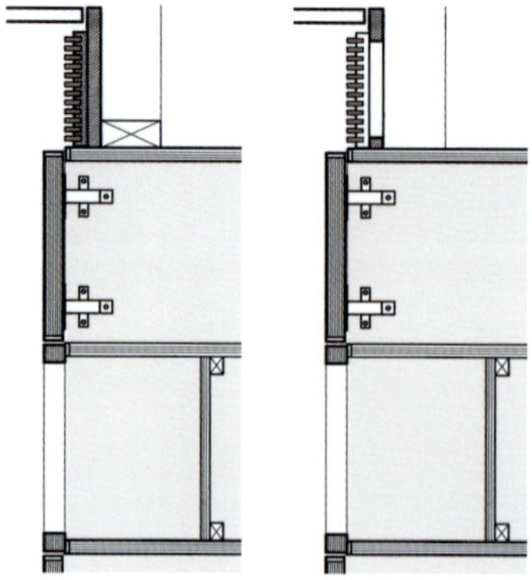

4.154

4.155

installations involved the ceiling, in particular the wiring for the extensive monitoring system, electric lighting and the sprinkler piping. Thus, if the core were interrupted, access from the mechanical room to the ceiling's distribution system would have been visible; nor would the sliding doors be seamlessly integrated in the core in this case, because they had to span floor to ceiling to separate the intended climate zones.

Instead of separating the core from the ceiling, we decided to articulate the transition in the form of a continuous grille located in the top 6 inches of the room (4.154). These grilles were removable, so that any possible wiring, for situations such as retrofitting or additional sensors, could easily be accommodated in the future.

But more important, the grille location and design was the result of integrating two cooling units into the core, functioning as fan-coil units. These units receive their cooling water from the cooling system located in the mechanical room, and work by blowing cold air into the room whenever cooling is required. The core had effectively become a mega-appliance, by unifying multiple cabinet units. The kind of grille we had envisioned to achieve a continuous appearance was not available as an off-the-shelf item, and those that were available were limited in width. However, a continuous element was required in consideration of the architectural context. Therefore, we had to either adjust an available product, or come up with our own design. We decided to go with the latter, after extensive research on the effort it would have taken to adjust an existing product. Instead, we used simple flat stock aluminum, and backed it up with grooved MDF (4.155). Wherever openings were necessary, this board was interrupted while the aluminum strips continued. With this method, we systemized an approach to accommodate multiple cooling units at flexible locations.

The integration of countless lighting fixtures, two unique color-morph traffic light systems (see chapter 4.10), running water in the kitchen, cooling units and sensors, and many outlets and light switches were all related to the same discipline of cabinetry and utility systems integrated into one mega-unit. The question remains: Could these elements be more effectively coordinated, streamlined and simplified to meet the conceptual requirement of a systematic approach? For example, could the cabinets configure the space, without any additional drywall and framing? The advantages are obvious, since a reduction of parts would result in a reduction of both cost and labor.

However, for our first project, this task would have required an enormous amount of planning. Furthermore, it became clear that refining such a multitude of design elements was too difficult for most of the relatively inexperienced students, and the related organization of teamwork.

Configuring the drywall first was a forgiving strategy, since any forgotten conduit, light switch etc. could quite easily be run after the drywall framing and before the cabinets were in place. The bathroom inside the core was also much easier to install with this framing in place, given that it was mandatory for the wet wall to conceal all integrated plumbing to enable the house to travel and properly function once in Washington. We decided that the wet wall, containing the plumbing rough-in, would travel with the mechanical room side of the house, without any need to disconnect any pipes. As a consequence, we took off the toilet bowl and the sink prior to transportation. This was the easier way of dealing with this issue.

4.156

4.154 Custom grille detail: closed (faux) and open conditions for ventilation/cooling.
4.155 Fabrication of grooved MDF strips at the table saw, serving as the carrier for the aluminum strips.
4.156 Installation of kitchen cabinets: Integrative thinking of plumbing, electrical and ductwork is required during fabrication; anticipation of any problems in the drawing phase save time in the long run.
4.157 Detail of integrated cooling unit: intake and outtake influenced the aesthetics of the core.

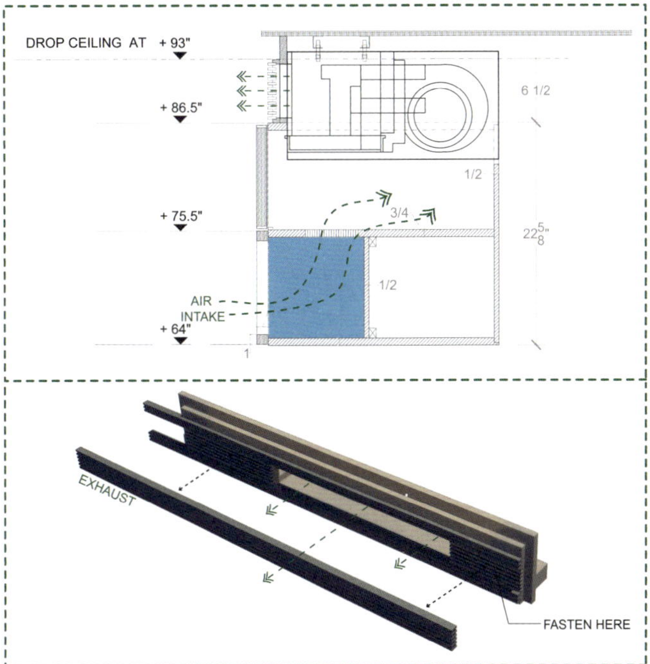

4.157

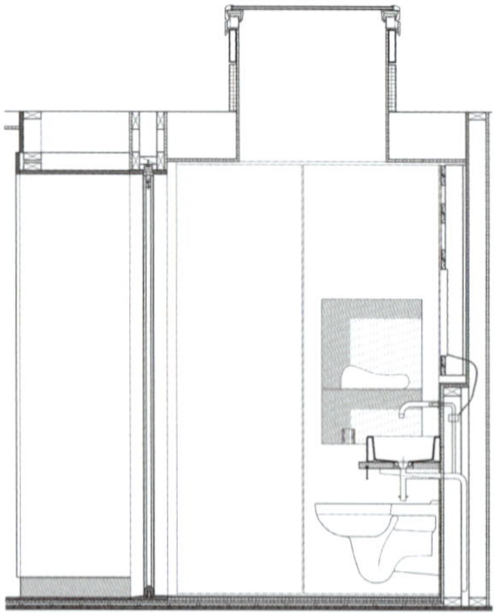

4.158

4.159

During the design phase, a critical concern was that the small bathroom would feel claustrophobic. On the one hand, we had to keep it as compact as possible so as not to increase the size of the core disproportionately, on the other hand, the experience in accessing the box-in-a-box bathroom should constitute a spatial climax.

Fortunately, the skylight brought this room to life, as daylight makes a room appear more interesting and spacious. Thus we also researched and specified white materials that reflect more light. For example, we installed a white glass-epoxy composite slab for the floor, which is commonly used for countertops. One piece measuring 4 x 5 feet perfectly covered the bathroom floor, with the exception of the laundry niche and the floor-integrated shower base (4.160, 4.161). The material's glass chips also contributed to a beautiful light effect.

We shied away from tile work for the walls, because we were concerned that it would be damaged during transportation, and we were looking for a more contemporary material that emphasized batch fabrication. Our research led us to plastic sheets made of high-density polyethylene (HDPE), which is made from ordinary recycled milk jugs. They are not reusable for milk according to the Food and Drug Administration (FDA). These jugs could either go to landfills where they decompose for a hundred years (1.47), or they can be recycled into other materials. These plastic sheets were beautiful and waterproof, and suitable for our bathroom.

We fabricated templates to manufacture the panels as accurately as possible (4.162), and laminated them to the water-resistant drywall (4.163). The edges of the panels we fabricated as tongue-and-groove, so that they would line up and continue

4.162

4.163

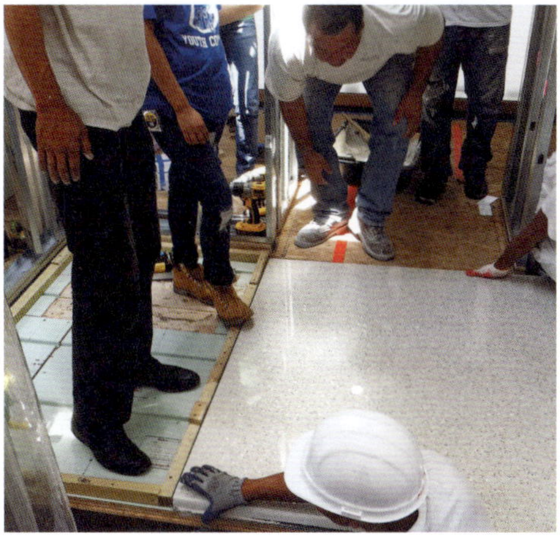

4.160

4.161

4.158 Bathroom section, with wet wall distribution; early design development.
4.159 Rendering, investigating spatial strategies.
4.160, 4.161 Laying the composite floor slab.
4.162, 4.163 Template preparation and lamination of the recycled wall panels.
4.164 Finished niche with wall paneling.

the smooth wall surface perfectly. During our design phase, we also discovered that coordination between the surrounding cabinets and bathroom enclosure was advantageous, specifically with regard to the clarity of cabinet functions. For instance, in one such case, we discovered that the flat screen TV did not require the 2 foot cabinet depth that was typical of the cabinets.

By reducing the depth of the TV cabinet, we had enough space to configure a niche in the bathroom for towels and other accessories (4.144).

We went through extensive template work (4.162) to coordinate between the different material systems, and arrived at a precise solution. To differentiate this interruption from the usual wall surface, we clad the element with recycled glass tile, also a result of our material research, and a nod to those team members in favor of traditional tiling. This juxtaposition contributed to an interesting detail in this space, too (4.164).

An even more important accent in the bathroom's materialization was the mirror wall, which was integrated above the sink and toilet, and included the off-the-shelf medicine cabinet. The coordination between the custom made mirror wall and the inexpensive prefabricated cabinet, was first processed by testing the fit in detail (4.165). The intent was to make the room appear more spacious than it actually was in lieu of the mirror's reflection.

Integrated into the mirror was a color-morph indicator to inform the bathroom's user about the surplus of water supply. The underlying idea here was to have a simple signal in the house to link the expenditure of resources, such as water, to the user's usage pattern. This indicator was connected to sensors in the garden's rainwater tank.

4.164

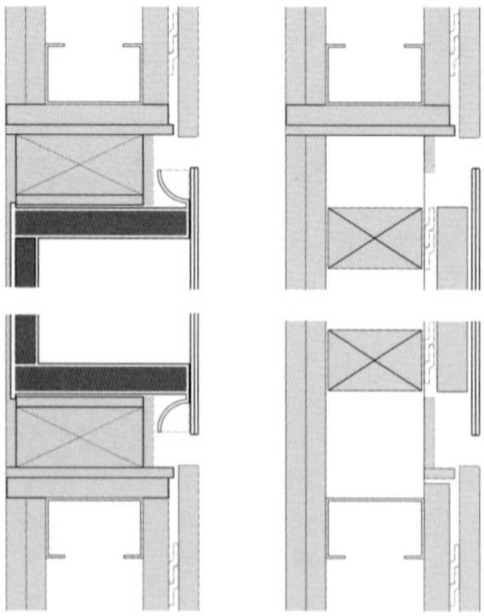

4.165

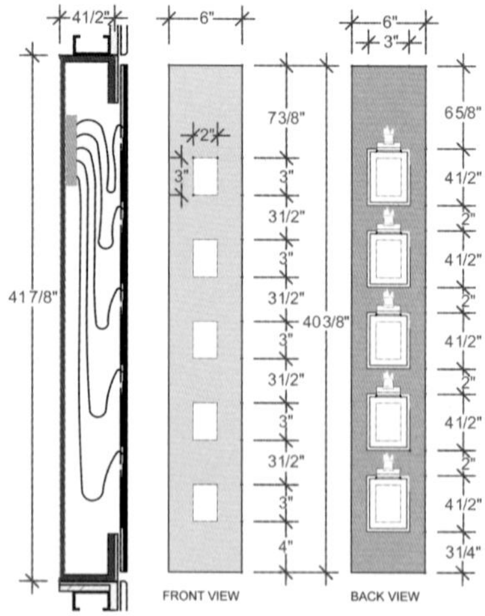

4.166

The control system would notify the indicator to change colors when the water level in this tank was changing. For lower water levels, the indicators would change to red, while upper water levels would be signaled in blue. Since the mechanical room was immediately behind the mirror wall, wiring this device was not difficult. However, one concern was that the color morph patches were susceptible to water damage and therefore had to be protected. After consultation with a professional glazier, the students experimented with delaminating mirrors by testing different techniques and chemicals (4.167). The space requirements for the electronics were also coordinated with the cabinet dimensions (4.166). The final result was an integrated custom design, wherein cabinetry and electronic technology were unified (4.168).

One of the most important items in the bathroom was the dropped ceiling. Since the house is designed to be transported in two parts to meet the truck/trailer dimensions allowed on highways, the task of reconnecting the house after transportation had to be approached systematically.

While the southern half of the house uses the mechanical room as an access point to rewire and reconnect all piping and conduits, this switching station for the northern half had to be established somewhere. The space most suitable for it was the dropped ceiling, which was directly connected to the mechanical room. We understood early on that this access panel had to be maximized; we thus moved the skylight out of the way, over the shower, and established substantial access.

Nevertheless, with the accumulation of conduit and piping, the ceiling cavity became quite crowded. This was not necessarily predictable, given that engineers' drawings are line diagrams without concrete dimensions. Furthermore, without the help of BIM modeling, this information is separated into several drawings, and these distinct systems needed to be connected into a cohesive plan. In order to not be surprised before it is too late, one must effectively solve the 3D puzzle. In an effort to anticipate these issues before they surface, intersections between trades are best established through hands-on learning. The advice of our electrical and plumbing professionals was crucial to still keep the organizational strategy intact. Since these pipes and conduits are not visible, the hands-on experience is crucial for awareness of this problem. The detailing and construction of the core was almost a project in and of itself, with a high level of complexity and necessitating integrative thinking. The approach was a highly customized unit, as fit with the idea of a modular house.

This kind of approach was necessary to address the specifics of the program and the requirements of modern comfort. We used this method of thinking as a point of departure for research on green materials, hardware applications and the integration of electronics, all related to thorough detailing skills.

A main interest in customizing the core lay in its economic cost efficiency, which could be established through the development of different tailored solutions within a systematic catalogue. This task was not approached in our project, rather the foundations were laid to competently process feasible custom solutions and develop a design strategy that considers different economic conditions. This is a typical challenge in professional settings, and directly relates to prototyping competency. The integrative detailing skills learned in this customized subsystem are thus paramount to realizing an architectural project.

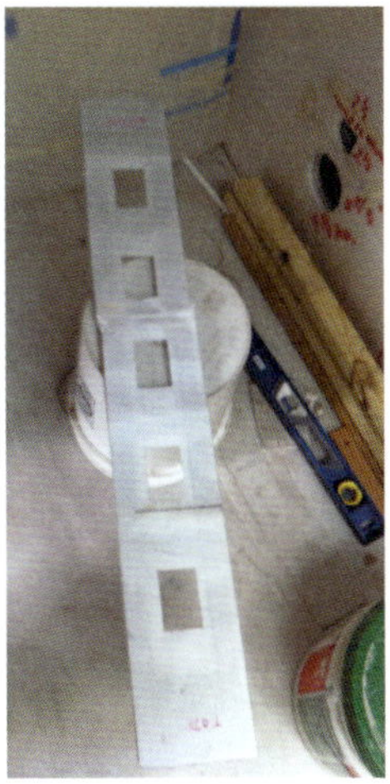

4.167

4.168

4.165 Detail: coordination between prefab medicine cabinet + mirror panels.
4.166 Custom mirror to incorporate the color-morph water indicator.
4.167 Template for mirror etching: water indicator.
4.168 Bathroom: Mirror wall with integrated water indicator; the reflection in the mirror shows the large ceiling access panel. This panel is necessary to reconnect the MEP systems after transportation of the house.
4.169 Bathroom access panel during construction. This space was host to a lot of traffic, with students becoming increasingly protective of their work, due to a high concentration of different trades.
4.170 Access space during transport to Washington DC.

4.169

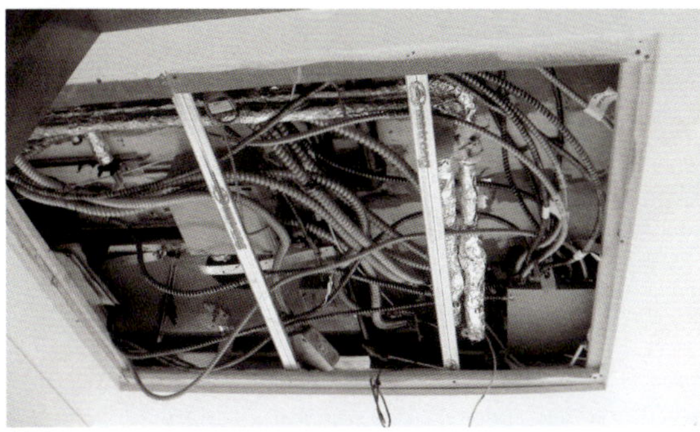

4.170

The 'roof trellis' fulfills several functions at once. By shading the roof and south windows, interior heat gain is avoided; photovoltaic panels are well ventilated, improving performance, and the mounting system is also easily retrofittable, when more efficient panels can displace current technology. The integrated jacks of the 'trellis' allow for orientation in different directions to ensure the best orientation due south in different urban settings.

A main concern of this subsystem was related to high wind loads. The 'trellis' would act like a sail in a storm. Thus, the angle towards south is kept low. The connection to the building is critical, too: six stanchions attach the trellis to the building, responding to the geometry of the space frame and building. The sequence of assembly was equally important. We assembled the full 'roof trellis' first, and lifted this light-weight structure onto its mounting points afterwards.

The 'roof trellis' is based on an industrially manufactured space frame system that was available in element sizes compatible with our planning module. The node of this system was especially beautiful. However, any generic system has to be adjusted to its specific requirements: the attachment of the linear racking tracks to the frame's tubes, and the attachment to the roof stanchions was also key. An engineer had to sign off on the design adjustments.

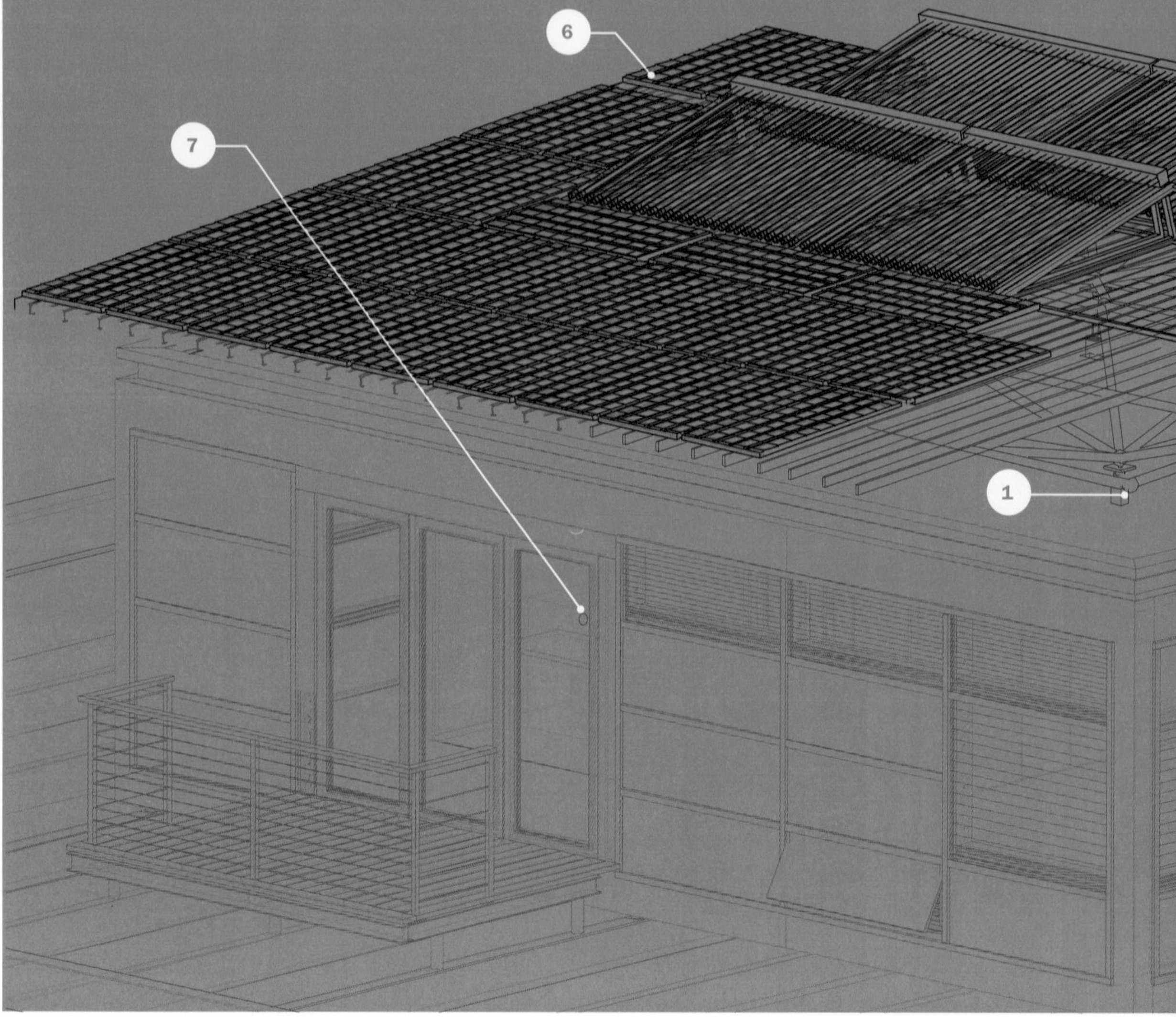

SOLAR TRELLIS

10

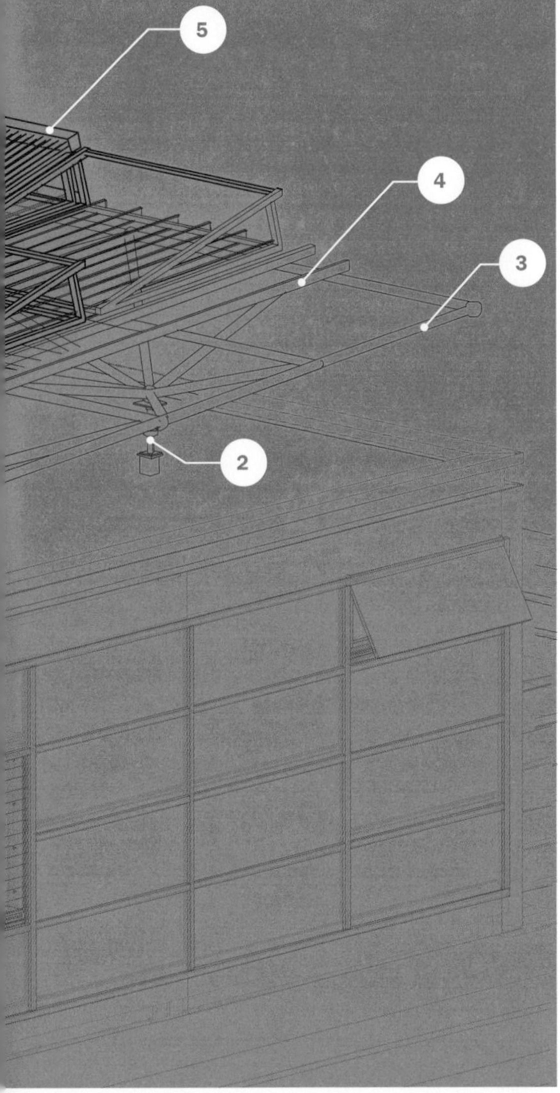

1. TRELLIS STANCHIONS WITH CUSTOM MOUNTING PLATES, STAINLESS STEEL
2. ADJUSTABLE JACKS; HINGED CONNECTION BETWEEN STANCHIONS AND SPACE FRAME
3. SPACE FRAME, 3" DIAM. TUBES, PROPRIETARY NODE SYSTEM, GALVANIZED STEEL/ALUMINUM
4. RACKING SYSTEM, SUITABLE FOR PHOTOVOLTAIC PANELS, ALUMINUM EXTRUSION (3")
5. EVACUATED TUBE COLLECTORS, 6 X 30 = 180 TUBES (~160 S.F.), ON PROPRIETARY RACKING SYSTEM
6. PHOTOVOLTAIC PANELS, 40 X 240 W = 9.6 KW, WITH MICRO-INVERTERS TO MINIMIZE TRANSMISSION LOSS
7. POWER INDICATOR DESIGN, COLOR-MORPH TECHNOLOGY INTEGRATED IN KITCHEN BACKSPLASH, SIGNALS POWER GENERATION SURPLUS/DEFICIT

10

SYSTEM 10: SOLAR TRELLIS –ENVIRONMENTAL HARVESTING AND SHADING

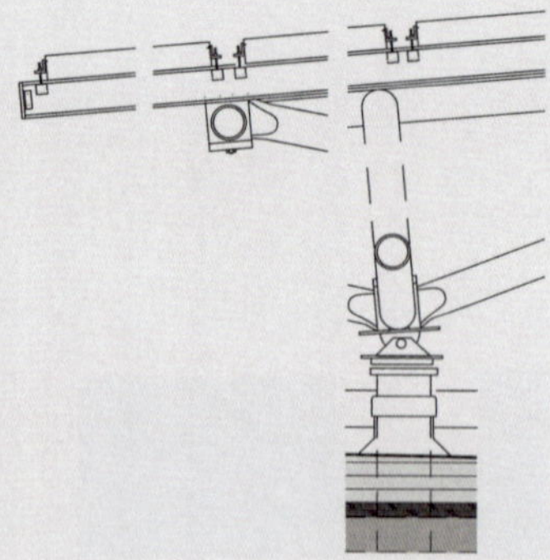

4.171

The building's original objective was to run on solar power. Therefore its architectural expression had to communicate the collection of solar energy. The house's solar trellis was more than a utilitarian device, and thus its materialization had to translate into architecture. However, a modular building with a repetitive pattern is at risk of losing its concise visual boundaries. Hence the floating trellis that we envisioned would complete the building's morphology by concluding it with a canopy.

Although this concept was easily translated in the drawing, the reality of creating the trellis's floating effect was much more difficult. The structural system required to transfer the loads of the solar panels and thermal tubes had to be considered, along with any dynamic loads of snow and, most critically, winds. Winds on the top of buildings can be fierce at times. In fact, winds nearing 100m/h have to be considered when designing a building in New York City, according to the building code. Because winds can come from all different directions, the structural system should employ regularly braced systems that are triangulated to distribute the loads axially.

Research done previously on dunnage systems showed that a prefabricated space frame was most suitable, being made from very thin members, which, coincidentally, echoed the modular look developed for our ingenious building block system. Additionally, all members of the prefabricated space frame could be prefabricated to the 5'-4" bay size we had developed.

One additional, and unique, load-bearing scenario appeared later for our consideration. Assembling the space frame system on the Roofpod's roof would have been more challenging due to safety precautions and restricted space. With this in mind, we decided to assemble the one hundred plus pieces on the ground and lift the assembly by crane once complete. Thus, the lifting points had to be determined structurally (4.172), and the load conditions had to be evaluated accordingly. During this evaluation we realized that temporary bottom chords had to be added during the lifting stage to keep the system stable (4.174).

In order to follow the strategy of reducing any necessary work on the Roofpod's roof, the connection points had to be minimized. The same objective was applicable to avoid thermal bridges that accrued in this structurally significant connection. The geometry of the space frame was adjusted to 6 bays, three in length and two in width. These bays were twice as wide as the regular module, meaning 10'-8" in dimension. With this geometry, the amount of stanchions could be reduced to six, and the members of the space frame could be reduced to 3" diameter tubes, rendering them relatively slender. Because all space frame supports were in the center of the building, they were virtually invisible. The perimeter of the space frame appeared as if it was floating, expressing the desired horizontal plane above the modular building.

In order to tilt the space frame toward the sun to make it more efficient, we identified that the best

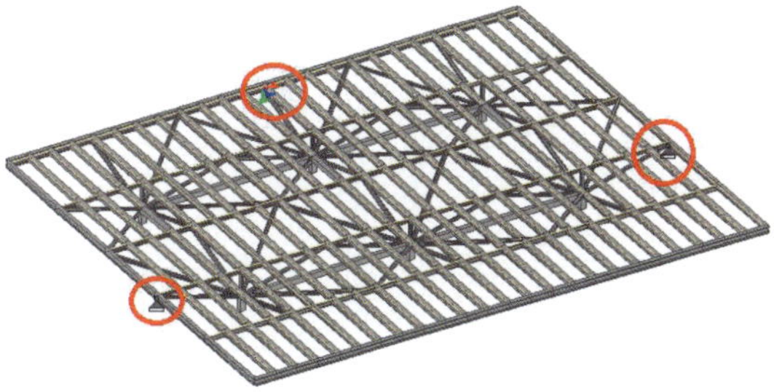

4.172

4.173

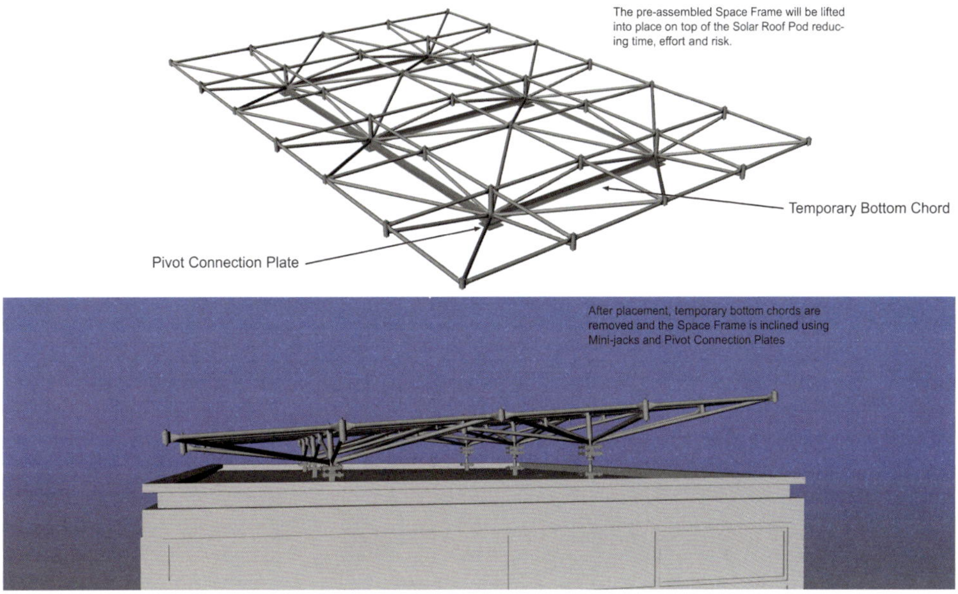

4.174

4.171 Detail of space frame and trellis edge.
4.172 Space frame design, with lifting points indicated and rails for photovoltaic panels.
4.173 CCNY: Assembly of prefabricated space frame: Discussion of assembly sequence in order to minimize tolerances.
4.174 CCNY: Space frame design, and anticipated installation strategy; student drawing.

way to manipulate the system was to insert additional pieces between the roof stanchions and the space frame. To change direction, all six stanchions received a pivot plate (4.178) with a steel pin strong enough to receive the cumulative weight sitting on it. This pivot plate was custom designed so that it fit our purposes exactly. The three north stanchions also received three mini jacks that could be adjusted in height, so that the entire trellis could orient towards the south.

With this tilt, the solar thermal collectors were able to operate on a buoyancy principle, and the efficiency of the photovoltaic panels increased, too. However, the slope of the trellis was passionately debated. The more slope the trellis received, the less of a visual floating effect could be achieved, which was an attractive design item. Models were built to study the Roofpod's overall appearance by testing different angles. Simulations were done in parallel to evaluate any quantitative results. It was clear that the greater the slope, the more solar energy could be harvested. However, if the trellis angle was too steep, wind loads would negatively affect the assembly. In the end, we settled on a solution that considered all these factors and synthesized them as comprehensively as possible.

The solar panels could not be directly mounted to the space frame, but needed linear aluminum railings holding the proprietary panel setting clamps. The connection between the space frame

4.175

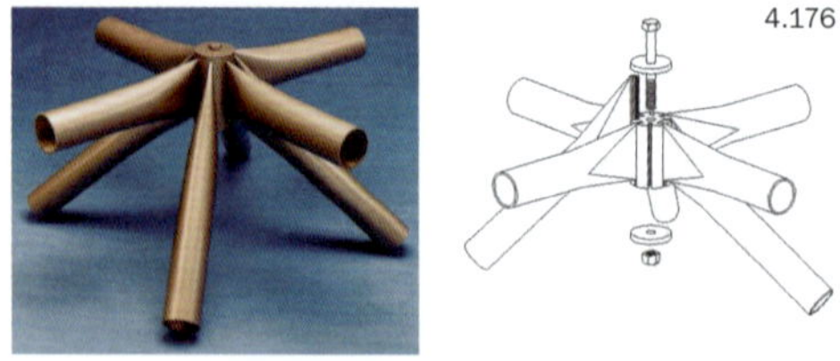

4.176

4.177

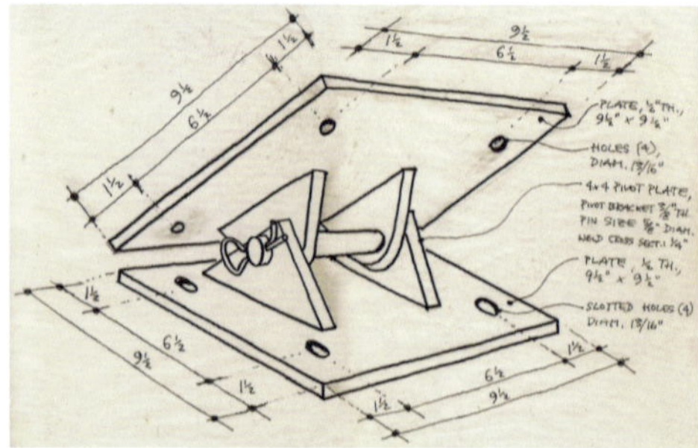

4.178

tubes and the railings was accomplished with simple U-bolts, which had to be structurally evaluated before we could mount them.

The railings ran north to south and were three inches deep. They were therefore able to cantilever five feet beyond the space frame. This added another benefit, since the trellis could now become a shading device for both the roof membrane and the south façade. Due to the sun's angle, the cantilevered panels could shade the upper windows in summer, while allowing the advantageous winter sun to hit the glass directly. The slope of the trellis was related to this passive control strategy, because it effectively lowered the panels' edge and increased its overall effect. This edge had to be carefully calibrated to work with the changing seasons' solar impact.

The space frame and railings were assembled on the ground, during which it became apparent that the final pieces were most difficult to fit together, despite the precise and repetitive geometry. When any system is expanded, a small dimensional discrepancy will ultimately become significant. Some of the last pieces had to be carefully fit with the help of a mallet (4.175). We also anticipated these tolerances in the design of the connecting plates between the space frame and the roof stanchions by using slotted holes to absorb the necessary adjustments (4.179). The greater number of pieces in a system, the more one has to plan for their expedient adjustment.

Before the space frame trellis could be lifted, final provisions were made to mount the micro-inverters (4.177). We had utilized a technology that avoids transmission losses by allowing every panel to be managed by its own inverter. If the panels are wired in series, as is usually the case, then the output of all panels will be affected by even partial shading of one panel. Therefore, in a situation where partial shading cannot be avoided, such as in an urban setting, the case for managing each panel individually will yield higher results. Also, with this emerging technology, each panel can be monitored individually. If a panel is broken, dirty, or covered by something, it is easy to identify each panel's performance via computer interface.

After all this work was completed, it was time to lift the assembly in place and install the panels. Again, the sequence of installing the photovoltaic panels was critical to ensure that they wouldn't obstruct other work. Fortunately, we had the help of a professional installer, who had laid out the sequence according to his experience. For instance, the wiring of the panels was important, since the cables were clearly visible from underneath. They thus had to be installed in a very orderly way.

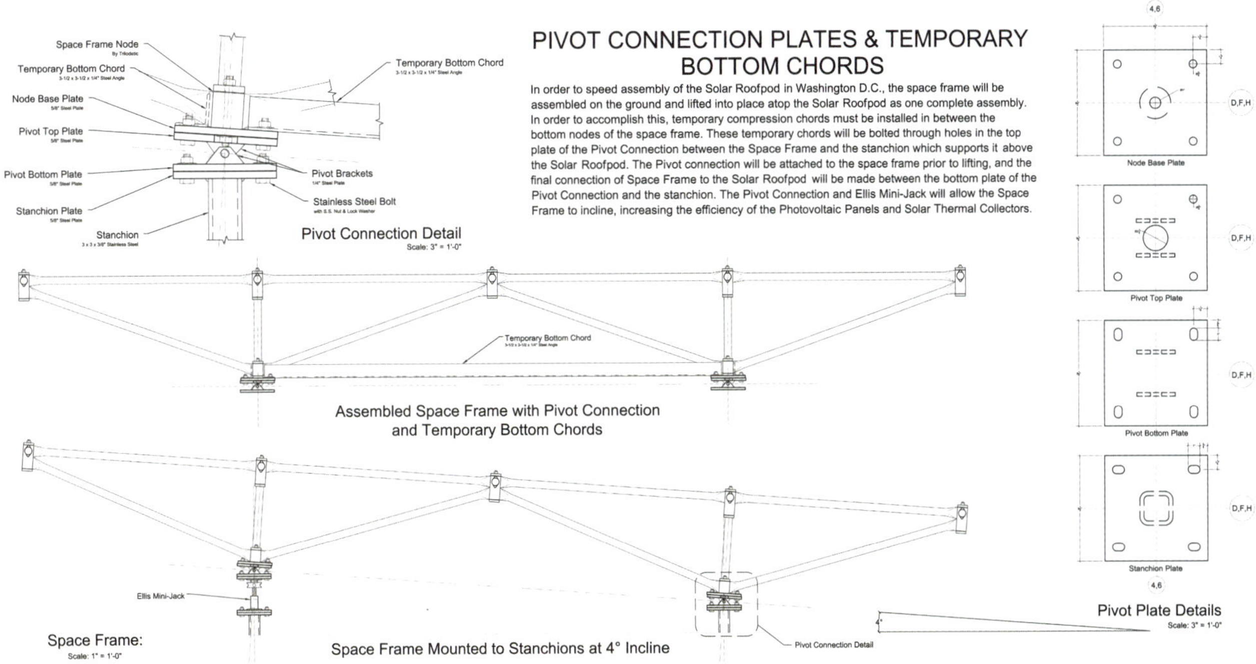

4.179

4.180

4.175 Space frame; Assembly tolerance. The final nodes in particular needed additional persuasion with a mallet at times.

4.176 Space frame node with proprietary slotted connections. After all tubes are inserted, two round plates finish the connector hub assembly.

4.177 Installation of micro-inverters onto the railing system.

4.178 Pivot plate: 3-D Sketch for customized fabrication.

4.179 Space frame connection to stanchions in order to angle the trellis toward the sun; student drawing.

4.180 After the original space frame's assembly, the railings were laid out. These tracks were necessary to hold the solar panels as specified. We took advantage of the system by cantilevering them over the south façade, therefore shading the Roofpod's upper windows in summer and thus avoiding unnecessary heat gains.

4.181

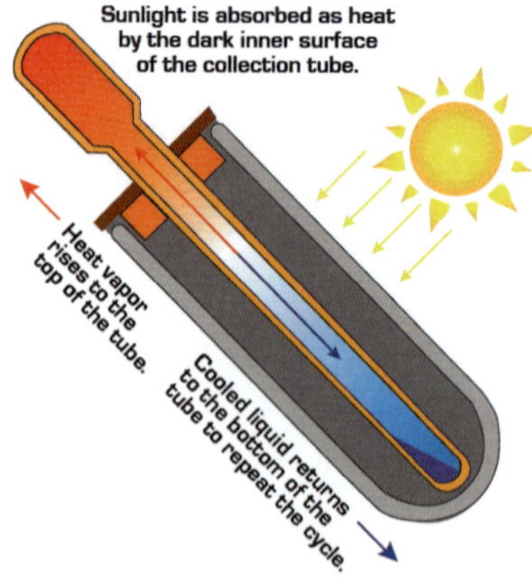

4.182

4.183

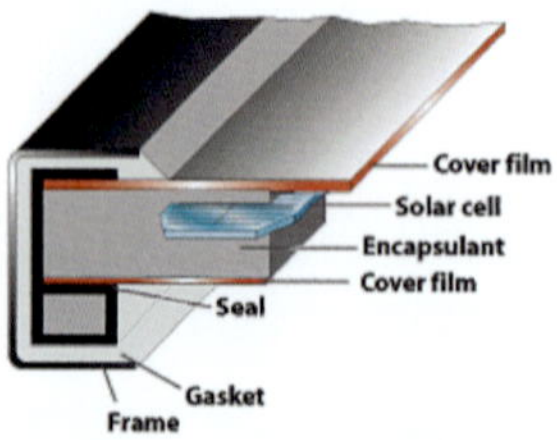

4.184

4.185

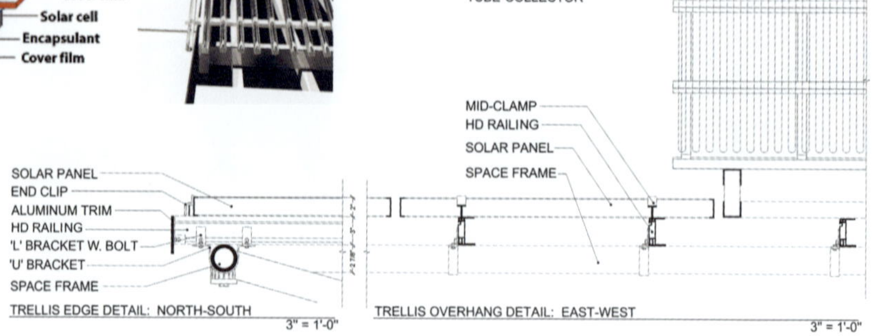

Each panel was held in place by only four clamps (4.185), which was initially a concern given that a hurricane was forecast for the last day of construction. But the proprietary connections proved strong enough to withstand hurricane wind forces after the storm had passed.

We were pleasantly surprised by the thinness of the mono-crystalline photovoltaic panel, which consist of very thin (0.2mm) silicon wafers, 72 of them (5"x5" squares), in one panel. These fragile wafers are imbedded in an 'encapsulant', which is typically a translucent plastic resin (4.183). The wafers are processed so that their corners are chamfered to avoid damage when being sliced. So the intersection of four wafer corners in the panel generates a diamond shape, which transmits light through the encapsulant. The south-side overhang featured countless gridded diamonds, which created, depending on light conditions, a beautiful translucent character, and tangibly communicated the materialization of the thin panel (4.194).

The solar thermal collectors consist of evacuated tubes. These tubes work on an ingenious principle, producing temperatures from the sun of about 180˚F. They do this by absorbing heat through a vacuum-insulated inner tube. A heat pipe is positioned within the inner tube, which is supported by conductive aluminum wafers.

The heat transports vapor by convection to the top of the tube, where it is released into a copper heat exchange manifold. The then cooled liquid returns to the bottom of the tube to repeat the cycle (4.182).

4.186

4.187

4.188

4.181 Solar trellis: U-bolt connection between space frame and secondary railings.
4.182 Principle of evacuated tube collector.
4.183 Composition of typical solar panel.
4.184 Computer model: overlapping evacuated tube collectors.
4.185 Detail: Attachment between space frame and solar-mount railings, and solar panels, resp. evacuated tubes.
4.186 Completion of trellis ground assembly, with micro-inverter boxes mounted for subsequent wiring on roof.
4.187, 4.188 Lifting of the pre-mounted trellis: the modular 'building blocks' receive the canopy.

In order to make this principle work, the tube needs to be pitched, yet another effective motive behind the trellis' tilt. Because there was a 10 degree minimum pitch required by the manufacturer, which exceeded our design intention of only a four degree incline, we had to coordinate the performance requirements with the appearance of the trellis yet again. After our previous research, it was categorically not feasible to change the overall angle of the trellis.

Since we understood that a steeper angle of the solar thermal installation would imply some shading on the photovoltaic panels, we had to optimize the location of this installation by moving the evacuated tube assembly to the north side, where shading would least affect the panels. Therefore, the cantilevered south side would feature the photovoltaic technology, while the north side would emphasize the solar thermal installation. We adjusted the proprietary solar thermal racking system to fit on the overall trellis, with an extra 8 degree incline.

Conflicts arose when coordinating the dimensions of both the photovoltaic panels and the evacuated tubes in order to fit them on our modular space frame. Even though our 5'-4" bay size correlated with the photovoltaic panel dimensions (5-'1" x 2'-7"; two panels = one bay), fitting in the evacuated tubes into that geometry presented a problem. The off-the-shelf tube assemblies of our sponsor company came as either 20-tube or 30-tube-units, which meant that no combination would fit in the space provided for the solar thermal installation.

4.189

4.190

4.191

4.192

4.193

4.194

4.195

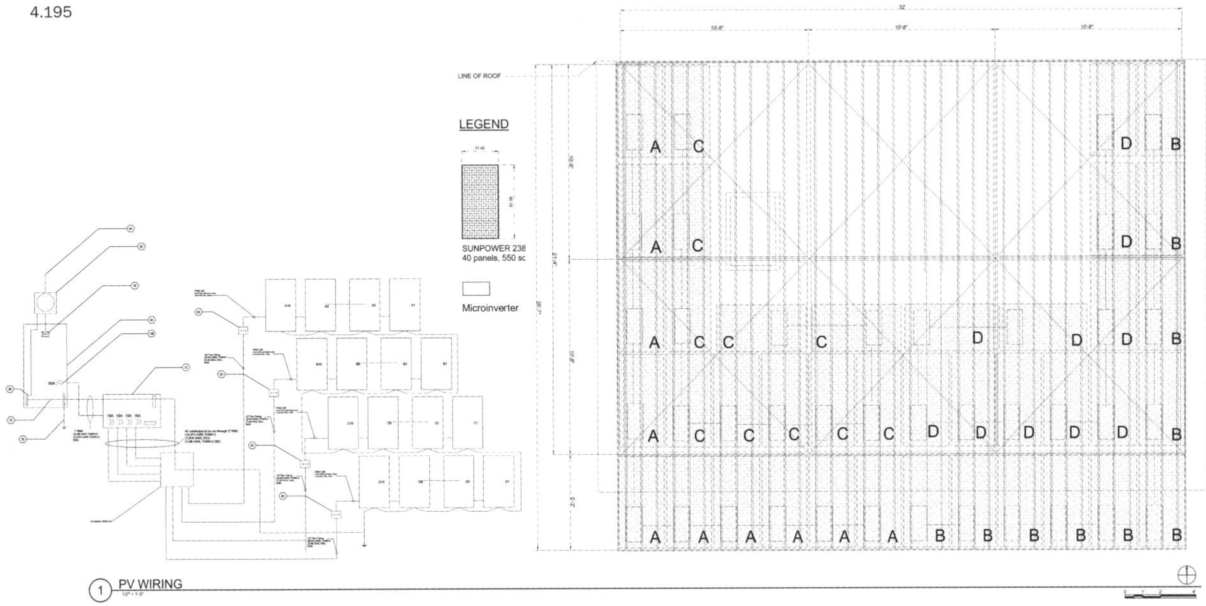

4.189 Solar thermal racking system ready for tube installation; the tube units overlapped the solar panels more than expected due to coupling issues. The overlap had been modeled in 3D to resolve any mounting issues and decrease in efficiency.
4.190 Evacuated tube installation: tubes were fragile and had to be handled carefully.
4.191, 4.192, 4.193 Photovoltaic panel installation. The sequence was key; the cantilevering panels were the most challenging.
4.194 South overhang of trellis: Shading the façade works well for the summer sun's steeper angle.
4.195 PV wiring plan and diagram; student drawing.

Ultimately, a 30-tube unit with 90" width would fit best, and three of these units could be arranged to closely match four bays of the 5'-4" modules. However, with the space necessary to connect between them, the evacuated tubes exceeded this arrangement.

There were two solutions to the problem: a) customization, or b) coordination within the existing geometry. Because the only element of the solar thermal collector that needed adjusting was the linear manifold header and its embedded copper heat exchanger, we attempted customization. We were able to convince the manufacturer to deliver units made up of only 28 tubes as copper can be machined relatively easily. However, our engineers were not happy with losing twelve tubes, given that they were necessary according to their calculations. Therefore, we had to revisit the problem and elaborate on a geometric solution, which would ultimately affect the photovoltaic panels, since the tubes had to be positioned above them (4.184, 4.185). Although this meant the four panels around the solar thermal collectors would at times be partially shaded, simulations showed that equipping these panels with micro-inverters resulted in a negligible amount of power loss over the course of a day. Thus, the panels would still yield enough to safely power the house. This was ultimately a compromise, and not necessarily optimal. With more time, or more experience to adjust the simulated predictions from monitored data, these decisions could be fine-tuned.

The trellis is conceived as a carrier system that could be retrofitted. Because solar technology is still changing, many compromises were made in an effort to make the product both economical and competitive. Thus the products developed in the future will replace our current technology, when more sophisticated techniques will prove to be more economical and thus more marketable. The mounting system fulfilled its purpose, but whether we can substitute the solar thermal units with a better fitting or a more efficient unit is inherently an open-ended design question. The trellis should be able to accommodate a multitude of different solutions, and the aesthetics involved will improve with more flexible applications and options.

4.196

4.197

When the students focused on designing the exterior landscape lighting they discovered that the trellis could be understood as an element in this environment. First and foremost, they were adamant about the need to emphasize the trellis at night. This element's nighttime presence related just as much to its immediate surrounding as it did to defining the Roofpod as a component of New York's night skyline. The trellis could be used as a reflector, distinguishing the Roofpods throughout the city. This thought took precedence over an initial idea to use the trellis as part of the ecological environment, possibly providing nesting places for bats, bees and birds. Instead, the group was in favor of this former concept for its far reaching visibility and lighting effect. The lighting was a selling point, emphasizing the presence and technological power to light the house, in particular at night and without the sun's presence. With the emergence of LED technology, this undertaking was not difficult to accomplish. The two fixtures were mounted so that the overhang was the emphasized lighting area, and only consumed a mere 12W per fixture. Thus, this was an affordable effect that made the Roofpod more attractive as a product. Furthermore, this effect emphasized the Roofpod as a repeated element in New York City's night skyline (4.196).

There was one particular concern addressed repeatedly in class when the topic of photovoltaic technology was discussed, and specifically in its design context. The concern was how to make this technology a tangible part of the architecture and not just a display of tiled panels on the roof, and ultimately to make this feature part of a "Net-Zero" lifestyle. Furthermore, it was key to understand the panels as a crucial part of the building's operation. We utilized the color-morph elements previously studied as a method of connecting the interior lifestyle of the house with the exterior photovoltaic array. The idea was to use this technology to indicate the performance of the panels in a simplified way. The best location to integrate this indicator was the core wall, and the best position was the kitchen backsplash. This location was selected because many kitchen appliances are primary power consumers, and the mechanical room was conveniently nearby, with ample space and flexibility for accessing the color-morph elements. The control system could easily evaluate whether more or less power was being produced in real-time as was being consumed. All appliances, lights and other power-consuming sources were already being monitored. However, the typical interfaces conveying this kind of information are mundane computer screens. Instead, we were looking for a more simplified and integrated display that functions as part of a cohesive design approach.

Color-morph material is interferometric display technology. It is a conductive sheet less than a millimeter thick. Pixels can be manipulated by an electronic impact almost like nanoscale-mirrors, absorbing different spectrums of incident light, and thus reflecting it back in a particular color. The technology is very energy efficient, as electricity is only necessary when switching the micro-mirrors to a new position. The technology is currently being considered for the next generation of computer screens. However, we were presented with several limitations. Currently, only small areas of these surfaces can be manufactured without defects, while in the future there is potential to produce wall-sized surfaces. For now, our indicators were limited to 3"x3" patches (4.197). Also, while the surface of this material had to be

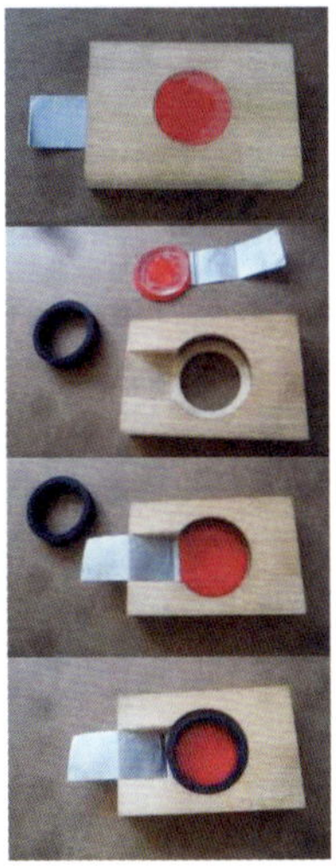

4.198

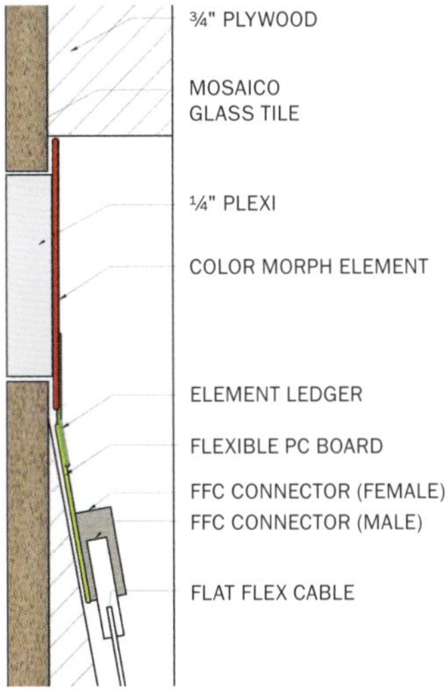

4.199

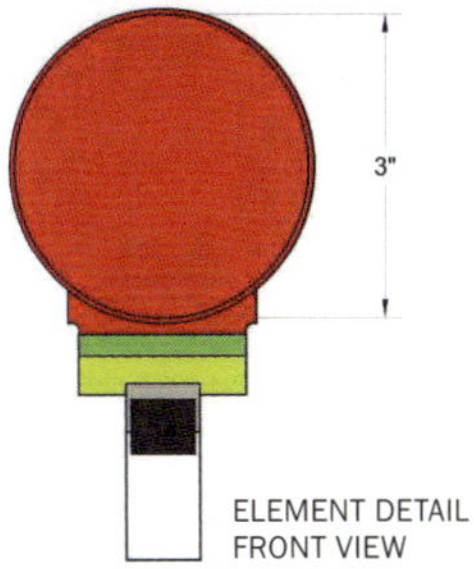

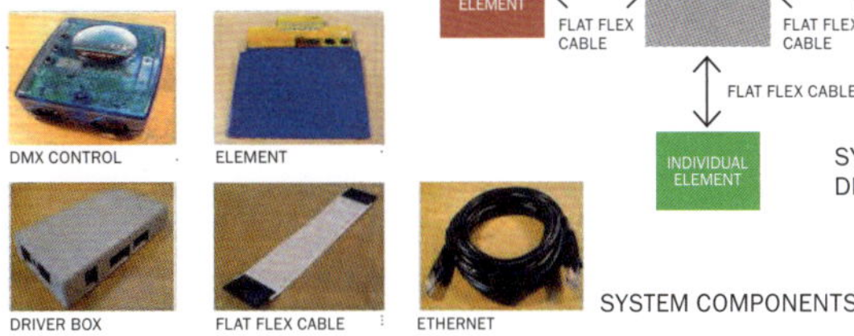

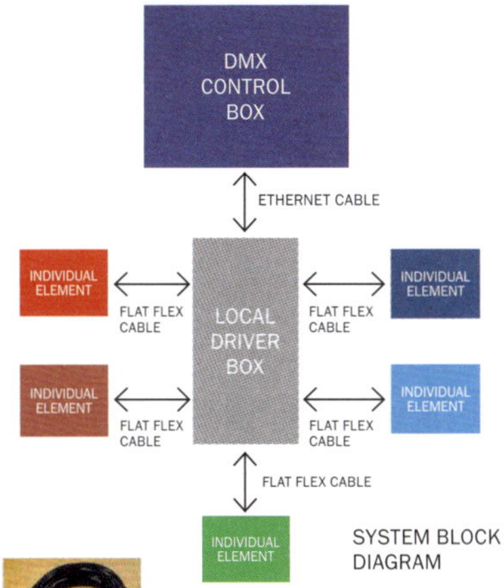

4.196 Trellis lights on: LED lights to utilize the canopy as a reflector.
4.197 Color-morph element with flat-flex cable attachment; prior to installation.
4.198 Mock-up for kitchen backsplash indicator: several technical and aesthetic criteria had to be coordinated. The mock-up was perfect for upcoming discussions.
4.199 Detail: Backsplash-integrated power generation indicator, custom design; student drawing.
4.200 Configuration of indicator openings; From 2-dimensional detailing to reality.
4.201 Operating indicator.

protected primarily against water and humidity, if this protective coating were too thick, the color intensity would diminish. We built several mock-ups (4.198) to understand how to address these limitations and anticipate how the assembly could be installed effortlessly, while resulting in a functional and attractive appearance. We realized that if the surrounding material was the same thickness as the protective lens, we could easily configure a rabbet to hold the disc in place (4.199). We also decided to use several of these 'traffic lights' in series in order to make them visually more appealing. Many patterns were drawn before a final design was chosen. This 2D-drawing was transferred to the epoxy-composite backsplash (4.200), then the holes were configured and the lenses inserted. When programming, we chose blue and green colors as the indicators for a surplus of energy generated, while different shades of red signaled that the consumption of energy was higher than the generated income. With this serene design we were able to communicate the house's performance in a cheerful way, which in turn would encourage the user to run appliances responsibly. The development of such a device required great attention to detail. Several areas of expertise had to be coordinated to realize the idea, namely the involved technology, buildability, and aesthetic appearance.

The solar trellis as a system was challenging. Its many small pieces, countless fasteners and yards of wiring all had to be combined into a whole. Precisely executing drawings and mock-ups was crucial to make the assembly work, from material research to the thread lengths of bolts. A student once mentioned that working on the trellis system felt "like managing a hardware store". The knowledge needed to combine large and/or small, prefabricated and/or customized items is very important, particularly when considering that meaningful incorporation of mass-produced items into a specific artifact is an essential design skill in contemporary architecture. The trellis was one of the subsystems where this kind of adaptive creativity was practiced extensively.

4.200

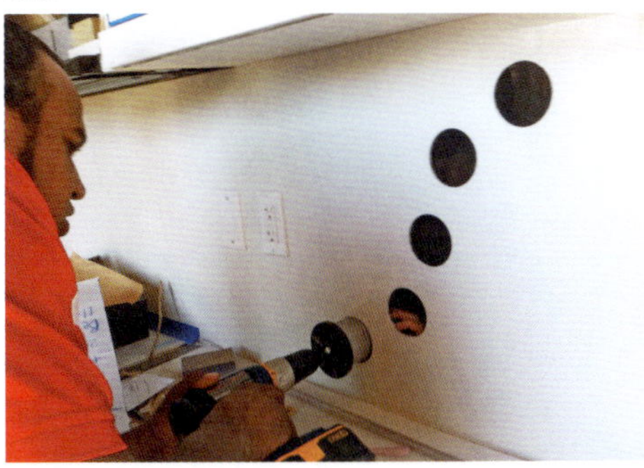

4.201

The structural members used for the dunnage system have enough additional strength to easily support the loads accumulated by an intensive green roof system, with planter depths of a foot of soil or more. This structural capacity is 'free', since these beams are already needed to transfer the loads of the Roofpod to the building. We used inexpensive solutions for our planter system, framing and surface materials. The modestly priced grey water irrigation system also keeps maintenance low. All these factors can contribute to low-cost operation while providing an attractive amenity.

Introducing green roofs to urban environments mitigates several common infrastructural problems, which impact our wellbeing. Adding extensive vegetation will reduce cooling load-related energy consumption. The soil's moisture absorption decreases temperature while remediating peak stormwater volume in the sewage system. The air quality is improved by increased vegetation, which mitigates airborne particulates and atmospheric carbon dioxides.

The complementary 'roof garden' is a vital element to develop the 'site'. Not only does it add an intermediate space between the interior and the skyline, it could satisfy several ecological needs, too. Urban revegetation of flora and fauna (3B: birds, bats, bees) as well as the opportunity of urban farming and organic waste composting (reducing waste stream) could flourish, besides extending the lifespan of the host building's roof membrane.

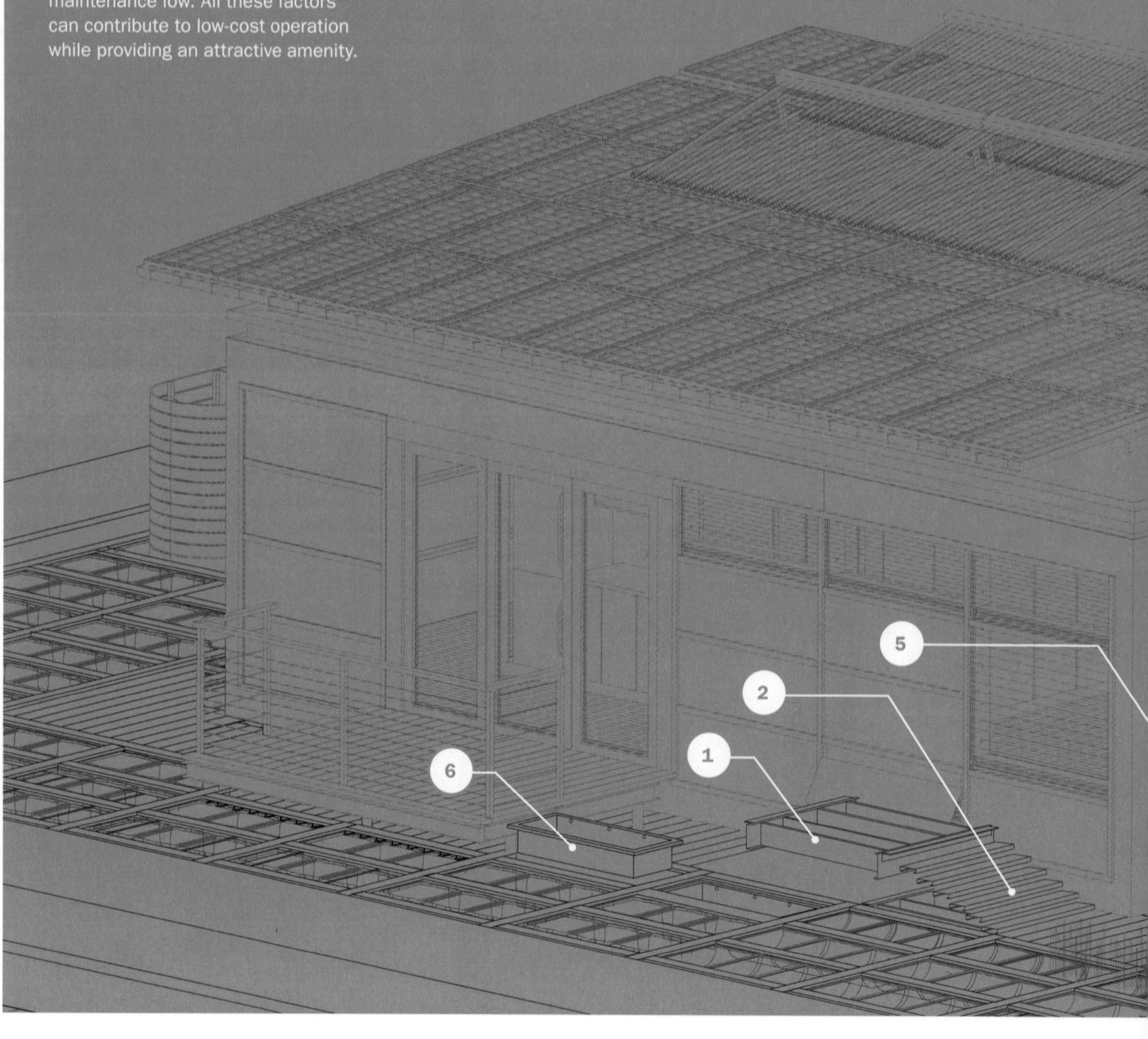

ROOF GARDEN –'LANDSCAPE'

11

1. LIGHT-GAUGE STEEL FRAMING, GALVANIZED, LOAD TRANSFER TO DUNNAGE; MODULES CAN CARRY STEEL PLANKING OR PLANTER UNITS

2. "HARDSCAPE": WALKABLE MODULES, PERFORATED GALVANIZED STEEL, INTERLOCKING PLANK SYSTEM

3. "SOFTSCAPE": PLANTABLE MODULES, INTEGRATED PLASTIC PLANTERS, WITH WOOD TRIMMING

4. PLASTIC PLANTERS, DIFFERENT DEPTHS, TO ACCOMMODATE A VARIETY OF DIFFERENT PLANTING CONCEPTS

5. LIGHTWEIGHT SOIL AND PLASTIC PIPE DRIP IRRIGATION

6. "WATER PLANTERS": STAINLESS STEEL TANKS, COLLECTING RAIN WATER FROM THE ROOFPOD'S ROOF

11

SYSTEM 11: 'ROOFPAD' –GREEN ROOF SURROUNDING

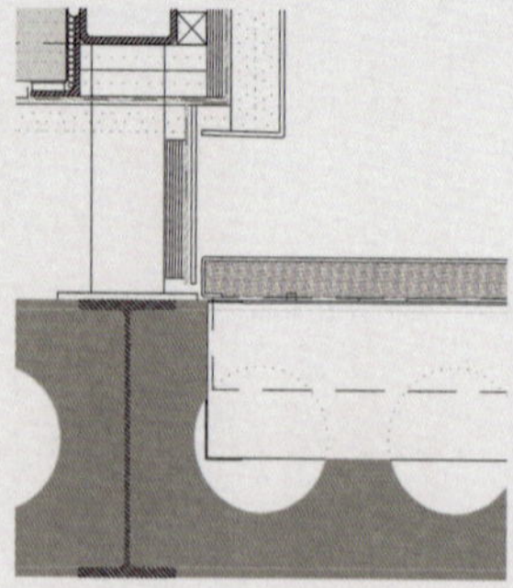

4.202

We understood early on that the interior space of the Roofpod would only bear its true experiential potential with the exterior landscape surrounding it. Here was an opportunity to activate and inhabit the entire underutilized, forgotten horizontal plain of our urban roofs, not just the real-estate gained for interior living space. This design step potentially had greater consequences in its urban context than for the building itself. Remediation strategies, such as the heat island effect, vegetation loss, stormwater management and air quality shortfalls are all effectively addressed by the large scale approach of greening urban roofs.

In this respect, the technical development and detailing of our roof landscape was the system that closed the circle of interrelating subsystems, and brought them into a cohesive whole. The landscape was mainly dependent on the dunnage system, which set up the planning grid for the project, and tied together both spatial movement and views. Furthermore, the landscape involved connecting and retrofitting pipes and conduits within the dunnage layer that would ultimately be tied to the host building via the configured crawl space. For that reason, this space had to be accessible and the landscape modules had to be removable. Thus, this landscape modularization strategy was quite significant, as they acted as access panels to the dunnage system's distribution layer.

The main advantage of the landscape system relied on the structural capacity of the dunnage beams. Since the beams required a lot of structural depth for the Roofpod's load transfer, they inspired and reinforced an intensive green roof concept of using heavy planters with depths of up to 12 inches. We used the dunnage beams as our scaffold for these drop-in landscape modules (4.203). We applied a similar technique as the location finder strategy (see chapter 4.3) to precisely connect the building modules. We attached socket head screws to the top flange of the dunnage beam and drilled corresponding holes in the structural attachment angles of the landscape modules (4.204). This was achieved by clamping all module support frames to the top flanges of the dunnage beams after they were installed, and then drilling pilot holes through both layers. Afterwards, the appropriate hole sizes were drilled separately. This method ensured that all modules were tied together and referenced to the dunnage beam assembly. Therefore we could be certain that all the pieces of the puzzle would fit in precisely the same way as when first installed, without risking the relocation of tolerances during the process of reassembly.

The support modules underwent a long evolution throughout the project. In the first mock-up phases (4.208, 4.209), we intended to build all the modules out of wood, including the module framing, to keep them economically viable. So our first design started with a trip to the local lumber and hardware store, where we bought 2x4 and 2x6 wood framing, some fasteners and some screws. The discovery we were most proud of on this trip, however, was that mortar mixing tubs seemed to fit well as planters in our initial scheme. We built two test modules that summer, one for the walkable surface,

4.203

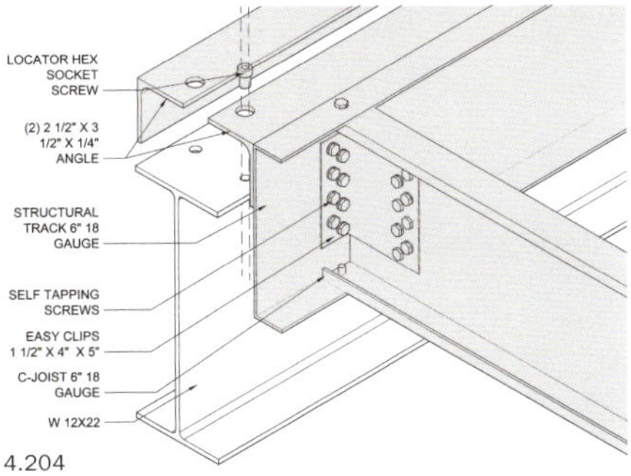

4.204

4.205

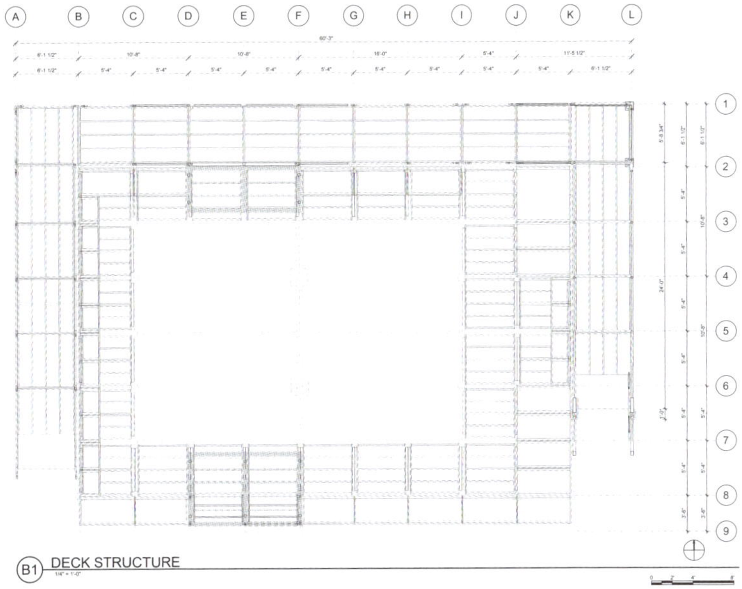

4.206

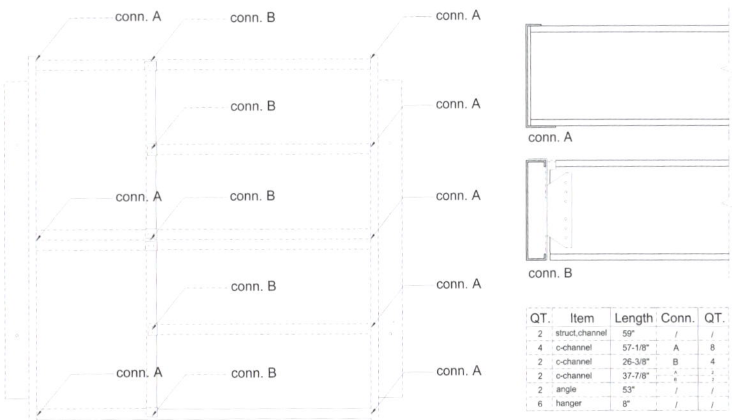

QT.	Item	Length	Conn.	QT.
2	struct,channel	59"	/	/
4	c-channel	57-1/8"	A	8
2	c-channel	26-3/8"	B	4
2	c-channel	37-7/8"	A B	2 2
2	angle	53"	/	/
6	hanger	8"	/	/

4.207

4.202 Detail: Landscape - Roofpod transition.
4.203 Installation of support modules into the dunnage system.
4.204 Connection between dunnage beam and drop-in module.
4.205 Production: Framing tips from experts.
4.206, 4.207 Roofpad: Framing plan: overview, detail and take-off.

called the 'hardscape', the other for the plantable surface, called the 'softscape'. Even though this mock-up resolved several structural issues and led to simplifications and aesthetic improvements, we also understood that the exclusively wood materialization had its issues. The most important objection was the fire department's regulation allowing only a limited amount of decking on roofs to be combustible, due to fire safety.

With this information, we decided that the support structure throughout the project could not be built in wood in order to meet this requirement. And so we changed our strategy and supported all modules with light-gauge steel framing (4.207). The students consulted their structures professor to correctly dimension the depth of the framing members, and practiced with a visiting professional builder who taught us the best methods to assemble the frames (4.205).

For the surface material of the hardscape, we decided to use perforated metal planks that also met fire safety requirements. Initially, we found lightweight aluminum elements that fit perfectly, but because they were too expensive for the project, we replaced them with galvanized steel planks (4.210, 4.211).

What we kept from our first mock-up studies was the height differentiation between hardscape

4.208

4.209

4.210

4.211

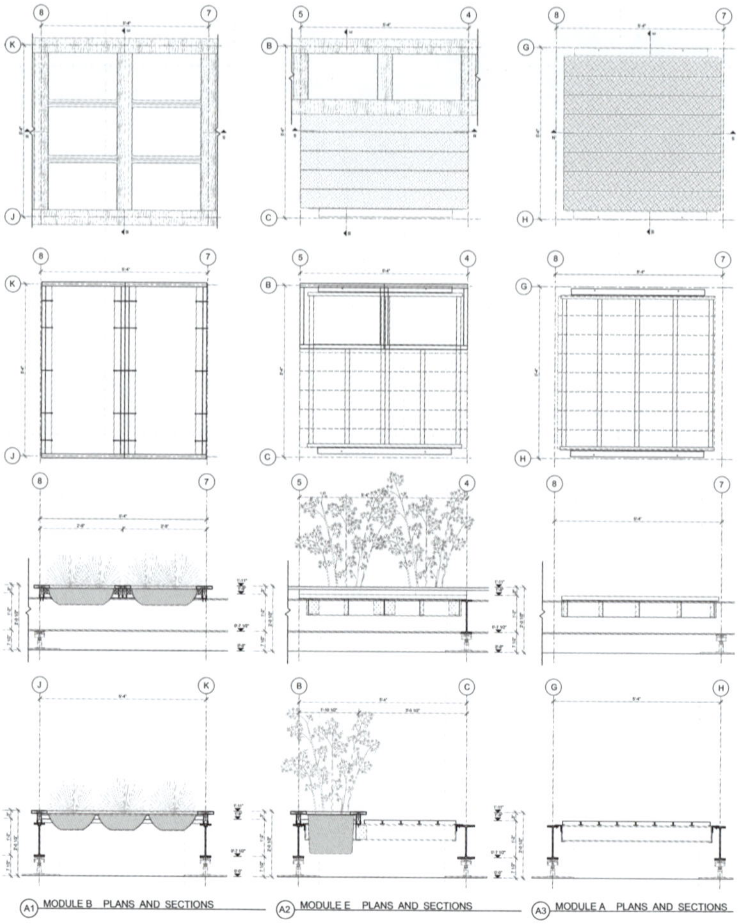

4.212

4.213

4.214

4.208, 4.209 First mock-up at early design development, entirely build of wood; ingenious use of mortar mixing tubs as planters.
4.210 Final landscape mock-up.
4.211 Transition between 'hardscape' and 'softscape'. A walkable path configured the transition between the house and the garden.
4.212 Landscape module catalogue: 'Hardscape' and 'Softscape' modules, and intermediate hybrids.
4.213, 4.214 Construction of secondary trim framing and insertion into dunnage.

and softscape, which looked good but made even more sense when articulated with alternating materials (4.211). This resulted in the trim of the planters being made of wood and the walkable surface of perforated steel. The height variations were a result of the different attachment methods between the surface materials and the substructure, which fit the purposes of aesthetics, too.

The wood species was cypress, the same material we used for the curtain wall mullions of the house, which visually tied the assembly together. We also configured a walkway around the house to emphasize the building. With a reveal at its bottom break metal edge, all three elements, garden, walkway and Roofpod were precisely differentiated.

Fortunately, the mortar mixing tubs, acting as ready-made receptacles for our plants, proved to be quite practical since they fit well into the modular structure. In addition, they were sturdy enough to carry the weight of the lightweight soil, which was easily penetratable for irrigation systems, and unbeatable in price. With their ten-gallon capacity, they were just the right size to be carried by one person (60 lbs.). Later in the design process, we found a deeper plastic container with almost the same width and length. We used both containers in different parts of the garden after our gardening team did extensive research on necessary planting depths. In order to take advantage of the dunnage system's capacity, we used the deeper planter for bigger and more interesting plants, while the shallower planters were filled with grasses, herbs and vegetables.

The choice of plants was influenced by the planting depth, our initial concepts of urban farming, and ultimately had to respond to the available precipitation. Even though the amount of soil and, thus, moisture storage was above average for green roof applications, we still had to be concerned with periods of drought as these planters dry out relatively quickly. Also, for heavy rainfalls, we had to ensure that the improvised planters would not drench the plants to death. The latter issue was easily resolved: we drilled holes in the bottom of these planters so that excess water would empty onto the host building's roof membrane and roof drain. The issue of providing extra water was more difficult, since we did not want to use tap water to keep the plants alive, given the availability of rainwater that could be managed more efficiently for these purposes. Thus, we decided to make use of the rainwater collected from the Roofpod's roof, as well as the grey water from the laundry and kitchen and bathroom sinks using exclusively biodegradable soaps to ensure that the plants would still thrive. We integrated two storage tanks below the ramp system to hold water that would serve the purpose of irrigation. In order to integrate an easy and accessible approach to irrigation, we had to come up with an automated system that would monitor the plants' moisture level and react with a controlled approach, something that could not be achieved manually with watering cans. Thus, we came up with a system that monitored both the moisture in the soil as well as the collected irrigation water available in the tanks.

4.215

4.216

Engineering and Architecture students worked together on developing this system, which was subsequently implemented during construction. In order to connect all the planters for irrigation according to plan, the system involved ½" diameter plastic pipes with one drip emitter and a coupling per planter run throughout the landscape (see 4.219, 4.220).

For that reason, the planters and the wood framing had to be perforated. We originally elevated the planter trim for aesthetic reasons, which also proved advantageous now for technical reasons as it eased the process of drilling holes through both the wood and plastic. Furthermore, penetrating the steel dunnage beams would not have been feasible. Finally, the pipes had to be laid out in such a way that they could easily be disconnected to separate the modules. The practicality of this sequence was also tested during construction.

While the functional operation of the system was resolved, most of its success was concealed within the planters, or buried below the dunnage system in the form of storage tanks and pumps. A more tangible approach was preferable, so that the design was self-explanatory. An additional interest of ours was thus to develop 'water planters' that would function as storage tanks while also containing water lilies, frogs or goldfish, etc.

Research on plants that could clean our grey water yielded a familiarity with living machines. This is an ecologically engineered technology developed to restore and remediate polluted water by replicating and accelerating the natural purification processes in several steps. With a series of these planters, we could even have used sewage to feed our irrigation water.

However, the implications of this research on the realized project were limited, since the Solar Decathlon rules restricted the use of sewage water. In the end, we integrated two customized stainless steel planters, a gift from one of our sponsors, at the front edge of the Roofpod's garden. These tanks came in handy, after we realized that our original greywater storage was not sufficient to irrigate all the planters. With the addition of these steel planters, which received rainwater directly from the Roofpod's roof, we were able to add about 150 gallons of water to the system. The effort to integrate the tanks into the landscape was quite complex, as it required locating accessible plumbing paths and cutting holes into the neutral area of dunnage beams to connect the system (4.217, 4.218). Physically realizing these water planters revealed that using various planning process tools, from renderings to details to

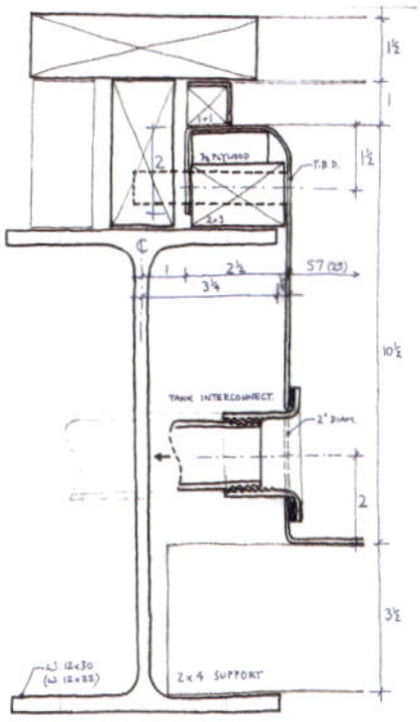

4.217

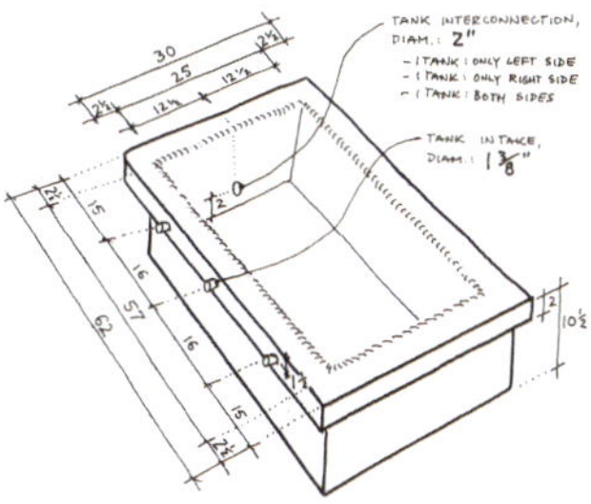

4.218

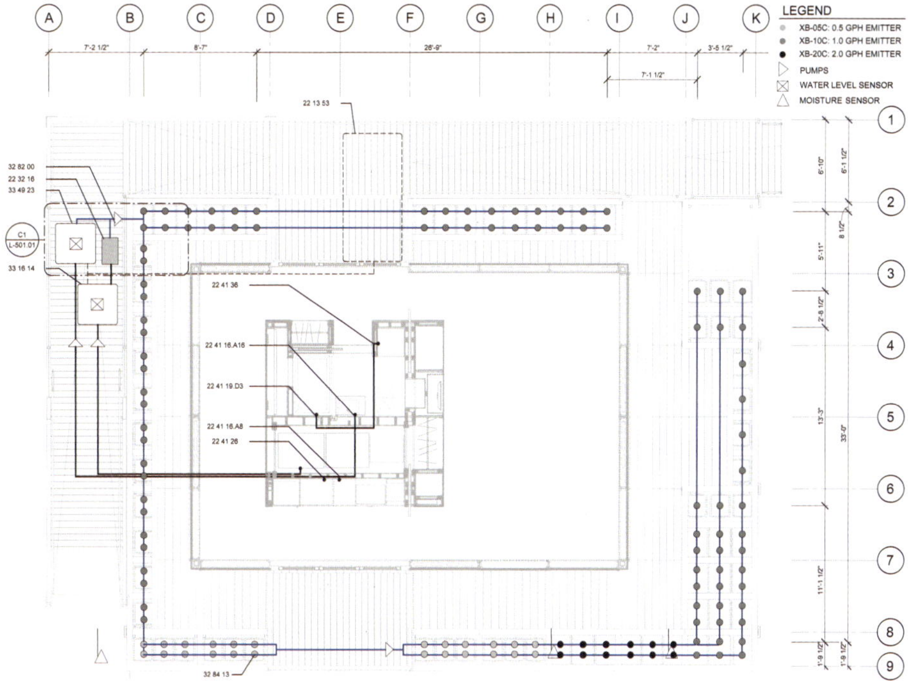

4.219

4.220

4.215 Installation of additional water tanks and the unifying trim that tied the landscape modules together. The coordination with plumbing and structural members was experienced first hand.
4.216 Rendering of the rainwater tanks: Geometry and material concepts could be processed here, while the coordination of conflicts that enabled the system to function had to be resolved in detail.
4.217, 4.218 Hand-sketches for coordination and fabrication purposes of the rainwater tanks.
4.219 Irrigation plan: rainwater and selected grey water is used to keep the plants hydrated.
4.220 Landscape: Installation of irrigation system within planter array.

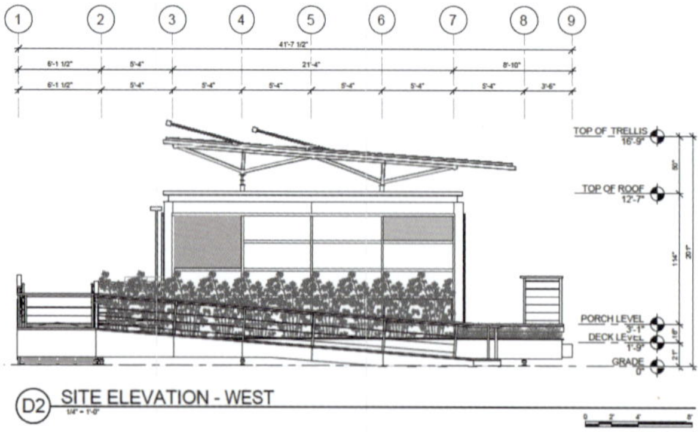

4.221

4.222

4.223

hand-sketches, was critical for integrating and coordinating all of the design overlaps into one functional overall system.

For the exhibit in Washington one of the most important landscape elements was the ramp system for access to the entrance level of the house. These ramps would not necessarily be needed for the Roofpod's generic application on an urban roof, which could otherwise be replaced by an elevator or stair. However, it was critical in resolving the difference in height between the bottom of the dunnage and the Roofpod's entrance level for the public exhibit in Washington DC (4.222). The slope of these ramps had to comply to handicapped accessibility regulations (ADA), and be capable of handling a constant stream of visitors.

Since the ramp was not a necessary part of the generic kit-of-parts strategy, we did not emphasize the 5'-4" module. We also focused less effort on requirements related to non-combustible materials, since there was no building underneath the Washington exhibit. However, there were a host of other reasons behind the precision of these pieces. Everything from walking surfaces to support girders, handrails and lighting had to be accurately thought through. Most importantly, even though it appeared as one continuous element in its finished assembly, it would also have to be transported to Washington in pieces. Thus, each piece had to be the right size and weight, so that they could be easily carried and assembled. Additionally, the ramp system covered approximately 750 s.f., which was

4.221 West elevation, ramp system.
4.222 Visitor ramp during the public exhibit in Washington
4.223 Installation of ramp module during construction; northwest corner.
4.224, 4.225 Detail and perspective representation of the ramp structure: coordination between subsystems.

4.224

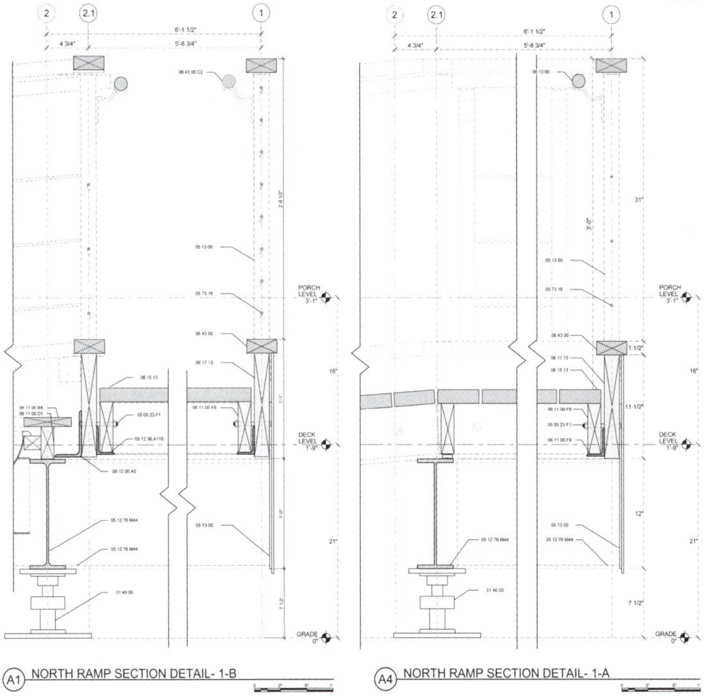

4.225

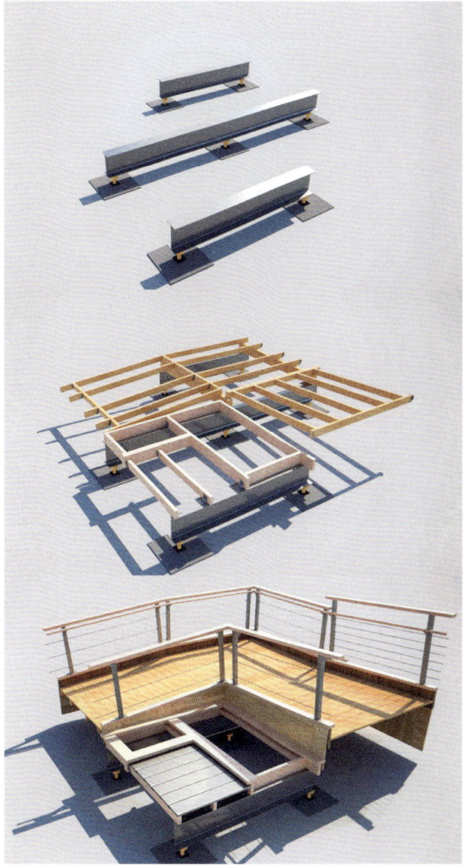

a significant amount of surface area; equating to about the same footprint as that of the building. Because it had to penetrate the dunnage system in order to reach the Roofpod's entry height, its structure was not related to the regular base layer.

Experientially, however, the ramp was a central element for this project, since it represented the viewing platform from which you initially got to know the house; with the change in height, the spatial sensation of getting closer to the building intensified the experience.

The strategy for materializing the ramp surfaces echoed the materials used for the rest of the project. The donated cypress became our preferred wood for the planter trims and Roofpod mullions. This wood species is plentiful and can withstand moisture well, since it originates in swamps. For these reasons, we decided to build the ramp walkways out of cypress, too. The substructure for it was simple 2x6 wood framing, in sections of typically 5'-4" by 6 feet, to accommodate the width required for the ramp by code. An interconnecting plank between ramp sections was typically mounted after two adjacent segments were installed, permanently connecting them together. Thus, during construction the spacing of these planks was very important for configuring the same edge condition for each transition. The modules were held in place by glue-laminated girders, which also supported the handrail posts (4.224, 4.225). These girders were, in turn, partially dependent on the steel dunnage system and additional extension beams, which had to be coordinated carefully with the changing elevation of the ramp structure. Both the dependency on the steel dunnage and the requirement of a 1:12 maximum slope (ADA compliance) were challenging. At one point during construction, the ramp design had to change when we realized that it exceeded the maximum allowable slope, otherwise it would be too long to fit into the assigned site perimeter. Again, this proved that conditions laid out in the drawings had to relate to actual conditions found on site. Earlier detailing of the ramp motivated the students to study prefabricated railing systems. However, most of these systems were either aesthetically unsatisfying, or too expensive. After weeks of discussion on this topic, a friend offered to teach a class in his metal shop, wherein the students were excited to design and build their own handrail posts. They refined the details with the input of their new teacher, enabling them to build beautiful stainless steel posts during the course of one semester, which were more slender and less expensive than the original options available on the market. These posts were combined with stainless steel cables for the horizontal elements, and with a cypress top rail to reflect the surrounding material concept, resulting in a code compliant and beautiful railing.

The top cypress rail served an additional purpose as a base for signage. We developed a jig for routing out an area of 2x8" at every foot, and laminated color-coded plastic signs within this area so that the sign's surface appeared flush with the wood surface (4.226, 4.227). These signs bore specific information on the Roofpod's environmental impact for approaching visitors. Since visitor lines were long, the waiting time could now be spent studying the Roofpod in detail.

Lights were also integrated into the ramp, beautifying the experience at night (4.228). Again, the logistics and consequences of running the wiring required for the execution of this design idea was addressed through first hand systems coordination during construction and reassembly in Washington. Even though these LED lights did not consume much energy (1.2W per fixture), they were a surprisingly beautiful addition to the overall appearance. The extra effort to integrate these fixtures and consider their wiring as transportable units was rewarded accordingly. The complexity involved in this design task was again best understood through the installation experience.

The transportation of the ramp system was a challenge in and of itself. The ramp was made up of support girders between 14 and 20 feet in length that were dependent on an efficient structural attachment to the dunnage steel system. Since these elements could not be broken down into smaller dimensions, their lengths determined the assembly method involved. We decided to keep the handrails and the wiring for the electric lighting intact on these elements, which significantly reduced the final reassembly efforts.

Again, negotiating between the conceptually ideal modular system intended as a kit-of-parts, and the concrete requirements of physical disassembly and transportation made it obvious that a categorical approach is not feasible. Rather, it required developing solutions that consider all specific factors comprehensively.

The assembly of our Roofpod project for the Solar Decathlon exhibit in Washington, fueled by the experiences and constraints of physical construction, was now complete. The described subsystems had emerged from negotiating a comprehensive logic, infused by many factors and scales. All subsystems had their strength and weaknesses. During construction, the students liked to discuss how they could have achieved a better result if factors such as time, tools, and money were different. The landscape system in particular revealed that its full inherent potential could not possibly have been realized given the time constraints dictated by the Solar Decathlon. For example, most of the planters could not be planted on the construction site, because it would render their transportation too difficult.

Thus, researched solutions related to urban farming and composting, and their evaluation based on real-life experience and practicality were left unresolved. Some remaining questions were: How difficult is it to maintain the plants? What species would work well? What efforts and benefits would crop production really yield? How could composting of food scraps from both the Roofpod and the host building provide nutrients for the green roof? What would be the resulting temperature difference and life-span extension from covering the roof membrane? Many more questions could be entertained, however definitive answers could certainly not yet be found. Even though simulations could yield preliminary results, the effectiveness of any system has to be tested, subsequently evaluated and then refined. The landscape system in particular is capable of being refined in continuous steps, since it is easily interchangeable and built of inexpensive parts. Future long-term installations and a focus on testing its efficiencies are thus essential for the successful development of this system for a broad application.

4.226

4.227

4.228

4.229

4.230

4.226 Routing out the sign area in the handrail to integrate information plates with the use of a jig.
4.227 Custom handrail design: The ramp rail along the viewing platforms edge provided additional information to visitors.
4.228 Exterior ramp lighting: wiring added complexity during construction and reassembly.
4.229 'Roofpad': Finished assembly in Washington, with a planted herb garden.
4.230 Glue-laminated ramp girders with preinstalled railings, ready for shipping; the cypress covers for the girders and the tension cables (see 4.224, 4.225) were added after the system was reassembled.

12

LOGISTICS: SITE CONDITIONS, DISASSEMBLY, TRANSPORTATION

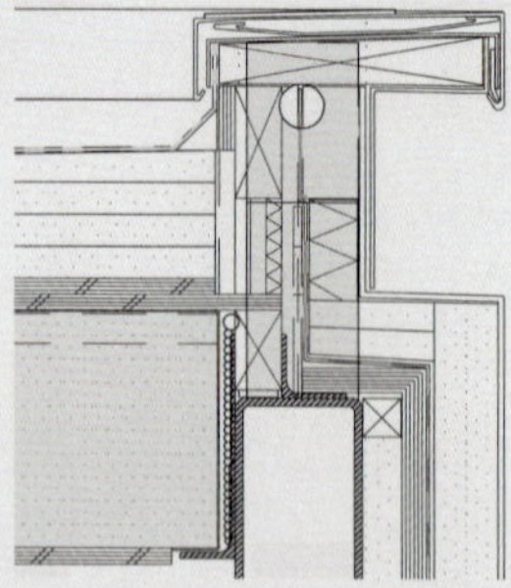

4.231

Throughout the end of the planning phase, we recognized how much the transportation requirements to Washington had affected the physical design of the Roofpod. Supporting the loads while lifting the house from the site onto the trailer had a particular influence on the integrated steel frame (see chapter 4.3), which made us especially nervous on moving day. Our concerns were related to the loads of 20 tons per module, and to the damages that might occur with both halves being separated and reunited after reassembly.

When planning for the logistics of construction, a topic of discussion was preventing the house modules from being accidentally connected during assembly. For example, carelessly applied screws or other fasteners could accidentally fuse the modules together, so we had to ensure the ease of both splitting the two elements apart, and fitting them onto their respective trailer beds. When building a house, one can easily forget that it will be taken apart and separated after completion. But that was exactly the challenge. One of our main objectives during construction was thus the protection of the transportation joint.

Another concern was of relevance, and related to this transportation joint. Due to curricular restrictions, we were permitted to start the construction site only four months before the Roofpod had to be transported to Washington. Paying attention to the tight schedule was a given, and delays, such as those from inclement weather, were simply not acceptable. The decision to use a tent for these bad weather days was accompanied with an awareness of additional costs. Consulting with tent rentals confirmed that costs increase dramatically with size. We were also concerned about the ensuing greenhouse microclimate, which would create uncomfortable work conditions in summer months.

We combined all these factors (rigging based logistics, transportation-related separation and weather protection) to generate our own game plan for construction, as follows: We would build the dunnage continuously, but position one of the two halves of the building 3 ½ feet away from its final location. By creating this gap we could place a tent over one half while the other half is exposed to weather, and, subsequently, be well ventilated. During rainy days, the tent provided the primary work space, while the exterior module was preferable during pleasant weather. Additionally, this physical gap ensured that everyone involved in construction activities understood at all times that the building was to be separated at this juncture (4.232, 4.234).

In order to move the preliminary placed half to its intended location, this split made a test crane run necessary. Even though this represented extra effort, we welcomed this as an opportunity to estimate the difficulty of operation for the days in Washington. We decided to merge the building together after most of the mechanical and electrical installations were complete, thus simulating the upcoming reassembly. Also, we would have a better understanding of the weight of the modules and the equipment needed.

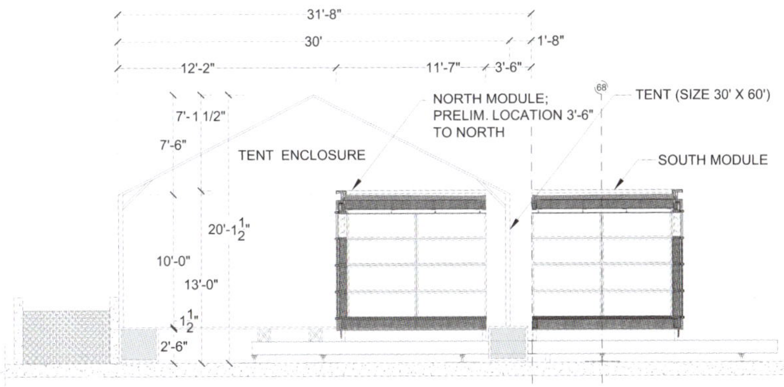

4.232

4.233

4.234

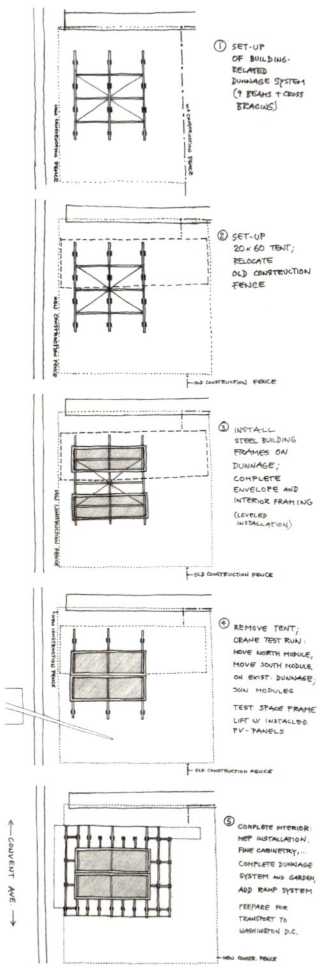

4.235

4.236

4.231 Detail: Roofpod-integrated lifting tab for rigging.
4.232 Tent set-up: Geometric coordination of elements. Host building's structure, tent size and Roofpod assembly.
4.233 Tent structure between Roofpod modules during construction.
4.234 Diagram of construction sequence, related to location and disassembly logistics.
4.235 Delivery to construction site rooftop on campus.
4.236 Location: Material deliveries and work conditions during bad weather had to be anticipated, in order to progress efficiently. For rainy days, a tent was permanently set up.

4.237

4.238

4.239

4.240

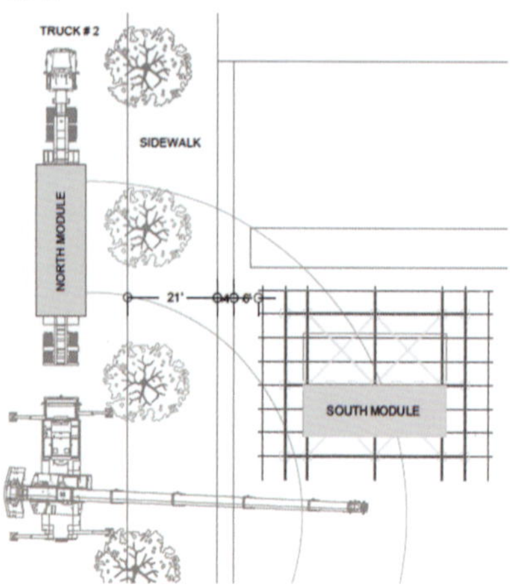

4.241

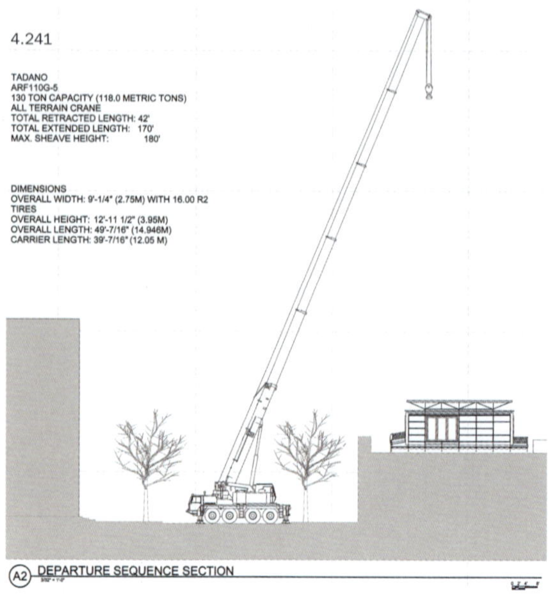

4.242

First, the tent would be removed. Then, during the rigging and connection of the two sections, any complications would be automatically addressed, it was especially important that the module locator (see chapter 4.3) be tested. Finally, the solar trellis would complete the building as planned, the solar panels and thermal collectors could be mounted and subsequently their performance would be tested during the last weeks of construction.

Several investigations had to be undertaken to make sure that the crane test was successful. This had become even more difficult, since the only available site on which to construct the house on campus was, appropriately, the roof terrace of a one-story building. This site (4.239) allowed us to investigate several aspects of a typical Roofpod installation, wherein preassembled building modules were rigged. The most important investigation was determining the turning radii, reach and weight capacity of the crane equipment. Early on, we met with crane operators to understand the 3-dimensional relationship between the equipment and the built environment, including the Roofpod, building underneath and available street space (4.240, 4.241). Several limitations became evident throughout this investigation. For instance, the location of the Roofpod was structurally dependent on the available load capacity of the multifunctional host building underneath. This building comprised different structural systems, such as a long-span structure for a gym that could not support the Roofpod's weight. The options for site locations were therefore limited. Also, street trees imposed limitations on the crane's reach, and the anticipated weight of the Roofpod had obvious consequences for the crane's tonnage. The supporting drawings had to be absolutely correctly dimensioned, so that predictions were accurate.

We found a site that was both suitable for a crane and commensurate to the existing building's structural system. The test run would also reveal any unpredictable issues. Since this test run was scheduled six weeks before the scheduled disassembly, we would have just enough time to adapt to any unforeseen conditions. In particular, we were curious about three specific aspects:

a) Would the location finder demonstrate how practical it was to the assembly sequence? We had not yet installed any finishes in the house, except to the adjacent building blocks abutting the east and west sides. These two seams made us nervous since the gridded nature of the blocks would make any inaccuracy immediately visible on the interior of the building. Double- and triple checking the building block geometry with laser equipment paid off, as the transition between the two interconnected halves worked out perfectly. The location finders corresponded with the dunnage beams, completing this major joint (4.248, 4.249). With this joint now accomplished, the completion of the interior of the house's cabinetry, dropped ceiling and finished floor could commence, since the Roofpod modules' juncture was permanently defined by this connection. During the entire project, an awareness of these two halves traveling separately was one of the main factors in finishing the interior.

b) Would any glazing crack during rigging? Our concern related to the deflection that occurs when the building module is suspended in the air from the crane's spreader bars. Initially, our design had four suspension points at the module's corners; later, due to this concern of extended deflection, we added two intermediate support points to the structural design in the middle of the span (see chapter 4.3). In order to effectively judge the risk of damaging glass, we had to first complete the glazing. This was challenging, given that we were already behind schedule at that point. Finishing only the northern module's glazing gave us enough information to assess the risks involved, since both modules were

4.237 Gap between house modules immediately after tent removal; the south half (right) had to be covered with a tarp when rain was forecast.

4.238 The two still separated house modules, immediately before test rigging.

4.239 Rendering of anticipated site during design development; student drawing.

4.240 Crane turning radius studies, defining possible construction site locations.

4.241 Crane reach: Anticipated weight and site location determined the size of the crane.

4.242 Spreader bars set-up at street level: preparation for the test rigging.

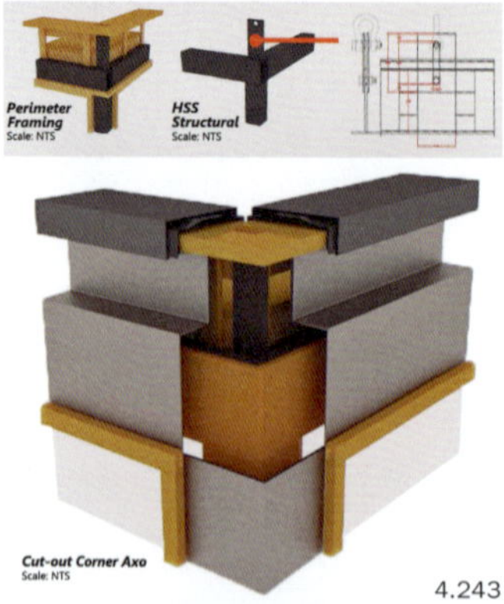

4.243

4.244

4.245

4.246

4.247

rigged the same way. No glass damage occurred, and the electronic equipment of the crane indicated only minor deflection.

c) Would the estimated weight be accurate? Since the six retractable lifting tabs were made of relatively thin plates (2x 3/16" per tab), we were concerned about any weight increase during construction. The test run would provide an opportunity to learn the modules' actual weight, since their tonnage could be read by the crane's electronic scale. Additionally, we could safely test whether the lifting tabs were strong enough. There was no significant risk involved during this test run, since the modules would be elevated just inches above the dunnage, so that if any complications arose, a structural back-up was in place. It turned out that the weight was close enough to the estimated 20 tons, and that the lifting tab connection was, indeed, strong enough.

The test run was perceived as very productive by the team, despite the additional costs. Many of the previously unanswered questions relating to the practicality of the Roofpod's moving operation and reassembly procedures, could now be assessed. This gave us confidence for the pending moving days and engendered the final strategy for the transportation of all the necessary pieces of the assembly to Washington. It was clear that the two house modules would fit on their flatbed trailers, as they were modified early in the design process for both the allowable transport dimensions for highways and their particular positions on lowboy trailers. Furthermore, all the other parts involved in the assembly had to be carefully packed, with a sequence and system in mind that would identify their location on the truck as they were needed. This is similar to organizing a tool box, wherein items are positioned in the most practical and handy configuration. In the case of the Roofpod, every specific assembly problem would have its corresponding tool box organization. For its reassembly, this was even more a significant

4.243 Parapet-integrated lifting tab; design study student drawing.
4.244 Early mock-up of the parapet with its integrated retractable lifting tabs.
4.245 Lifting tab: Test on half-finished south module.
4.246 Test run: Preparations to connect the module's lifting tabs to the spreader bar system.
4.247 Test run: Rigging of north module to its originally intended location.
4.248, 4.249 Building module before test rigging and at the moment of abutting the north module; the location finder plates were essential to accurately align the building block façade. All other finishes were installed later.

4.248

4.249

4.250

4.251

4.252

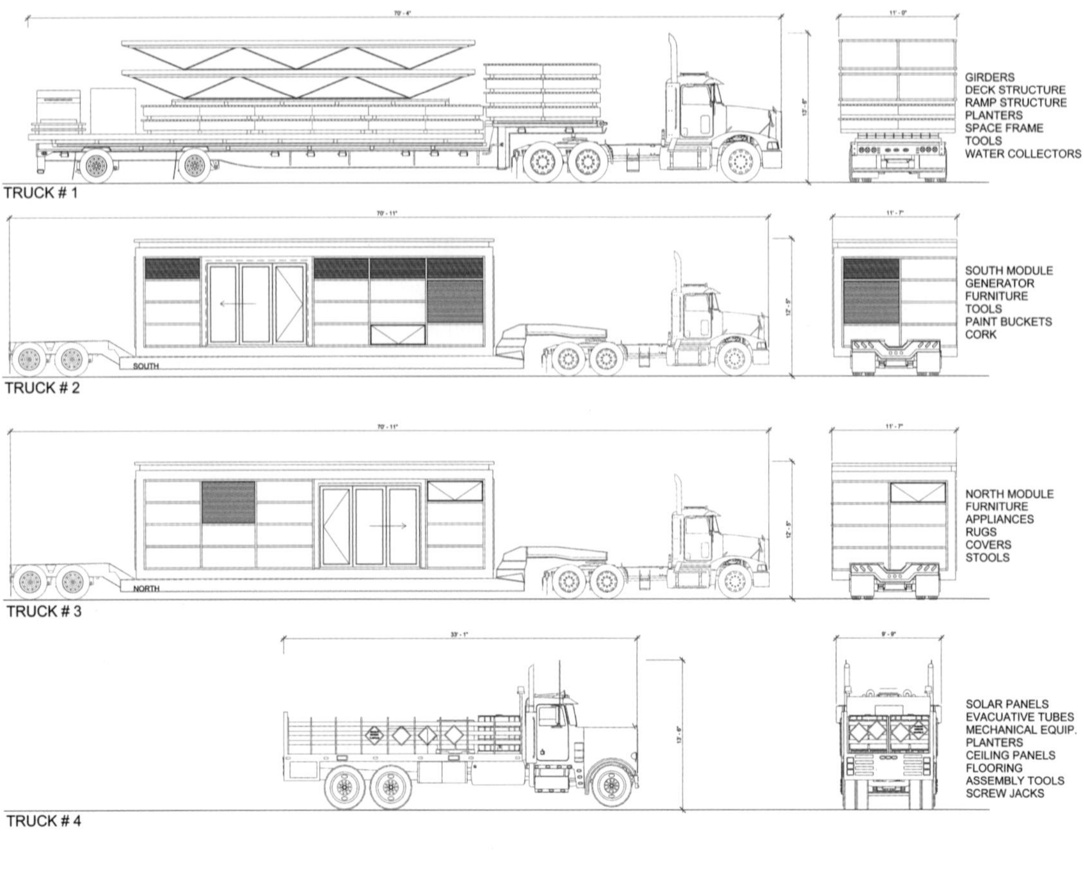

4.253

4.250 Loading of truck #4 (enclosed trailer) with landscape modules, tools and smaller items; sequential thinking was key to finding everything later.
4.251 Arrival of truck #1 (flatbed) which carried the dunnage beams, space frame and cooling tower. Unloading by telescopic forklift worked perfectly.
4.252 Transportation logistics: How many trucks are needed, and how to organize the loadings? It turned out that certain elements, particularly the space frame, had to be disassembled and put back into crates to make all the items fit.
4.253 North module, almost ready for transportation to Washington: the modules were shrink-wrapped to protect them during highway travel.

problem, since not all trucks would have access to the building site on the National Mall at the same time. Assembly items that related to each other had to be on the same truck, while elements whose installation was dependent on rigging had to arrive early. Finally, the nature of the transportable elements themselves also determined their location during transportation. For example, the long and heavy dunnage beams were transported on a flatbed trailer where they could be picked up by a forklift. This order was also reflected in the cost of the necessary trucks. It became clear that compromises had to be made: In order to fit everything on only four trucks (2.252), the space frame was disassembled to create enough space. Despite the additional work required to reassemble the system in Washington, the benefits far outweighed the effort, making it an easy decision. Reassembling the space frame was a familiar process to us, and, therefore, the efforts were quite predictable. It became apparent that we underestimated the equipment and building element's bulk volume. The fourth truck we intended to rent grew according to our requirements. We used an enclosed trailer to contain the many smaller pieces.

The final distribution was as follows: Truck #1: dunnage system, the crated space frame, and the cooling tower with its custom enclosure; Truck #2: South module Roofpod; Truck #3: North module Roofpod; Truck #4: Landscape modules and ramp system, solar panels and evacuated tubes, tool chests, water tanks and plumbing, generators, furniture, floor and ceiling elements for finishing the transportation joint of the module, and the screw jacks and plates for the dunnage system's transition to the ground.

Truck #4 arrived first to set up the site, position the plates and screw jacks, so that the dunnage system on truck #1 could follow. Using a telescopic handler, the dunnage beams were set while the space frame was being assembled. Then the crane arrived to set the Roofpod, starting with the north module, which was further away, then the closer south module, and finally the reassembled space frame. The sequence between setting the modules had to be anticipated well ahead of time, otherwise the location finder would not have worked.

To anticipate the way the building had to be transported and re-erected was part of the design challenge, even though it was not visible in the end. Much of the design time was spent studying the relationship between the various 'mechanisms'. Even though these conditions were more complex, they exist in any project. Anticipation of such logistics, in particular under challenging conditions, is thus paramount for realizing a successful building assembly.

The site atop the apartment building, depicted here as a case study, represents only one of countless possible applications in New York and other urban environments. Any Roofpod prototype design is thus influenced by the specifics of the host building, by time and budget constraints, by environmental context, by building codes and of course, by access to technical means of fabrication. The conditions under which a first one-off is produced are different from a streamlined fabrication process for multiple Roofpods in series. The next logical step would now be to collaborate with specialists in fabrication, real estate and policymaking in order to develop this building as a feasible contemporary typology.

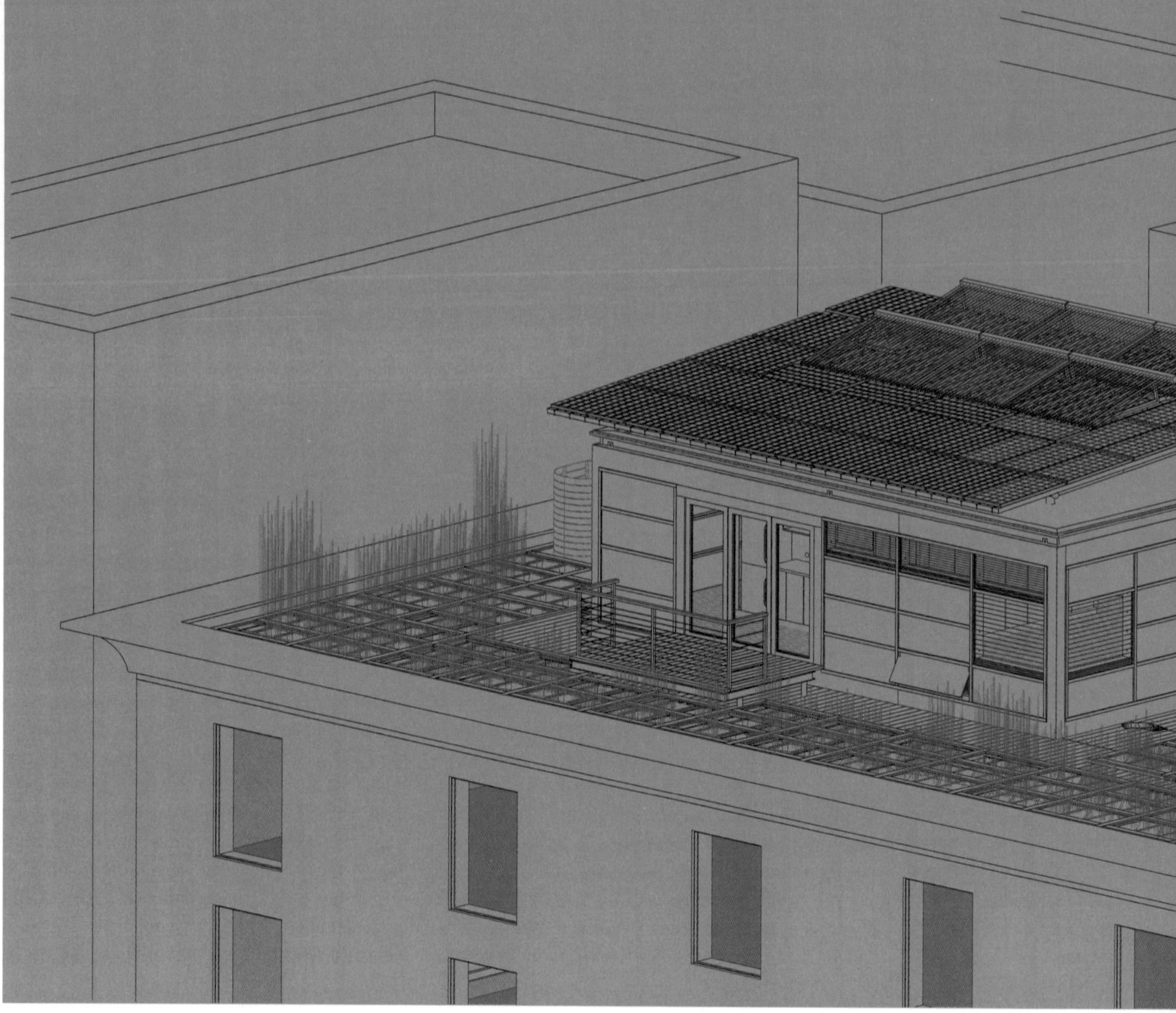

COMPLETED ROOFPOD ASSEMBLY

12

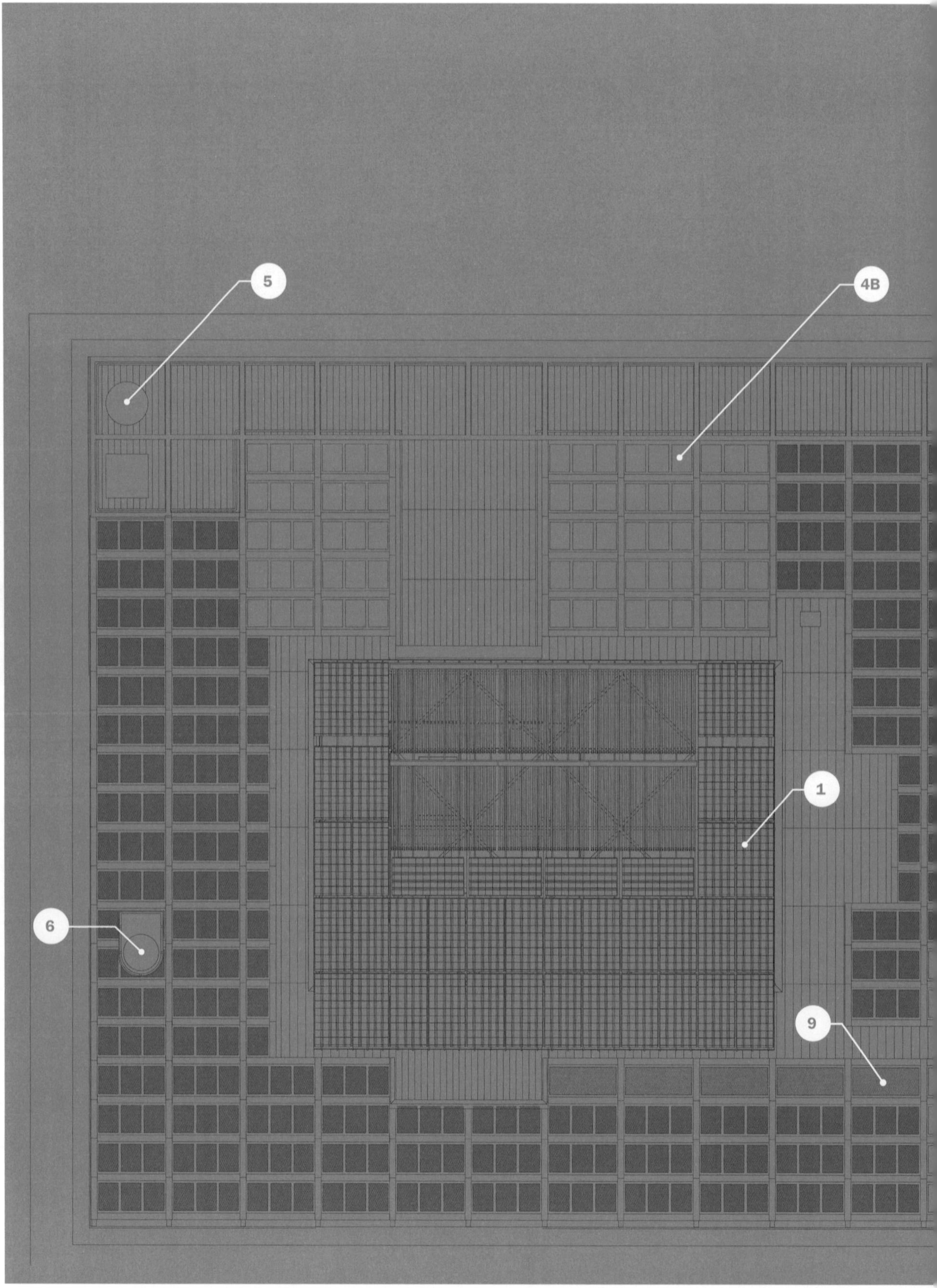
5
4B
1
6
9

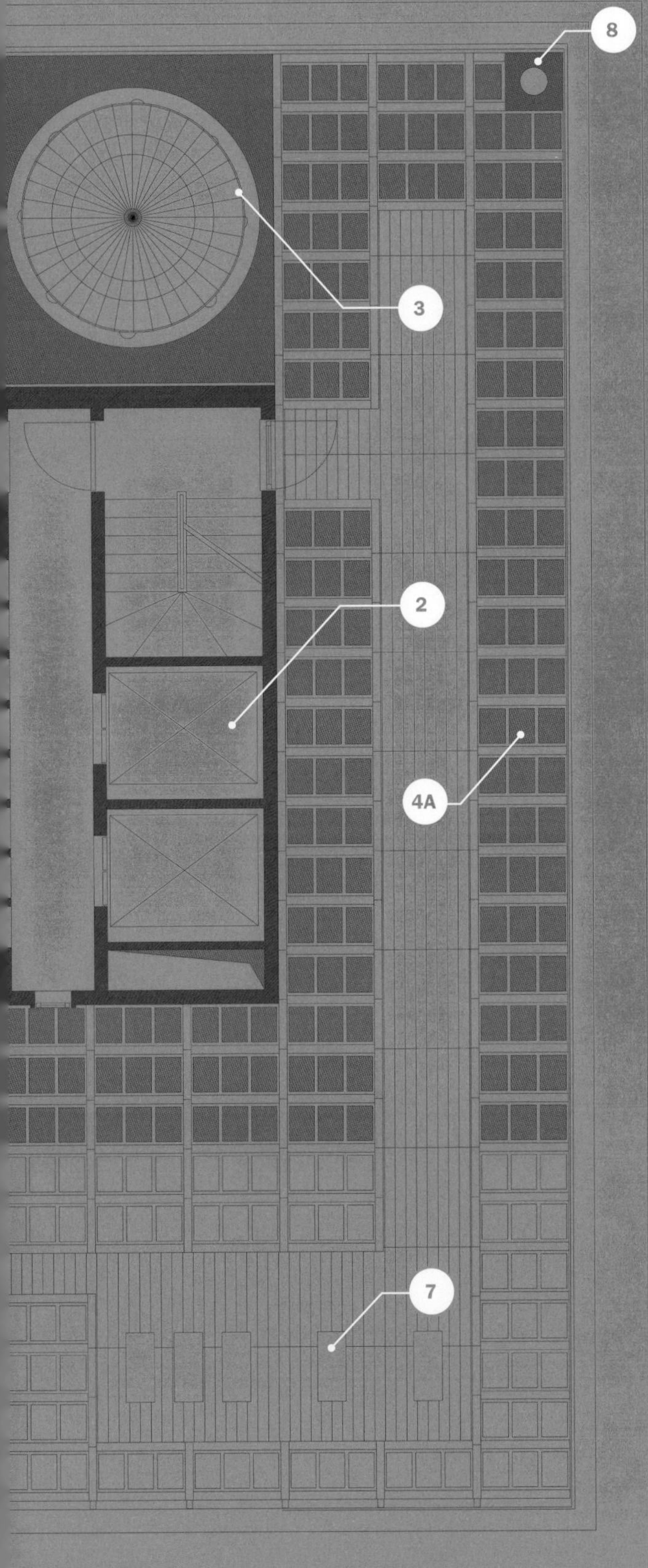

PLAN, PROPOSAL

1. ROOFPOD
2. STAIRCASE AND ELEVATOR ACCESS
3. WATERTOWER; AND EXTRA GREYWATER IRRIGATION SYSTEM
4. ROOFPAD
   A) INDIGENEOUS PLANTS;
   B) URBAN FARMING: HERBS, FRUIT, VEGETABLES
5. COMPOSTING AREA FOR ORGANIC WASTE: FOOD SCRAPS, ETC.; SOIL PRODUCTION
6. COOLING TOWER, SOLAR COOLING SYSTEM
7. COMMUNITY GARDEN
8. WIND GENERATOR
9. INTEGRATED RAIN WATER TANKS

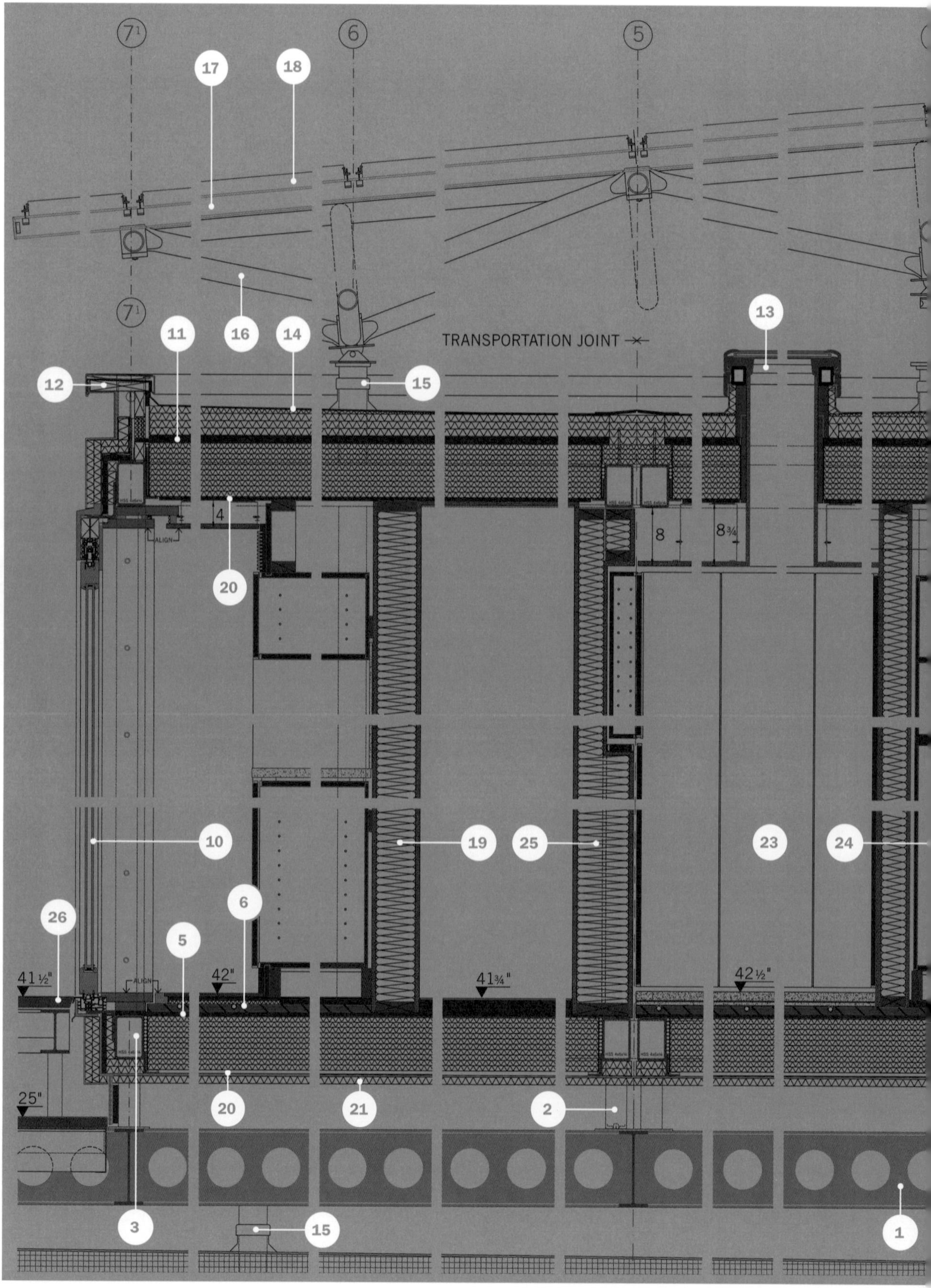

TRANSPORTATION JOINT
ALIGN
ALIGN
41 ½"
42"
41¾"
42½"
25"

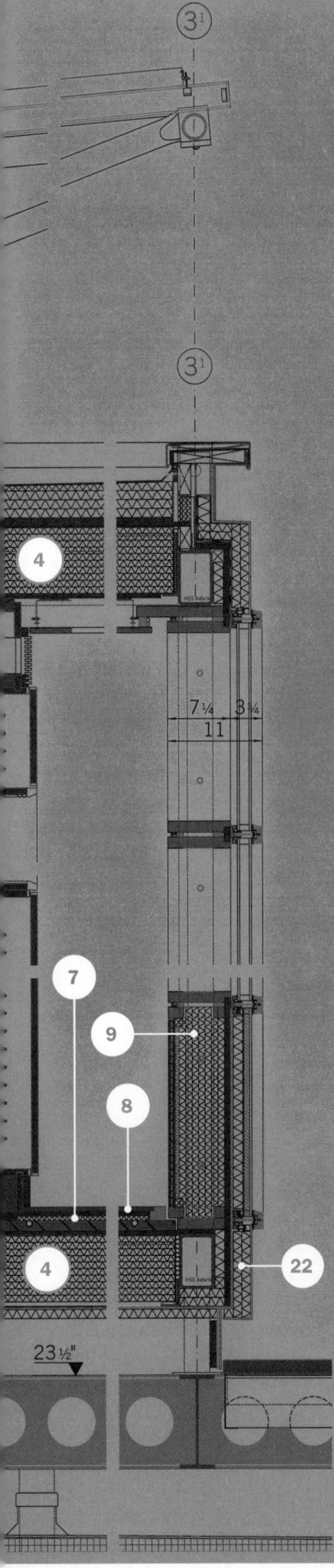

DETAIL WALL SECTION

1. DUNNAGE SYSTEM
2. MINI- COLUMN, HSS 3" X 3"
3. STEEL FRAME, HSS SECTIONS
4. COMPOSITE INSULATED FLOOR AND ROOF PANEL (EPS + LIGHT GAUGE STEEL), 7.5" THICK
5. ½" OSB FLOOR SHEATHING
6. 1-1/8" PLYWOOD W/ GROOVED ALUMINUM FOIL AND RADIANT ½" PEX TUBING
7. INSULATION, POLYISOCYANURATE, ½"
8. CORK FLOOR, TUNG+GROVE, 3'X1', ON ½" OSB SUBFLOOR W/ SUPPORT STRIPS
9. BUILDING BLOCK: POPLAR FRAME+PANEL; INSULATED W/ CLOSED CELL FOAM, ¾" OSB SHEATHING, WATERPROOF MEMBRANE, INSULATED GLASS UNIT; STEEL FIN REINFORCED.
10. BIFOLD DOOR, WOOD: DOUGLAS FIR, 3 LEAF, W/ TEMPERED GLAZING
11. 2X ½" OSB ROOF SHEATHING
12. PARAPET FRAMING, WOOD, LIGHT FRAME
13. SKYLIGHT, MOTOR-OPERATED; W/ TEMPERED GLASS
14. TPO ROOF MEMBRANE, 80 MIL., ON TAPERED XPS INSULATION
15. SPACE FRAME STUB, 4X4 HSS
16. SPACE FRAME, 2–7/8" DIAM. TUBING
17. PHOTOVOLTAIC RACKING SYSTEM, ALUMINUM, 3" DEEP
18. MONOCRYSTALLINE PHOTOVOLTAIC PANEL, 238 W/PANEL
19. LIGHT-GAUGE WALL FRAMING, 5/8" DRYWALL, ACOUSTIC INSULATION
20. ½" BOARD, FIRE-RATED MATERIAL
21. RIGID INSULATION, XPS, WATERPROOFING
22. ALUMINUM CLADDING, BREAK METAL PANEL
23. MDF + BAMBOO CABINETRY, POLYURETHANE SEALED; MISCELLANEOUS HARDWARE
24. BATHROOM FINISHES: POLYETHYLENE (HDPE) PANELING AND COMPOSITE CONCRETE SLAB
25. BATHROOM WET WALL, PLUMBING ROUGH-IN
26. PORCH: WOOD DECK ON STEEL FRAME

THE SOLAR ROOFPOD

PHOTOGRAPHIC DOCUMENTATION

Solar Roofpod
TEAM NEW YORK / CCNY

1

ROOFPOD: ON CAMPUS AT THE CITY COLLEGE OF NEW YORK

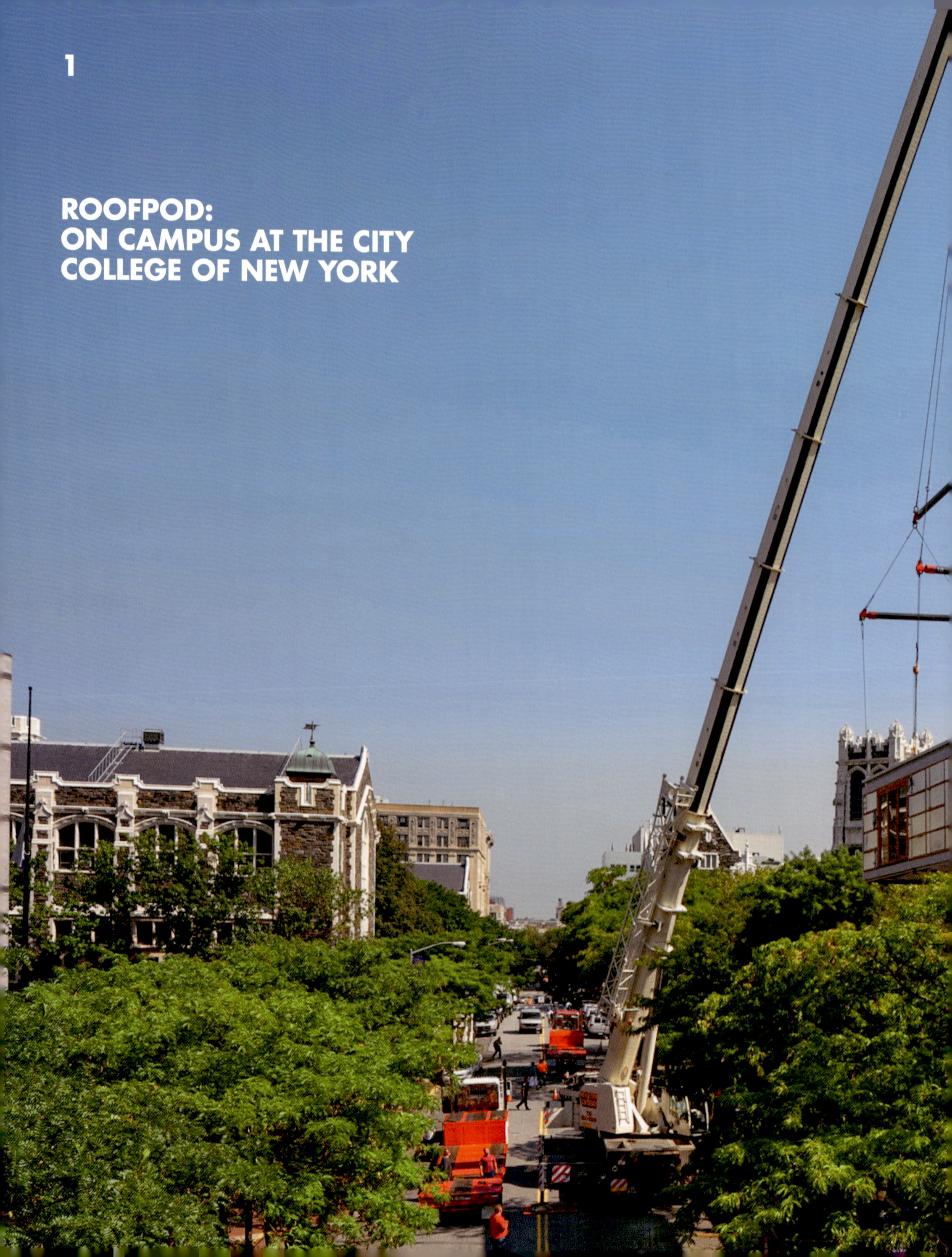

JCB

2

ROOFPOD: REASSEMBLY AT THE NATIONAL MALL, WASHINGTON DC

GROVE
WIDE LOAD

SAFETY
FIRST!!

CCNY

WDF

the city college
of new york
solar roofpod team ny

the city college
of new york
solar roofpod team ny
PENSKE

403
Team New York
the city college of new york
solar roofpod team.ny

the city college
of new york
so ar roofpod team ny

United Rentals

United Rentals
450AJ

CCNY

SOLAR DECATHLON
403
Team New York

3

ROOFPOD: FINISHED PRODUCT

NY

NY

EDUCATION – COMPREHENSIVE DESIGN

0

THIRD APPROACH: "WHAT?" 'KNOWLEDGE'

5.1

5.2

After having designed and built the 'Roofpod' with a team of students and having them seen change in their intellectual appreciation of what being an architect might mean, I have thought much about the ways architecture should be taught in school. Architecture is not, as many of us initially thought, and countless characters in novels and movies suggest (5.1), a profession based on isolated, sometimes unrecognized genius of drafting and thinking, but a task that needs at its essence the hands-on knowledge of making; it requires steady attention to multiple criteria, not only based on two-dimensional drawings, but on three-dimensional processing, necessary to materialize the imagined (5.2).

WHAT?

It is undisputed that many skills are needed to conduct architecture in a meaningful way. In the professional world, architecture has been constantly shifting from hands-on knowledge towards being administrative, servicing the building and real estate industry. While at the beginning of this century many architects still came from crafts backgrounds, and knew very well the techniques necessary to process material, today this perspective is increasingly rare.

Architecture is predominantly taught in classrooms, not in craft shops. The main tools with which to convey knowledge to students are words and especially drawing; this means of expression has developed considerably due to the progress made in computer-supported programs and illustration methods. The way we understand artifacts has changed with it: Often surface appearances are emphasized, while the structural qualities of a design stay concealed. The comprehension of how to actually build the two-dimensional image is seldom seriously approached in depth in academic teaching.

Simultaneously, studying in general has become more popular. Today, more people study than ever before, instead of, for example, entering apprenticeships in the craft shops that contribute to the production of tangible architecture. Nowadays, working with the mind is more highly valued than working with the hands. Career opportunities and career decisions go in the same direction. A disconnect between planning and making is apparent.

We are forgetting about craft. In contrast to study requirements in a handful of local areas, where architecture students have to do internships in building trades, the ability to learn concrete materialization skills is – at least in many industrialized countries – largely underdeveloped. However, a comprehensive materialization alphabet is essential to develop an efficient, productive and creative designer skill set.

What is the subject of design, as we culturally understand it? Is it the stimulating articulation of an abstract idea, or does it include assembly and operating instructions?

Are we equipped to present abstract ideas well enough for non-ambiguous realization, if they are based on drawings? What makes these ideas better than those of hands-on specialists in fabrication?

The profession thrives with comprehensive thinking, and thus has to include design processing:

5.3

5.1 "The Fountainhead", Warner Brothers, 1949 (after a novel by Ayn Rand, 1943): The architect is portrayed as a misunderstood genius, isolated in his own intellectual world.
5.2 Jean Prouvé, on site, (Maison du Peuple, 1938), retouching the stairway railing, damaged in transportation.
5.3 Comprehensive design factors, activities and intellectual output to design the 'Solar Roofpod'; the students were involved in all aspects of design-building. They had to reposition their output according to progress made.

All aspects, from idea to product, have to be considered equally.

Similar prototyping skills can be found in industrial design and also in areas of applied engineering: Good design justifies itself by the comprehensive opportunities it develops during its prototyping phase.

Any kind of architectural education should suggest a field of professional application. In architecture, – the art and science of building –, often the skills you pick up in school are not sufficient to prepare you for the holistic processing capabilities of making an artifact. Although history or social sciences are all significant components to intellectually understand and improve architectural projects, one of the supposedly mandatory elements, – to materialize the imagined – is fairly absent in the curriculum.

How can we, as teachers and professionals, develop design teaching strategies that make students aware of an intertwined skill set of intellectual and practical components, and change the perspective early on, so that engaging in hands-on methodologies becomes more attractive and throughout one's professional life.

How can we bring back practical skills to an academic field, neglected in the past decades in favor of a top-heavy "higher" world of art, politics and philosophy?

1

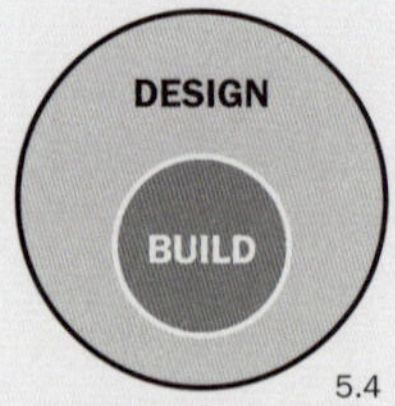

5.4

PROTOTYPING: LEARNING TO MATERIALIZE

Design is – at least partially – a backwards process: While in assemblies form grows by adding pieces, in design, form often has to be deconstructed. The surface conditions are the starting point for exploring the deeper structural conditions, in order to make them realizable.

The image of a chinese puzzle (5.6) demonstrates an area of intellect that most likely cannot be developed by abstract studies, but needs the involvement of the hands, to investigate its structure and joinery. It explores the interplay between the overall object and the way it is put together.

Architectural structure has to be approached from both angles: Starting from surface conditions, but also from assembly strategies.

Joinery, construction sequences and tolerances can most likely not be anticipated by discussing or drawing them, but must be experienced by motoric skills, to positively affect the assembly strategy. The intrinsic characteristic of drawing doesn't include tolerances or sequences. Thus, three-dimensional hands-on skills must be developed in parallel with the two-dimensional graphic skills (5.4).

This could also be explained with such objects that do not reveal significant parts of their nature by just looking at them, for example toys, tools or utensils: they have to be used to be decoded.

When, in the 50s psychologists such as Benjamin Bloom identified categories of different ways of learning[61], they confirmed that its output is based on developing an efficient combination of interrelated domains. The domains identified were: 1) cognitive, 2) affective, and 3) psychomotor learning.

While the first two domains are related to mental and emotional development, thus mainly mind-oriented, the third area included manual or physical skills related to intellectual processing. Arguably, different professions need different combinations of intellectual processing, therefore different involvement of the above domains.

Architecture is related to physical assemblies, since this is how form is constructed. This task comprises, however, not just the repetition of common construction details, but is based on invention, specificity and ingenuity. Thus, it must be intellectually processed, in order to generate results.

The process of materializing a formal concept is thus a constant comparison between the means of assembly and the predications of form. The negotiation between them must be based on experiential learning.

In the 70s, David Kolb laid out his experiential learning theory (5.5), which idealized the involvement of two related approaches to gain knowledge: Concrete experience and abstract conceptualization[62]. Both of them use different methods to process intellectual content through reflection: cognitive/emotional, and testing: psychomotor[63].

While cognitive ways of learning are well explored, the research on the psychomotor domain[64] is still underdeveloped. The explanation for this oversight might be based on the lack of experience in teaching manual skills in academia.

However, these skills are vital for comprehension of what a 2-D detail drawing expresses: the way architecture is put together.

PROTOTYPING

If architecture is, among other activities, sophisticated fabrication, then the process of prototyping must be practiced. The word proto-type derives from the Greek words 'proto' = first, original, and 'typos' = impression, form. A prototype is thus an early impression of an idea, or a materialized concept. It is a concrete impression: thoughtfully materialized, with the intention to learn from it, and avoid future mistakes. It is usually used for complex mechanisms like cars, or electronics that have to be practical. It relates to industrial prefabrication, often used to mass-fabricate an item.

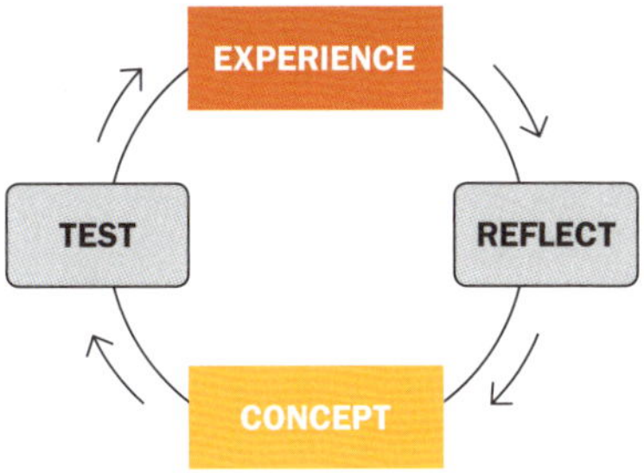

5.5

5.6

5.4 Building as an intrinsic part of design research, particularly in developing anticipation skills.
5.5 "Experiential learning cycle", David Kolb, 1975
5.6 Chinese Puzzle: The assembly cannot be understood without physically investigating its joinery by taking it apart. Similar in construction: Intellectual access to sequences, tolerances and joinery can best be experienced hands-on.

In that sense, the prototype is an investment – as a learning process and fabrication challenge – while the post-prototype production becomes – as a consequence – intellectually effortless. However, the intellectual process to get from point A to point B, from a first impression to a refined method of materialization is what builds the actual competency.

In that respect, it doesn't matter much if the item will be in fact mass-produced: the fabrication challenge is the essence of the problem; the prototype is used as a materialization learning tool.

In an architectural application, comprehending the material dimension is key. Testing of performance factors: gravity, moisture, light, or stability and joinery – of course related to appearance – develops anticipatory thinking and intellectual transfer capacities. This facilitates further abstract process conclusions. Experiments are learning experiences.

MATERIAL THINKING

In architectural education, this perspective is significant. While teaching Construction Technology, I often observed in the students' early large-scale details that the physical dimension of materialization was missing.

The anticipation of an assembly sequence, or access and ways of fastening were often non-existent, because they actually are meaningless for the students' practiced craft of drawing. How then to teach this material-related understanding?

The common academic way is to redline the students' drawings and make them redo these over and over again, until the result is credible. However, it is questionable whether this is the most effective approach to intellectually develop the ability to anticipate physical problems of material joinery and fabrication processes.

Another way would be to visit construction sites to learn by observation. Most contractors are, however, hesitant to allow this, due to liability issues. Also, what often arrives on the sites is in large part already fabricated: in an industrialized world, where – from drywall to window profiles to entire facade elements – many items originate in factories, the act of fabrication cannot easily be experienced anymore. Production methods stay concealed. Fabrication in architecture has taken on a different meaning, more focused on assembly of semi-products, which again takes training to combine them meaningfully.

The standardized assembly of industrial systems, however, cannot be the professional center of our attention, otherwise architecture would become administrative.

Instead, we have to develop our own 'fabrication thinking', and teach assembly as part of that greater problem. System thinking skills make the architectural profession viable. The ability to customize, considering a set of comprehensive factors, is a vital value of our present-day profession; we cannot compete with industrialized processing. The architect can however be the artisan of construction.

SYSTEM THINKING

The main occurrence of Architecture is in fact represented in one-offs, or to put it another way, single attempt projects with unusual, complex sets of intentions.

In that respect, the use of prototypes seems obsolete, since the same fabrication will most likely not occur. However, the capacity to organize and interrelate the many parts of a system to each other is, where complexity must be generated.

With this organizational challenge, the necessity of developing skills in prototyping processes, grows more rather than less important. There is only one chance to get it right. The materialization takes place only once.

Accurate presumptions about the outcome of the design, what creates the more attractive product,

5.7

5.8

by estimating the effort and benefit within every design move, are like creating a comprehensive puzzle.

If this puzzle – which defines the materialization of the project – is not considered appropriately, most likely, in one form or another, failure will strike, either through physical deterioration of the product, or worse, the user's dissatisfaction and subsequent neglect (...not to mention the headaches of law suits and compensation claims). Therefore, the architect's main task must be to get the prototype right.

SYSTEM PROCESSING

The reason why we set up our own 'production line' when fabricating several components of the Roofpod ourselves, was surely not to start an industrial endeavor. It was to refine the skills necessary to process design towards materialization. While usually time and budget are limitless in theoretical projects, here, we had to negotiate in a much more complex way between design intent and the possibilities of construction. The difficulty of this negotiation is that the conditions are initially dynamic: What tools are available?; How much money can be spent?; What connections are feasible?; What level of precision can be achieved?; What materials are available? How much labor is necessary to realize the intended detail? There are countless factors.

Only constant research that asks the right questions and responds properly while refining the design, will achieve the resolution that is necessary for the start of production. In that sense, the same questions are asked that would also be addressed in industrial processes. However, the balancing criteria are different. Here, the balance between time, money, quality and beauty bears more weight in favor of a more specific and probably superior product.

The design can be more responsive to the specific conditions. Therefore, a deeper and more refined way of processing is achieved.

The issue is of course the proper balancing of applicable factors influencing the decision. An avalanche of information can easily bury the good intentions. Prioritization is paramount. Thus, the basics that have an immediate impact on haptic and sensual perception should be primary factors.

When discussing the factors that shape the design, opinions were initially diverse. While some students took the industrial challenge seriously, others wanted to produce the most beautiful appearance imaginable. During the research phase it became clear that it is often impossible to compete with the advanced manufacturing techniques supported by millions of dollars of investments. The strategy therefore had to concentrate on more attention to detail and greater creativity to integrate subsystems holistically.

We realized that the specialty of our profession is anchored in the capacity to invest more specific thought, practical and theoretical and therefore develop more quality towards the final product. The negotiation methods needed to produce the product might be similar to more industrialized areas. However, the 'production line' allegory served mainly for investigating the conditions of how architectural systems can be better processed.

In the course of this discussion, we compared our task with the profession of a chef (5.7); the artisanal qualities of fabrication are essential to learn, while one has to take advantage of technologically advanced methods of prefabrication, but always with the aim of producing a superior product. It is questionable, whether we meet the objective of teaching better systems processing skills to design students. Can design thinking be comprehensive without engaging in three-dimensional materialization skills?

The priorities in design representation happen on paper, small-scale models and in verbal form. Serious material-focused research, i.e. on tangible properties and methods and strategies of joinery, improving constructability and attractive appearances, are rarely pursued in an academic setting.

The evaluation of a student project is usually academically judged in terms of conceptual thinking – which often removes the project from concrete realization issues. In the past decades, an intellectual tendency has become apparent that suggests that the act of construction is inescapably a banalization of an original intellectual idea. This position, a capitulation to the concrete realm of making, doesn't acknowledge that a superior product in demand of a more original and immediate realization would need more – and not less – research and interdisciplinary attention (contractors, fabricators, etc.) to produce it. Instead, in school the drawing becomes the endpoint of development. It functions as the final product, without the intention of materialization. The drawing, originally the materialization processing tool, masquerades as the intended artifact.

Material processing – why and how we use certain materials for certain purposes and expressions – is withheld from the aspiring design intellect, reserved for maturation in one's professional career. There are not many other professions, where academic study is as different to the actual professional profile as architecture is. Problem-solving in school and problem-solving in professional life seem to be different tasks.

One of the main criteria for the academic evasion of teaching what is needed for concrete tangible materialization, is grounded in a lack of opportunity to translate abstracts into materialized real-life application. It is difficult to get intellectual access to that field if a confusing overload of multiple abstract criteria blocks access to these basics. However the synthetization of concrete and abstract is necessary to reach a product quality superior to banal, often mass-produced, common building items.

A design philosophy is needed that accommodates the processing of comprehensive design factors, materialization as its intellectual center, in a pedagogic step-by-step approach. A design isn't done when represented by a computer model or drawing. The choice and refinement of techniques are part of architectural thinking:

An applicable metaphor, once offered to me by one of my teachers, understands design as an acrobatic act, like juggling balls: You start with a few, and incorporate more of them during movement. The more consistent the movement and the more balls in play, the more attractive the acrobatic act. It is the same with the topics of design investigation: You start with the primary, most basic items, incorporate more, and control them simultaneously by overlaying, comparing and ordering them. The essential and skillful movement is the act of designing. I like the 'playfulness' of this metaphor: often haptic skills are best built by playing and experimenting.

In this respect, being the deeper and more considerate systems designer, we have to embrace methods to convey how three-dimensional fabrication happens. We have to develop better approaches how two-dimensional and three-dimensional methods are connected, and how both of them can be used to develop a better product.

A more refined product can only be generated by comprehensive thinking. The architect has to wear different hats to consider a design holistically. One of these hats, the one of the fabricator, is a very important one, to master the coordination of trades and material know-how.

It is clear in the juggling metaphor that there can be no end to this feat: refinement is time related. As soon as you stop moving and the hands-eyes-brain coordination ceases, the balls will drop, and the process ends.

Thus, the longer and more efficiently one can sustain movement, with coherence as a result, the more refined the product. The strength of true architectural design in our times is a higher grade of comprehensiveness and refinement.

5.7 What kind of chef would you want to be? Artisanal qualities necessary to become a superior chef are comparable with the skills needed to become a more creative architect. Prefabrication has to be integrated into holistic system thinking, otherwise it has limited value.

5.8 Last design review before construction of the Roofpod started: The pin-up was conducted under the construction site tent. Students had to explain how anticipated construction sequences and details were related to the envisioned appearances. Both hats were needed: The contractor's and the architect's.

2

TEAMWORK AND PROFESSIONAL ADVICE

5.9

Several crucial contributions helped the students' learning of realization competencies.

ADVISORS

Early in the process, we reached out to faculty and alumni to join a voluntary technical advisory board, which was present at most of our early reviews. The members of this board were specifically selected to represent and develop structural, envelope, and energy-efficiency expertise. They were able to communicate to students essential strategies to develop a feasible and organized realization approach, i.e. how to set up drawing sets and specifications, but also – more importantly – how to control progress in a way that the drawings wouldn't separate themselves from real-life requirements, the students had no experience with.

But, these experts were recipients as well, by acting as the students' counterpart for the students' customary efforts to push the envelope. While often in this process the solutions proposed were quite ingenious, the professionals quickly identified real-life obstacles. An ongoing exchange of positions, with the development of more detailed drawings and improving argumentation skills, was the outcome of this early stage.

TESTING

The second learning factor was related to testing the discussed materialization ideas, translating the drawings in haptic full-scale mock-ups. While in the early stages of design the students were forced to compare their goals: "What is conceptually necessary?" to their capacities: "What is realistically possible?", they later discussed regularly as a team what mock-up versions they wanted to realize: "What can be learned?" and "Is it worth the effort?" Many of the sub-systems involved in this project went through several materialized re-iterations, large-scale models or mock-ups, with a methodic analytical process and the gained confidence in concrete materialization skills. The mock-ups functioned as the necessary training phase to later build the house in full. Craftsmen and contractors offered their expertise repeatedly. During all the design phases it was clear that the students were responsible for physically constructing most of the house themselves. Thus, the advice was taken seriously.

While in the beginning, the students' judgment on materialization processes was sketchy, the mock-up experiments as well as the discussions with the professionals made them much more confident and balanced. In the course of planning advancement, not only were the advantages of the preferred solutions emphasized, but now also disadvantageous viewpoints were addressed. This was crucial, since the design had to stand its reality test. In contrast to the drawings, where you can cheat to present a seemingly working result, with the mock-up method there was no space for excuses. The solution had to work!

Due to this process, students were aware of crucial production steps and necessary check-ups;

5.10

5.11

5.9 Construction as teamwork – framing the walls of the Roofpod core.
5.10 Reviews with professionals and faculty: Technical advisors explain and listen to guide the design development.
5.11 Feed-back on 'building block' mock-ups: Refining the design with the insights of specialists in related fields.

it is clear that they also perceived the finished result very differently than if they had been uninvolved bystanders. In the coming decades of their professional activities, they might have a stronger oversight and communicative potential within a construction team.

However, these skills didn't come from the motivation of learning technology alone. It was a by-product of wanting to build an extraordinary house for New York's future.

TEAMWORK

One of the most valuable achievements of this process was based in the discovery of the students teaching themselves, instead of being taught by their teachers. The discussions among the students were, in hindsight, at times more productive when they were alone, since the level of knowledge was at an equal level of development. Thus, they felt more comfortable in expressing their opinions.

There were always follow-up discussions at official studio times, so that unfavorable decisions could be corrected. Their sense of competency and social identification within the team grew clearly. The culture of explaining how and why decisions were conceived, what factors were evaluated, and in what hierarchy and interconnections those factors were ordered, became more sophisticated through the repeated group interactions.

While during the beginning of this phase, the students followed instructions to work on different research components, after a while they came up with their own assignments and distribution structure of work. They quickly realized that the scope was too extensive to attempt the design individually. Leaders crystallized according to their level of interest in certain topics and based on the expertise of professional backgrounds. Social skills were important, too. Friendships grew, often related to teamwork activities within the project's infrastructure.

The self-organization indicated the growing identification level: the passionate discussions about design priorities demonstrated that the students took their design task seriously. They were highly motivated, mainly due to the prospect that what was drawn would eventually be built. They wanted to get the materialized version as right as possible!

RESEARCH

The sense of place within the team triggered follow-up research: students were eager to mark their territories by developing expertise. They often engaged in their private, self-initiated investigations. After weekends, students reported on their findings: One student, who was assigned to draw an initial distribution proposal of the radiant heat system, had studied several books in her spare time. When discussing the layout, she explained why certain solutions were preferable: The maximum run length for pipes to transmit heat evenly, possible ways of reconnecting the system's two transportation modules, and the related size and location of the manifold. There was a clear sense of responsibility for her part of the project.

5.12

5.13

5.14

Another student studied our sponsor's curtain wall catalogue; hundreds of pieces and order numbers. He memorized everything, in order to come up with a drawing that would not be "made up", as he commented, but was based on real conditions.

There are countless instances that could be listed to show the students' active role in material- and product-based research. Even though the pace of progress slowed sometimes down due to this research, the eagerness to get the design resolved without flaws, no matter how long it takes, was a productive attitude. It is one that you usually do not develop in academic environments but that is common to offices. Complex problems have to be processed in an organized social setting. It is usually based both on team interaction and individual research. This vital part of production must be practiced in school to prepare for the challenges ahead.

After spending several months on the project, approximately 50 students (architecture and engineering) had figured out how to best work together. They were more frequently joined by professionals supporting the school, and from our college's facilities department: electricians and plumbers, carpenters and experienced all-rounders. A general contractor, an alumnus and long-time friend of our school, helped by providing regular visits from his best foremen, and later on, assisted with the logistics of procurement and construction coordination. Other contractors, electrical and plumbing particularly, attended regular meetings, to help transition theoretical drawings into practical strategies. Similar to the earlier support from our technical advisory board during the design phase, the construction tandem of students and professionals worked well: first we got hands-on instructions if best practice, then the students took over and completed assignments. During the construction, we depended heavily on the MEP-related installations. It became clear that what is visible in architecture is in the end only part of bringing the design to a functional resolution. Management skills, especially related to ordering a smooth construction sequence, were experienced first hand.

SUBSYSTEM ORGANIZATION

The strategy to define early on the most important building subsystems was not only helpful in understanding the assembly in its composition, it also delineated the responsibilities of student teams: The building was thus several times drawn in phases of its construction (4.2). By not only showing the end result of an intended appearance, but the path of how to get there, tasks were identified, and distributed – with shifting priorities – to prepare for the crucial logistics of construction.

The main teaching coordination to manage the specialized groups, was related to time management and coordination, buildability reviews, procurement assistance and integration between the systems interfaces. The time coordination was especially crucial, in order to have necessary parts in place for construction. The students' processing capacity to develop strategies in producing shop drawings and details, spreadsheets and order lists that had to work in real-life contexts became much stronger over time, especially with the input of professionals.

The students identified strongly with the subsystems they developed. The pressure was apparent, especially during construction, if for instance related progress of another group made their own work cease. The need to understand the sequence and the interdependencies, and develop step-by-step strategies for a smooth process, was recognized early on. It surely added a new dimension to the way the students looked at architecture.

Interdependent work had to be coordinated. Fitting the design pieces back together to anticipate the construction sequences was thus part of the teamwork, too. The skills learned were clearly different to those in ordinary design studios; so was the structure for the teaching and monitoring of progress. The outcome was particularly satisfying, since the growth of skills present at the project's completion, was related to the real-life requirements of comprehensive design thinking.

The typical prejudice of professionals: that students usually don't learn in school what it takes to fit into professional contexts, can be proven wrong by adapting teaching methods, implementing materialization and project team work in the everyday curriculum. Combining the common design skills, usually taught in school, with the techniques of hands-on building and coordination is advantageous for a real-life design process.

Competent communication skills based on anticipation make the transition into a professional workforce easier and more efficient.

The experience of touching and perceiving the logistic path of construction, based on what had been drawn before, were pivotal incentives for most students to engage in this project. The improvement of collaboration and communication were the main consequences, both of which are vital to developing architectural processing skills in a professional manner.

5.12 Weekly meeting of 'sub-teams' involved in interior finishes, lighting and MEP installations to discuss the required design integration and assembly strategy with their professional supervisors.

5.13 Investigation during fabrication hours: Different kinds of methods were studied hands-on, involving the experience of professionals. Sketches complemented these studies; afterwards, these sketches were the basis for correcting the set of construction drawings.

5.14 Design team after phase 1 (construction documents and mock-ups): Completed large scale model based on the preliminary construction drawings, with a 'building block' mock-up study in the background. By combining different media and research scales (drawing, building, product investigations), the design result was more and more refined, considering comprehensive factors and ensuring constructability.

3

'ENDOSCOPIC' AND 'ECTOSCOPIC' DESIGN FACTORS

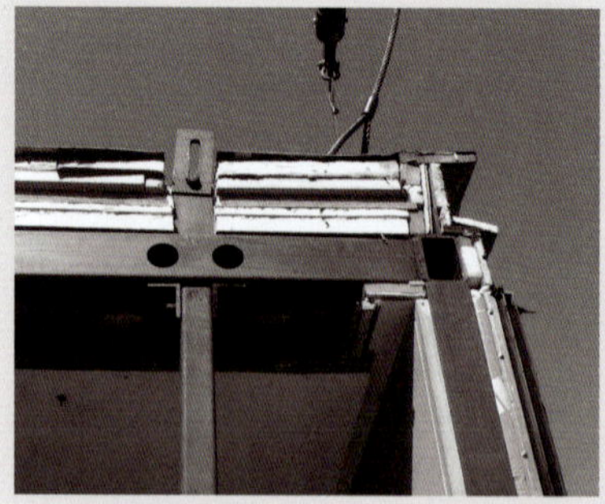

5.15

Compelling design pedagogy always consists of various factors. An extensive set of socio-cultural and political circumstances surround our daily lives, sometimes barely noticed, but with a great impact on our self-conception. What is normal, or acceptable? When traveling to other countries, or when comparing our lives to the lives of our parents, one realizes that certain conditions are significantly different. When confronted with difference, one will often take a stand on what seems better or worse. When better ways are discovered, the next question asked is: What should be normal?

It is noticable that students are receptive to these questions, due to the fact that they are on the look-out for how to fit in this world, and how to make a difference in it.

'ECTOSCOPIC FACTORS'

I call the every-day-life conditions surrounding us 'ectoscopic' factors: they shape part of any design narrative, such as that for our Solar Roofpod project. For example, the city's PlaNYC agenda was well known by many of our students before we started. They were interested in the strategies developed by NYCity agencies to transition our urban lives for a better future. The everyday-life factors are helpful to develop an intellectual vision: guidelines for a design undertaking related to how we understand and wish to change our world. However, they stay – like most visionary approaches – rather daunting and abstract, since they are dependent on complex and uncontrollable economic, cultural and political mechanisms. These are diverse: neither easy to analyze, nor to influence instantaneously. They are also interpretative and subjective.

Can design change the way we live? It is debatable every artifact can have an impact on socio-cultural conditions. However, if design embraces the complex nature of the 'ectoscopic' world, and if it can meet the requirements of superior product design (3.3), it will change our culture. This has been the case with transportation and communication devices. And, it might be the case for energy-related inventions, which are currently being discussed more and more.

However, an idea cannot change the world without its competent implementation. The skills needed to create something that is beautiful, practical, efficient, inexpensive and long-lasting are related to understanding artifacts in detail.

'ENDOSCOPIC FACTORS'

A second set of factors, identifying the hands-on prototyping aspect of design-build projects, I call 'endoscopic' factors: Intertwining requirements of aesthetics, available material sourcing, assembly methods, load calculations, structural conditions, material properties and performance criteria, integration of mechanical and electrical subsystems, budgetary and time constraints of the physical realization: a completely different realm of architecture, seen almost as though under a microscope. These factors are dependent on pragmatic research, material-based and mechanical thinking and the skills to synthesize and intertwine available methods and systems.

Even though there is also a subjective aspect to it, the possible options of solutions are more limited due to the many concrete immanent restrictions based in materials. Even though this realm is

5.16

absolutely necessary to materialize the imagined without the echo of the larger-scale human perception, of putting the experience into a larger context, the 'endoscopic' detailing process would be useless.

SYNTHESIS

When both of these parameters come together, especially when the ephemeral 'ectoscopic' and the immanent 'endoscopic' design factors in our project were reflected against each other, a learning process was started that stimulated real life design processing, based on practical and intellectual, cognitive skills. Students responded very well to this overlap, because it gives the power to physically articulate a picture of a better world. The fantastic but at times daunting world of vision develops the capacity to become real. Many students were attracted and extremely motivated by this particular prospect.

The project calibration within the comprehensive forces and intentions improved the project in an academically unusual way: While in a more common design studio project a linear process ('this is what I wanted') is articulated by means of a narrative, here, in a design-build context, a rather negotiative set of rules ('this is what is after consideration of the circumstances feasible, or manageable within our capacities'), is transferred into a real life setting, common to the practiced profession. In our repeated detailing process, more and more comprehensive reiterations moved the design towards the realizable materialization strategies.

The teamwork aspect in the refinement for a more feasible solution was one of the deciding pedagogic tools. Although the same starting position was presented to the students, the subjective interpretation of priorities triggered heated discussions of, for example, superiority of aesthetics over performance – or vice versa –, or of neglecting budget, code and time constraints. The friction between viewpoints contributed to refinement of the design and to problem-solving skills. The students were, for instance, forced to take a decision either to fundraise, or invest their own labor to achieve the aspired level of quality and detail. They understood very quickly that it takes both. The learning process of combining the means of possibilities empowered them more and more to come up with prudent decisions.

Equally, they were visibly motivated by the conjunction of visionary narrative and necessity of materialization. The fascination that the abstract part of the idea contained within the project, manifested as tangible evidence of their creativity, drove the process of sometimes cumbersome material learning. This learning process was of course physically more straining than any common studio design project. However, the final tangible result and the retrospective knowledge of the process were the intellectual rewards. The outcome contributed to a can-do attitude and confidence in processing skills of the multi-headed nature of design-build projects.

The pride of achievement was most detectable at the public exhibit in Washington DC (5.15, 5.16). The students' explanations to visitors combined both 'ectoscopic' and 'endoscopic' content. They were aware of the fact that they now were in a position to explain their vision by means of their own materialized artifact.

5.15 Rigging day: Interior composition of the Roofpod revealed; construction as a means to an end.
5.16 Public exhibit in Washington DC, National Mall: Students explain their reasoning for materialized design decisions; interior.

5.17

5.18

5.19

5.17 - 5.20 Several team-building activities during construction.
5.21 Proud 'owners' of the Solar Roofpod.

5.20

5.21

4

CURRICULAR IMPLEMENTATIONS AND LOGISTICS

In the beginning it was difficult to enthuse many of the students about the Solar Roofpod design-build project: too many uncertainties were apparent, and there were no design-build precedents to study, at least, in our school's history.

The insecurities were closely related to concern about investing an entire year of study in a high-risk project; high-risk, because the scope of work was so large that one certainly was dependent on other people's manpower. It was also clear that many extra hours would have to be spent. And many students were concerned about exposing their lack of expertise in Construction Technology, and the consequences that would have in the building's appearance.

DESIGN STRATEGY

Often design students solve realization issues, – if they have to come up with a drawing that shows how something could be actually built – by copying standard details from one of the established technology books, without addressing the specificities of their individual design. An in-depth investigation of how materials and methods could be used to help the expression of a building in detail is thus avoided. This is unfortunate, since there is a lot of creative potential in full-scale design investigations. Materiality can best be discovered in full scale.

The neglect can lead to a mindset built on the assumption that someone else will solve the issues of materialization, which have not been addressed in the drawing, in the real world. The students' explanations of their projects in reviews sometimes suggest something similar: the opinion is every so often that the drawing is indeed already a project in itself. There is no intention of thinking about its translation into the material world. However, during actual builds, this very material interpretation is one of the main challenges in architecture.

The students who decided to join the team from the beginning were generally attracted by this perspective. They wanted to study in a learning environment closer to the professional world. "It's like starting your career early – while you're still in school", one student said.

The Roofpod students increasingly understood the essential challenge of the architectural translation problem, from its 2-D design to its 3-D built condition. They overcame their technical anxieties by addressing and processing builder's issues themselves through experiential learning. They were in charge of defining what is feasible for a build. Drawings became more realistic and technical, and models and mock-ups complemented this process over the time of the project. In the beginning, when I asked the students to explain what their decisions were based on, often their answer ended with: "But, it's up to you!" However, I insisted: Since we will build our house together, we will also have to decide on the effort and methods together.

One of the essential strategies was to use drawing primarily as an abstract tool for the preparation of construction. However, in order to closely tie together the technical means with the aesthetic appearances, these drawings always had to be complemented with the elevations and renderings of the same detail in a spatial context. By building mock-ups based on the detail drawings, diverse

5.22

references between intention and outcome were available as an interpretation tool. With this method, explained in chapter three in terms of prototyping the 'building block', the design was processed both as a Design exercise and a Construction Document workshop. At its center was comprehensive negotiation, not only between the multiple design factors, but also between the academic disciplines of Design and Technology.

ORGANIZATION

Over the course of an entire academic year, the schematic design concept, initially submitted to the Department of Energy after 5 weeks of work, was translated into the finally built project.

Due to the complexity of the task, several courses were involved in that translation. At the center of activities was a 16-student-strong Design Studio, meeting at least twice a week, which acted as the coordinating link to the other courses.

In that studio, general project deficiencies were monitored. We tasked research assignments, and developed technical solutions and coordination strategies for other courses. The studio had constantly changing sub-teams of about 2-4 students working on separated design problems and subsystems. Initially, overview drawings (mostly 1/4"=1'-0"), with extensive dimensions, and related to a common coordination line grid system were drawn. This grid system was important to coordinate between the different teams, to be able to tie the separated tasks back together. Later on in phase 1 (fall semester 2010), key details were developed. These details became increasingly interconnected: the façade with the trellis system, then with the landscape system, including the access ramps. The wall section drawing (see: p. 278) was the summary of this investigation.

Simultaneously, each student had certain duties related to spreadsheet takeoffs, order lists and preparations, schedules and cost estimates, which I double-checked regularly.

A BIM (Building Information Model) computer model was built by designated students. That information platform tied most of the previously separated

5.22 Design-build 'living classroom': During construction, the detailed drawings were pinned up in the house, so that students were able to compare the technical representation with the final result.

tasks back together, and generated the Construction Documents. The 1/4"=1'-0" scaled drawings we initially started with, became more complex during the course of this process, and were increased in scale (1/2"=1'-0").

The students were also involved in in-kind fundraising efforts. This was necessary not just to secure some of the necessary capital; often, they fell in love with certain (not easy to pay for) products: fixtures, hardware, linoleum, cork, specialty glass, bamboo paneling, etc. In order to make them realize the monetary constraints, they had to contact manufacturers, obtain quotes, ask for installation details, and eventually also for donations. With this approach, we were able to receive interesting products for free, or at reduced costs.

At the same time, in an additional elective, they were shown how specifications are written and how project manuals are prepared. They researched material information regarding its environmental impact, they learned to understand the ways projects are usually organized, and had to deliver their own project manual as one of the Solar Decathlon deliverables. Also, in coordination with the college's development office, major monetary donors were approached: the students had to give convincing presentations. In total, about forty presentations were given this being a good opportunity to prepare the students to guide their house visits during the Solar Decathlon event.

PHASES

The entire project can be differentiated into four phases (5.23): The initial phase (0), was dedicated to developing the schematic design, with the finished presentation board and a ¼" scale model as the final result for submission to the Department of Energy's Solar Decathlon (5.23).

The following three phases (1-3) were then related to the design-build pedagogy described, with the final construction and the commencing exhibit in Washington. In phase 1, the fall semester, mockups and materialization strategies in initial technical drawings were developed, finishing with about 90% of a preliminary Construction Document set and a clear understanding of how to build many subsystems of the house.

Phase 2, the spring semester, was dedicated to completing the missing details, and improving the weaker design elements until the Construction Documents were sufficient to support the upcoming construction.

Presentation drawings were also continuously updated. Renderings were done to connect the technical drawings to the appearance envisioned in a spatial context.

Now, an additional 16 students worked in a parallel studio on explaining the project in graphic representations, emphasizing particular aspects of the design (see illustrations in chapter 2). These drawings connected the now more refined Roofpod back to the original conceptual intentions. It was possible to observe how the design had improved, and had become more efficient in terms of its performance and its constructability.

Most of these drawings were used for a brochure, which was distributed at the public exhibit in Washington to thousands of visitors. The communication design was a vital part of the project. It represented the public relations that are necessary to launch an idea in a broader scale. Many subsequent publications and websites used these images to feature the Solar Roofpod.

At the same time, in early 2011, we started production on some of the do-it-yourself prefabricated items we had developed in the semester before: the cabinets, the ramp railing and other steel custom elements, such as the two porch platforms, a canopy to the main entrance of the house, and an NY subway door to be fit into the mechanical room doorway. Also the landscape modules for the 'Roofpad' around the house were begun at that time. Due to the large amount of modules, this work continued well into the summer.

SHOP - LEARNING

Prefabrication shop activities operated in parallel with the studio: I was able to convince two friends and professional long-time craftsmen, owners of a steel and a wood shop, to teach the students their crafts in two electives during phase 2 in spring 2011. These classes were held in their shops in Brooklyn, in the middle of the week.

For many students, these classes were the high point of the project. They were exposed to the real techniques of fabrication, which are not often taught in school. These experiences were also important to prepare the students for the upcoming construction phase. Most significantly, the students came into contact with materials. They developed their own know-how about fabrication. They were able to transfer the knowledge gained into their drawings. The details improved immensely: the students had made a connection between the fabrication they had experienced and their own craft of drawing.

SOLAR ROOFPOD

0 STUDIO: CONCEPTUAL PHASE

SCHEMATIC DESIGN

COMPETITION BOARD
SMALL SCALE MODEL

---

1 STUDIO: PROCESSING PHASE

DESIGN DEVELOPMENT

90% CONSTRUCTION DOCUMENTS
MOCK-UP PHASE
- —BUILDING BLOCKS,
- —ROOFPAD, RIGGING

LARGE SCALE MODEL

ELECTIVES

—LANDSCAPE 1 CONSTRUCTION DOCUMENTS
—METRICS RESEARCH
—MATERIAL + SPECIFICATIONS
—ENGINEERING*

---

2 STUDIO 1: DESIGN-BUILD

100% CONSTRUCTION DOCUMENTS
DETAILING
FABRICATION PHASE
- —BUILDING BLOCKS,
- —ROOFPAD

RENDERING REDO

STUDIO 2: EXHIBIT DESIGN

RENDERING REDO
GRAPHS/BROCHURE DESIGN
COMMUNICATION / WEBSITE

ELECTIVES

—LANDSCAPE 2 Mock-Ups And Framing Roofpad

—STEEL: SHOP-DRAWING + FABRICATION
Railings
Porches
Subway Door
Entry Canopy

—WOOD: SHOP-DRAWING + FABRICATION
'Core' Cabinetry
'Building Blocks'
Bathroom Panels + Vanity
Interior Doors

—ENGINEERING*

---

3 SITE CONSTRUCTION

BUILDING ASSEMBLY

*) "SENIOR DESIGN" (ENGINEERING CURRICULUM)
TEAMS:
- —Electrical Design + Photovoltaics
- —Mechanical Design + 'Solar Cooling'
- —PCM Tank/Thermal Storage
- —Monitoring + Control System

5.23

5.24

5.23 Time schedule divided into four phases: Phase 0 was a typical ca. 5 week studio assignment; Design-build phases 1-3 integrated hands-on activities into design thinking, from full-scale mock-up research (1), to prefabrication (2), to site construction (3) during the summer.
5.24 Initial model, schematic design phase.

5.25

5.26

Many of the students were amazed by the possibilities they observed during shop fabrication. They were exposed for the first time in their lives to professional fabrication methods.

In the woodshop class, the students had to develop their shop drawings for the cabinets forming the 'core': they were advised by the master cabinetmaker on what to change (and why to change it); when these explanations were not clear, the three-dimensional explanation followed in the form of hands-on demonstrations in the shop.

Different material properties and techniques were explained. For example, a piece of solid wood was measured at the beginning of a class, put in a water bucket, and measured at the end of the class again. The measurement demonstrated not only that the piece had expanded, but also that the expansion varied in different directions of the grain. This demonstration was of course more interesting than the dry theoretical explanations usually done in Construction Technology lectures.

Most popular was the 'What is it?' game, where a bucket was passed around with many different pieces of spare hardware, found in the shop. Students had to explain in what application hinges were used, and how the pivot point would work. They then had to do quick sketches of its installation in a cabinet context (5.25).

They also had to come up with their material orders and execute cutting lists to use the sheets of plywood and MDF as efficiently as possible. Later on, after the shop drawings were completed, they witnessed how their drawings materialized. Not all the work was fabricated by the students. Not only was there not enough time; the missing experience would probably also have slowed down the process. The students were responsible for all the rough cutting of the door and cabinet panels. They also laminated the bamboo paneling onto the cabinet doors. They helped to build three sliding doors, one of them with custom glass cut-out. And they fabricated the plastic paneling for the bathroom and the 'building block' elements in the shop in Brooklyn. They were also involved in the cabinet installation on site, later in the summer.

The students did not fabricate the cabinet carcasses, since too much work had already accumulated. Most of the finishing was also done by professionals, so that the result came out as desired. However, all phases of fabrication were at least demonstrated to the students, so that they were able to comprehend the full scope of getting from their initial drawings to the final result.

Students started the steel shop class with shop drawings but not to the same extent as in the woodshop class. Instead, the students learned many of the different material techniques to build i.e. their custom railing system, for example. They learned of the necessary fabrication steps, which are sometimes not visible in the drawings. After they were shown the processes, they had to separate into groups, and form their own assembly lines.

Since the sixty posts were based on repetition, jigs were used in order to facilitate production. First tubing, plates and attachment angles were cut, then screw holes and the holes for the tension wires were drilled. In the meantime, the first welding training had taken place. All the post pieces were finally assembled, grinded and polished. Again, the experience of materializing what had been thought on paper, by comprehending all the necessary steps through the physical process, was what fascinated the students. Most of them would agree that their understanding and appreciation for the crafts had grown through this activity. The lines in their drawings had now more meaning.

5.27

5.28

5.25 'What is it?' Hardware studies at the woodshop in Brooklyn. Preparation for cabinetry shop drawings.
5.26 Woodshop, Brooklyn: Hardware installation for the 'core'-integrated murphy bed cabinet.
5.27 Woodshop, Brooklyn: Cutting list and order list preparations; shop drawing coordination.
5.28 Brooklyn woodshop: cabinet panels and bamboo veneer had to be cut to size.

5.29

5.30

5.31

5.32

5.33

5.29 - 5.32 Woodshop, Brooklyn: The bamboo 'core' veneer had to be laminated panel by panel. Students learned the techniques involved in producing what they had drawn.
5.33 Opinions regarding 'true' detailing were passionately discussed during woodshop classes.
5.34 Woodshop, Brooklyn: Plaining handmade ventilation grille for 'building blocks'.

5.34

5.35

5.36

5.37

5.38

5.39

5.40

5.35 - 5.40 Steel shop, Brooklyn: The initial "assembly line" production tactic for the railing posts became more and more a material investigation into developing customized design and fabrication skills. Welding was of particular interest.

In addition, an additional elective for the preparation of the landscape around the house, called the 'hardscape' and 'softscape', was created. Colleagues specialized in landscape design taught this class. About a quarter of the students were also enrolled in my Design Studio. Any design changes were thus always reported immediately, and integrated into the overall drawings and in the complementing documents.

Different detail approaches for the transportable modular 'landscape units' were developed. Later, this elective produced their own mock-ups to evaluate the best option. Small mistakes that wouldn't have been obvious in the drawn format, were caught before production started. The entire area surrounding the house, dubbed the 'roofpad', was produced with the help of our contractor's foremen. Here too professionals gave the initial hands-on instructions until a workable solution was derived. Then, the students were put in charge of production based on the refined mock-up prototype and the remaining units.

Most of the design studio students eagerly joined these electives, which were restricted for enrollment to just that particular year. Waivers had to be signed to ensure that the college wouldn't be liable for any mishaps happening off-premise, based on the unusual scope of work. Fortunately, no injuries happened. We also completed a 10-hour OSHA (Occupational Safety and Health Administration) training course for injury prevention and safety during construction. Every student involved in constructing the house had to complete it. There were three student safety officers on site and they took their responsibilities seriously. Whenever they detected risky behavior or missed safety precautions, they addressed them right away, so that everyone present would learn to avoid future mistakes.

This was at times not easy, since in summer, with temperatures of up to 100 °F and high relative humidity, wearing hard hats and construction boots was not necessarily very comfortable.

When construction time had finally arrived, in mid-May of 2011, we were well prepared. The students had developed better team working skills, and knew how to apply an efficient organizational structure. Also, expedient progress of construction was somewhat secured by the many prefabricated items that had been produced previously.

Many of these items were stored in an old firehouse on campus that the facilities department had allocated to us. That space became both our 'factory' and 'warehouse': All the envelope 'building blocks' and the landscape modules were assembled in that space. In late spring, many additional items that we had ordered ahead of time, to make

5.41 Finishing of the 'building blocks': An air-conditioned environment was advantageous. Therefore, the blocks were finished where their development had started, in the design studios.

sure that we would not be held up by pieces missing in the supply chain, were arriving, too. The glazing, the solar panels, the thermal collectors, the bi-fold doors, the curtain wall system, and cork flooring, to name a few, had to be stored in a way that would allow for easy access.

In the meantime, the delicate 'building blocks' took posession of the studio spaces in our school building. Not only were we concerned that they would be damaged in the crowded firehouse, also the air-borne dust and humidity of the firehouse did not contribute to the finish that we desired.

This meant that in the end the classroom, which was the starting point of the 'building block' prefabrication project when experimenting on the mock-ups, had also become the place of the last action before they found their place in the Roofpod.

PREFAB - CUSTOMIZATION

The idea of prefabricated customization as an academic concept in design-build projects is advantageous, since it helps to prepare for construction long before it is supposed to start: Initial units can be developed in a timely manner, as both drawings and mock-ups. The mass production of the final design can contribute to the training of hands-on skills as a second step. The work on the construction site then becomes, as the third step, less overwhelming. However, the accuracy of these prefabricated items is essential. The management of dimensional tolerances is not fully understood early on, so that the outcome of inaccurate work can really jeopardize the entire production. The coordination of this aspect is thus essential.

With the described approach, we were able to physically construct at least ¾ of the scope of work ourselves, in the summer of 2011, under the weekly supervision of our contractor and coordination meetings at the start of each week.

The work executed by the students included basic framing and drywalling, all layers to configure the building envelope (walls, floor, roof), metalwork, woodwork, fine cabinetry, window and door installations, cork flooring, railing assembly, lighting fixture installation, and countless other tasks. The engineering students were able, under supervision of master electricians and plumbers, to do significant parts of the wiring, plumbing and mechanical installations by themselves. Where it was for safety reasons mandatory, the supervising professionals completed the tasks. Afterwards, all students knew not only the anatomy of their building in and out, they also knew what skills and efforts are necessary to get from the drawing to the built artifact.

With the deep knowledge the students accumulated during all phases of construction, it is clear that they understood this building differently than if they had only be involved as bystanders.

They experienced the structure of subsystems as an ordering principle. They knew that the coordination between trades and systems, in order to reintegrate these planning segments into a whole, is one of the key considerations of any project. Sequential planning and anticipation is needed to be in control of the assembly of an artifact.

The intellectual progress in anticipation skills had been extraordinary. To me, it was an eye-opening experience. Design has to be approached from both angles: the big picture and superficial form that has to be analytically deconstructed to understand it better; and also in terms of the countless number of pieces as a structural composition that has to be physically assembled to find an accurate formal explanation.

Therefore, architectural education should include tangible materialization as an inherent part of the curriculum.

5.41

INTEGRATION

How 'design-build' projects should be integrated, at what place, and in what sequence and distribution in the overall structure of architectural education, to make it a viable addition to the curriculum, is difficult to say. Our Roofpod project, although innovative as an approach for design teaching, was, in its scope, costs and involvement of faculty and resources, too big and too extensive.

The students that joined the project when it was already in the schematic design phase, spent about one year working on this one singular design. Practice on different programmatic and typological design problems – housing, institutional buildings, landscape and urban design is however also necessary to rediscover design processing from different viewpoints.

Furthermore, later in the curriculum, the combination of these different viewpoints in one design problem was also advantageous. It might even be possible to integrate material thinking in several of these approaches. Similar design-build projects approached in the future should be limited to a shorter development period: An ideal sequence would foresee one semester, preferably spring, for the preparation of construction documents, mock-ups and procurement, while the following summer would be used to do the physical construction, prefabrication and the completion of on-site assemblies.

As an alternative strategy to intertwine the design-build project with other research, for instance on an emerging fabrication technique or even the evolution of fabrication over its history, a complementing elective would prove advantageous. The objective is again to develop different perspectives on interrelated topics.

The main question of broadening the impact of 'design-build' projects is related to its interrelation in the full curriculum: What is the right year to integrate this kind of material thinking, related to psychomotor learning? At what stage of the students' intellectual development should the exposure happen? How can one develop a holistic picture of courses, coordinated with the school's faculty, bringing the message of the importance of craft skills across within a connection to other classes?

The initial discussion at a faculty level should focus on the organization of theoretical and practical design processes as a sequential problem. In what way could theoretical and practical skills complement each other in the students' education, involving faculty from different teaching fields? The strategy of 'ectoscopic' and 'endoscopic' factors could negotiate between the theoretical and practical influences, since the positive friction between them seems to generate a particular motivation in students. Critical thinking becomes more performative by combining different thematic and personal inputs.

Haptic realization skills should be acquired as often as possible in an architectural curriculum. I feel that the preferred spot for a design-build project, so that it doesn't become a dry construction management exercise, would likely be at the end of the academic sequence, when advanced comprehensive thinking abilities have evolved. The students must be able to process diverse inputs simultaneously, without losing track of design qualities. Many generative factors, from an artistic approach to the research on available technology, structural requirements or performance criteria must be included in this investigation.

Critical thinking evolves after processing multiple generative factors. The development of the students' design skills should start with basic design drivers, addressing space, light and context conditions. In later years more and more comprehensive criteria can be integrated, related to

5.42

5.42 Website: Virtual presence. With two students from the art department, our students were able to come up with their own website design. Updates were posted frequently.

5.43 + 5.44 CCNY events: At the end of construction, a farewell party was hosted, inviting all sponsors, the press and students with their families.

professional design practice and the ramifications of everyday processes, i.e. client input or budgetary constraints. Within a multidirectional framework, students should be able to define their specific architectural interests in later study years, and trace them against the experienced design-build reality. They would therefore be exposed to a similar professional skill set that they will need to adapt to in their future careers.

Practitioners preferably alumni from many diverse backgrounds, such as construction management, real estate, policy making, sustainability, history and theory, industrial design, etc., could participate in combinations, which of course would have to be tailored to the particular topic investigated.

The purpose is to have those reviews and advisory meetings identify design weaknesses by connecting the conceptual parti and realization proposal to all sorts of real requirements and refine the approach by offering mature expertise from the professional world.

This would generate a way of thinking, where design-build is understood as a negotiative process, in which the answers given are compared to the reality-checks brought in by these advisors. The final resolution of the project, interpreting, questioning or even provoking these real conditions would still be in the hands of the students, with the same motivation of developing their power by relating to the 'ectoscopic' realm by means of materialization.

The conscious intellectual confrontation of idealistic and practical ideas, traced against a concrete design-build artifact, would surely develop self-awareness of the role and skill set necessary to be instrumental in the field of architectural making.

As in other professions, i.e. the cabinetmaker's 'Meisterstück' in a traditional apprenticeship context, the testing of the comprehensive use of knowledge facilitates the transition into a creative and prolific professional occupation. In our particular case of the Roofpod project, it simultaneously prepared for interdisciplinary collaboration and social engagement.

Finally, it is important that the students realize their own shift towards a different (more physically productive) processing capacity. One of the main reasons why many students shy away from more practical efforts is the anxiety of failure, which is more apparent in built than drawn projects.

Instead, teachers should apply encouragement to overcome insecurities in the field of concrete materialization. 'Prototyping' – as material-based designing – should be considered research. This research clearly depends on originality: errors are part of the risk-taking process necessary to discover facts. This challenge should be approached

5.43

5.44

optimistically. Confidence leads to student momentum and future engagement. In the end, the students (alumni in spe) could hold the greatest power to change the hands-on culture of education, by transforming the attitude to create and preserve the built environment. In order to secure long-term success, permanent presence of the artifact as well as continuous social involvement are the best drivers to achieve cultural leadership.

FINANCING + PR

The high financial burden of design-build projects has often been criticized. It has also led to many cancellations before well-planned projects have even started. The pressure during the procurement phase of our project was severe. Without the generous support of the entire College's community, we wouldn't have succeeded. To establish a permanent program, several infrastructural provisions would have to be met:

The integration of PR strategies within the process of the program is mandatory. The fact that the 'Solar Roofpod' was advertised and sponsored with $100,000 by the U.S. Department of Energy, was highly advantageous. Its guaranteed public presence on the National Mall w/ 350,000 visitors, was persuasive for professional sponsors. Endorsements of the project by renowned organizations, such as the Clinton Climate Initiative and the New York Building Congress, were significant steps towards sponsor commitments, too. The involvement of the public is paramount to arouse interest with possible sponsors and shape long term commitment.

The scope of a design-build project does not have to be high-tech, but can be much smaller and simpler in scale. It is more important that the program is receptive to existing needs – and possibly existent funding sources – and is fit into a particular social context. In order to secure its long term success, it must develop a strong presence within its particular community. The anticipation of how the structure will be managed after completion is thus vital for the project itself, and for the program's overall reputation.

What all initiatives must have in common is a precisely defined communication concept, with a website, public events, exhibits and a publication presence as part of the design planning process. The efficient involvement of the college's administration for PR and the commitment to long-term funding must govern the efforts: more endowments dedicated to design-build pedagogy would be essential to secure a broader long term perspective.

5

OUTCOME

After completion of the Solar Roofpod project, I was convinced that it had changed the participating students' understanding of their future professions. It was obvious that their responses to questions of materialization and project management had evolved; but, I also suspected that with it, their general design attitude had changed, too.

SURVEY

However, my subjective viewpoint lacked proof: To get a better account on the students' preferred education goals, I conducted a survey within my department: A reference group of ca. 150 students, among them 25 of my Design-Build students, were asked about the most important skills a professional architect should have. Subsequently, they were then asked to compare these skills against their own academic preferences.

I set up the survey (5.45) in a way that 22 possible criteria were predefined. For each question, a maximum of seven answers was possible. The response could also be processed graphically, by taking advantage of a circular diagram, organizing the priorities.

The students also had to explain, why they thought professional and academic skill sets should align or differ.

At the end of the survey, I also asked what initial motivation started their desire to become an architect. I wanted to understand whether their initial understanding of this profession had influenced their priorities.

The surveyed students were mostly undergraduates. They had about 15 minutes to complete the form. The survey was conducted about four months after the 2011 Solar Decathlon had ended.

It became apparent that both my Design-Build students and also the reference group decided that professional and academic skills should actually differ, predominantly due to their interest in experimentation, and their identified need for more intellectual stimulation during academic studies. However, the two groups developed very different areas of emphasis in both their professional and academic foci.

The areas, in which the differences between the two groups were most pronounced, were of course most interesting, since they are most informative about different intellectual approaches and trajectories.

In the professional area – being an architect – the reference group emphasized 'Conceptual thinking' and 'Technical drafting (BIM, CAD)'; the Design-Build students identified 'Buildability' and 'Spatial thinking + form-making skills' as the far more crucial items (5.47).

In the academic area (being an architecture students), again 'Conceptual thinking' and 'Digital rendering' were emphasized by the reference group with a comparatively high percentage, while the Design-Build students defined 'Coordination among comprehensive factors' and again 'Buildability' as their preferred criteria (5.48).

Other items, where Design-Build students excelled, were 'Construction methods and related appearances', 'Implementation of environmental factors' (for both the professional and the academic field), while the reference group liked the subjects of 'Socializing skills and networking' and 'History of architecture' as their more favored targets. It is obvious that the Design-Build students were strongly influenced by their experience.

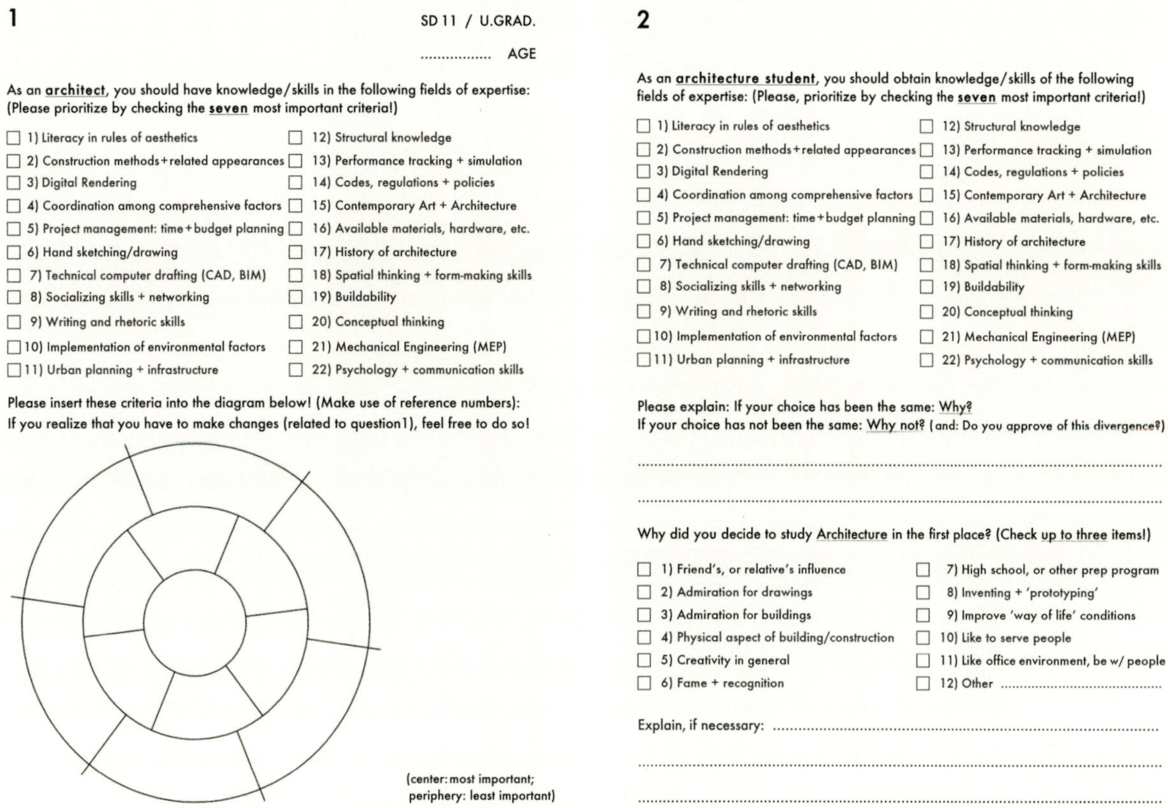

1

SD 11 / U.GRAD.

.... AGE

As an **architect**, you should have knowledge/skills in the following fields of expertise: (Please prioritize by checking the **seven** most important criteria!)

☐ 1) Literacy in rules of aesthetics
☐ 2) Construction methods+related appearances
☐ 3) Digital Rendering
☐ 4) Coordination among comprehensive factors
☐ 5) Project management: time+budget planning
☐ 6) Hand sketching/drawing
☐ 7) Technical computer drafting (CAD, BIM)
☐ 8) Socializing skills + networking
☐ 9) Writing and rhetoric skills
☐ 10) Implementation of environmental factors
☐ 11) Urban planning + infrastructure
☐ 12) Structural knowledge
☐ 13) Performance tracking + simulation
☐ 14) Codes, regulations + policies
☐ 15) Contemporary Art + Architecture
☐ 16) Available materials, hardware, etc.
☐ 17) History of architecture
☐ 18) Spatial thinking + form-making skills
☐ 19) Buildability
☐ 20) Conceptual thinking
☐ 21) Mechanical Engineering (MEP)
☐ 22) Psychology + communication skills

Please insert these criteria into the diagram below! (Make use of reference numbers): If you realize that you have to make changes (related to question1), feel free to do so!

(center: most important; periphery: least important)

2

As an **architecture student**, you should obtain knowledge/skills of the following fields of expertise: (Please, prioritize by checking the **seven** most important criteria!)

☐ 1) Literacy in rules of aesthetics
☐ 2) Construction methods+related appearances
☐ 3) Digital Rendering
☐ 4) Coordination among comprehensive factors
☐ 5) Project management: time+budget planning
☐ 6) Hand sketching/drawing
☐ 7) Technical computer drafting (CAD, BIM)
☐ 8) Socializing skills + networking
☐ 9) Writing and rhetoric skills
☐ 10) Implementation of environmental factors
☐ 11) Urban planning + infrastructure
☐ 12) Structural knowledge
☐ 13) Performance tracking + simulation
☐ 14) Codes, regulations + policies
☐ 15) Contemporary Art + Architecture
☐ 16) Available materials, hardware, etc.
☐ 17) History of architecture
☐ 18) Spatial thinking + form-making skills
☐ 19) Buildability
☐ 20) Conceptual thinking
☐ 21) Mechanical Engineering (MEP)
☐ 22) Psychology + communication skills

Please explain: If your choice has been the same: Why?
If your choice has not been the same: Why not? (and: Do you approve of this divergence?)

....

....

Why did you decide to study Architecture in the first place? (Check up to three items!)

☐ 1) Friend's, or relative's influence
☐ 2) Admiration for drawings
☐ 3) Admiration for buildings
☐ 4) Physical aspect of building/construction
☐ 5) Creativity in general
☐ 6) Fame + recognition
☐ 7) High school, or other prep program
☐ 8) Inventing + 'prototyping'
☐ 9) Improve 'way of life' conditions
☐ 10) Like to serve people
☐ 11) Like office environment, be w/ people
☐ 12) Other ....

Explain, if necessary: ....

....

....

5.45

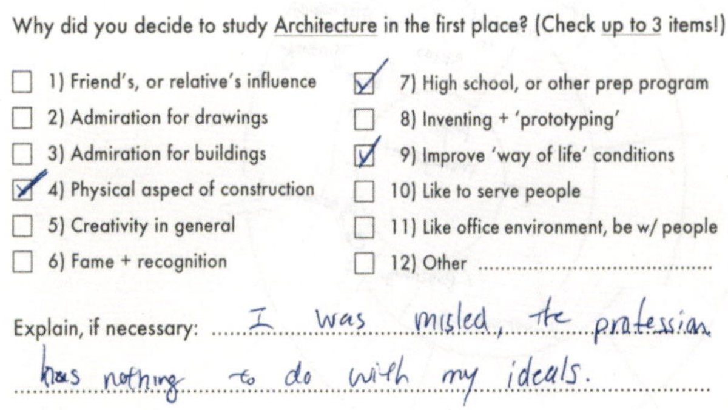

Why did you decide to study Architecture in the first place? (Check up to 3 items!)

☐ 1) Friend's, or relative's influence
☐ 2) Admiration for drawings
☐ 3) Admiration for buildings
☑ 4) Physical aspect of construction
☐ 5) Creativity in general
☐ 6) Fame + recognition
☑ 7) High school, or other prep program
☐ 8) Inventing + 'prototyping'
☑ 9) Improve 'way of life' conditions
☐ 10) Like to serve people
☐ 11) Like office environment, be w/ people
☐ 12) Other ....

Explain, if necessary: I was misled, the profession has nothing to do with my ideals.

5.46

5.45 Survey with 150 CCNY students: What priorities of skills and knowledge should you have as an architect, or as an architecture student? Why did you decide to study architecture in the first place? Comparison of 25 Roofpod Design-build students with 125 'regular' students, who studied a more common academic track.
5.46 Survey with 150 CCNY students: Why did you decide to study architecture? Friction between the ideal 'way of life' conditions and the physical aspects of construction. How can 'ideals' be realized?

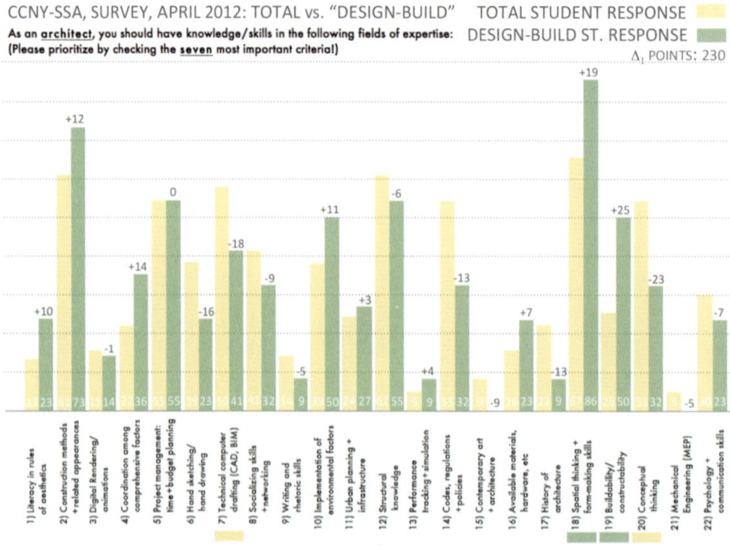

Table 1.1: Survey: Skills of an Architect; Reference group vs. "Design-Build Students"

5.47

It is also obvious that the disciplines preferred by the reference group are more theoretical, often emphasizing abstract means of representation. Their perception of the design process seems predominantly based on describing matters by two-dimensional, virtual means.

The subjects preferred by the Design-Build-students were mainly practically oriented, towards concrete production of three-dimensional artifacts, techniques learned as they developed the Roofpod project.

With respect to the materialization skills needed to 'realize the imagined', these students chose more design-transformative criteria that utilize both two-dimensional and three-dimensional means of investigation.

The processing methods lead to a more literate and creative practitioner. The more diverse the tools for analysis and processing are, the more complexity can be achieved. It becomes easier to connect the abstract world of drawing with the practical world of construction.

If design-build projects were instated as a mandatory component in architectural education, this might have in fact significant impact on the creativity level in the field of built architecture, where these transitions between drawing and building must take place every day, but not necessarily with the intention of creativity in mind. If the abstract conceptual world could connect to the world of materialization in a more sophisticated way, so that the transition between both is understood from either end, form and assembly, more customary systems would come into being. This would transform our built environment, which currently features more and more industrially stream-lined items. It would make our built surrounding more interesting and inspiring.

To promote these competencies, capitalizing on practical problemsolving and hands-on materialization, the implementation of design-build projects within the curriculum will lead to better results in built architecture. The impact would be much greater, since it is not the top percent of production that is targeted, but the broader culture.

It is clear that certain architecturally related skills can only be learned if experienced haptically, and motorically. The psychomotor domain associated with acquiring these skills must be integrated in architectural education.

It is necessary to process the complexities of assemblies both two- and three-dimensionally. The creative puzzles of assemblies cannot be learned by drawing, but only by doing.

CRAFT

There are many techniques for fostering a more material-driven architectural profession, but much of the work must be empirical and based on intellectual negotiation with craft. Opportunities in gaining these experiences must be given at an early development stage in education so that they can be seminal. They must start in school!

After our project, the students were very aware of the need to develop their material know-how, in order to process creative thinking and to materialize what one dreams of.

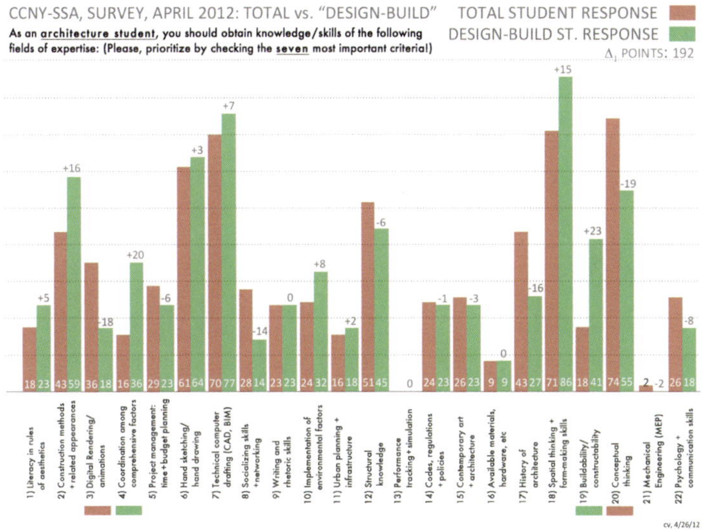

Table 1.2: Survey: Skills of an Architecture Student; Reference group vs. "Design-Build Students"

5.48

5.47 + 5.48 Survey results: Even though both Design-Build students and reference group think that competencies developed as an architecture student should differ from the ones of a professional architect, the focus on different fields of expertise is significant.

They changed the way they looked at built artifacts and how to communicate their observations. When for example looking at buildings while traveling in Washington during the Solar Decathlon, students noted details, and referred to them by describing the build processes or required tools. This materialization intellect, which is not visible on its surface, but instead within the deeper structure and process of the building, demonstrated their newly learned abilities best.

They were able to look behind the curtain, into the nitty-gritty of their profession. These skills are absolutely necessary to process architecture in a conscious and creative way. The more competently these transferences can take place, the greater the possibilities to create genuine buildings, beyond what is available on an industrial market.

Contemporary architects, if they want to be more than administrators of the building industry, should embrace artisanal qualities, be interested in craft and material processing.

It is necessary to communicate with all trades involved in creating a tangible product. In that sense, the market niche for architects lies in a more refined and more attractive product approach, with more knowledge and background information to make it happen.

COMPREHENSIVE DESIGN

Individuals are interested in comprehending the world around them. Students are especially motivated in that respect: They constantly try to explain the just experienced world to themselves. They are not interested in isolated product processing, but in the greater coherences of our culture. Design lenses[65] therefore have to be changed, and different ways of seeing the world have to be combined, to not oversimplify something that is complex in nature. The change of scale, of perspectives, of abstract and concrete resolutions, is inherent in the process of designing. It is a dynamic process of negotiation. Therefore, the system laid out in the survey has its weakness: the students were only allowed to check a mere third of the possible criteria, even though all of them are important to get to a proper synthesis.

The unevenness of the results is interesting, since it shows that certain criteria are overemphasized, while others are not addressed enough in our everyday teaching. The survey could thus be a permanent measuring instrument that shows what neglected areas need to be pushed forward.

Besides that, do all students really have to develop the same direction of expertise? Or, should the upper years of study programs be given space to build-out one's strengths, according to the particular talents that were individually discovered? Could different architectural tracks be articulated? There are many different but related scales and topics involved in implementing an idea. Due to the complexity of the architectural discipline, the field is dependent on teamwork anyway. Different angles on comprehensive thinking can expedite a process, if it is grounded in a common basis.

The survey also asked, what particularly brought about the decision to study architecture. In many statements, a friction between an envisioned way of life conditions and the learning of the physical aspects of construction is detectable. Both lenses, the big picture analysis and the small but significant processing abilities, need to be integrated in learning the skills of the profession. All information gathered must be processed holistically to generate a comprehensive design proposal. The processing abilities to do that efficiently, in a way that promises superior realization, must be one of our main teaching objectives.

Architects, in the movies and the media, are always known as the ones that lay claim to the intellectual property of a new building. They present themselves as the creators of the whole: not just the idea but, the materialized building itself. That is what fascinated me when I decided to become an architect. How, I wondered, is it possible to be so efficient? How can one develop organizational and artistic abilities to realistically claim such an enormous scope of work? It is not possible! The image is twisted: Although as an architect you are responsible for a lot, one of the main ways to be successful is to collaborate and continuously try to expand the horizons in 'how to make things work'.

This is not at all to belittle the architects' impact as an artist and a socio-cultural entrepreneur. Understanding the bigger picture is a significant talent. But, one cannot implement it with a solitary mind. Implementation according to changing cultural benchmarks is dependent on collaboration in order to successfully generate tangible paradigms. The most convincing response is a complex well-conceived artifact that is able to demonstrate its practicality, viability, feasibility and beauty.

It becomes clear that many factors must be involved when considering a design problem holistically, as an environmental problem, to improve the built and beyond it the cultural environment. However, the tangible beauty and elegance a building conveys is immediately related to the insights of know-how to produce it. The skills we have to foster are inherently craft-related to develop a superior quality in our work. If you want to claim a building as your intellectual property, you have to know what mistakes to avoid and what strategies to use when approaching the next iteration. The ability of refining a prototype is a long-term project.

If you don't know how it can be done, how can you claim it is yours?

The emotional rewards craftsmanship holds out for attaining skill are twofold: people are anchored in tangible reality, and they can take pride in their work. But society has stood in the way of these rewards in the past and continues to do so today. At different moments in Western history practical activity has been demeaned, divorced from supposedly higher pursuits. Technical skill has been removed from imagination, tangible reality doubted by religion, pride in one's work treated as a luxury. If the craftsman is special because he or she is an engaged human being, still the craftsman's aspirations and trials hold up a mirror to these larger issues past and present.

Richard Sennett
"The Craftsman", 2008 [66]

EPILOGUE

After the Roofpod was entered in the 2011 Solar Decathlon, where it was an inspiration to thousands of visitors, it has been dormant in storage for more than two years. It has thus yet to realize its terrific potential, since it was operated for only weeks in Washington. Our interest is in relocating it to a site where many visitors can experience its architectural qualities. We also intend to further monitor and measure real results, and subsequently tweak its performance.

For that reason, we are in the process of preparing the return of the Solar Roofpod back to the Spitzer School of Architecture's rooftop. We intend to install it in early Spring 2014 and make it a Center for Sustainable Projects, connecting professional and academic spheres.

In this setting, as a living classroom and a learning center, the Solar Roofpod will continue to communicate its idea. It is a one of a kind, self-sustaining prototype, which generates power but it offers even more: it stands for an educational model of advanced environmental stewardship at CCNY, for the surrounding community and New York City and likely as a role model beyond that context.

The building can bring to life CCNY's mission to access of excellence, as one of the first typological solar installations in Manhattan and a living example of our sustainable future.

In this book, the connection between three-dimensional and two-dimensional architectural processing has been one of the key concerns. The Roofpod is a prototype, a beginning from which has to be learned. The Roofpod's presence is thus an educational tool. It fosters the idea that design is not understood as a problem of superficial appearance, but as a calibrated composition of integrated systems and their performance. We must continue to learn with it, not only for its prototypical qualities, but also with the aim for an improved architectural methodology for any future projects to come.

In that sense, the Roofpod could become the three-dimensional compliment to this publication, the tie-in edition of the house along the book.

APPENDIX

ACKNOWLEDGEMENTS

by **Christian Volkmann**

A project of this scope, both the design-build realization of the Roofpod itself and the preparations to compile all the information into the format of this book were dependent on a large team of individuals – probably so large that not all will be named in this section. However, I am very grateful to everyone, named or unnamed, who understood the necessity of teamwork and collaboration for projects of this scale, and larger... My thanks go first and foremost to:

STUDENTS

(Always passionate, never quitting, even though tired at times; you are my heroes!)

Farah Ahmad, Ozgur Akkas, Francisco Arias, Yinery Baez, Adra Bubesi, David Cano-Bedoya, Michael Catalano, Mike Cheng, Steve Cummings, Chetram Dasrat, Mary Doumas, William Eng, George Fahim, Joseph Fisher, Karl Francis, Analisa Gerken, Giovanny Giraldo, Yelisa Grullon, Alex Gurevich, Andriy Herasymchyk, Christopher Hopstock, Joann Lee, Qi Fu Li, Tiffany Liu, Shwuyi Liu-Ma, Melissa Lopez, Juan Luengas, Astrit Maksuti, Carlos Matute, Samuel Mikhail, Javier Montesino, Wallace Ng, Selton Ng, Kyle O'Connor, Maulesh Pathak, Pablo Patino, Ross Pechenyy, Kurt Peters, Iskra Petrova, Wennian Qu, Nisu Quayum, Rajeevan Ratnanandan, Joileen Richards, Edwin Rivas, Asher Salzberg, Billy Schaefer, Mohsin Shabbir, George Sulley, Relbin Thomas, Zaman Tushar, Polina Tymoshenko, Ivan Uquillas, Fredy Uruchima, Teicha Villegas, John Vlahakis.

FACULTY

Prof. Hillary Brown, Prof. MT Chang, Prof. Alan Feigenberg, Prof. Jorge Gonzalez, Prof. Holly Kallman, Prof. Susan Kaplan, Prof. Nicholas Madamopoulos, Prof. Dominick Pilla, Prof. Ali Sadegh, Prof. Lee Weintraub, Prof. June Williamson (my special thanks to June Williamson for her lead on communication design).
Schematic Design: Prof. Alberto Foyo, Prof. Fabian Llonch;
Construction Assistance: Louis Hernandez, Jorge Plazas, David Seto, David Wright.

ADMINISTRATION

CCNY President Dr. Lisa Staiano-Coico, Provost Dr. Daniel Lemons; Spitzer School of Architecture: Dean Prof. George Ranalli, Chair Prof. Peter Gisolfi, Vice-Chair Prof. Gordon Gebert; Camille Hall, Jessica Maktal;
Grove School of Engineering: Dean Prof. Joseph Barba;
Development and Institutional Advancement: VP Shelly Butler, Rosemarie Weiss, Niamo Muid, Margaret Ambrosino, Leslie Timothy, Edward Schaefer, Karen Wenderoff;
City College Fund: Elena Sturman, Exec. Director;
Public Relations and Communication: VP Mary Lou Edmondson, Ellis Simon;
Facilities and Campus Planning: VP Robert Santos, George Varian (Physical Plant Services Supervisor), Holly Kallman;
Alumni Association: EVP Donald Jordan, Olga Mikhlina.

FRIENDS OF THE SCHOOL

Bernard + Anne Spitzer, Robert B. Catell, Andy Grove, P. + R. Fenster, Floyd Warkol; City College Architecture Alumni Group, especially Carol Kurth, Bill Stein, Joseph and Carolyn Fleischer, Laura Osorio, Frank Sciame, Howard Schlesinger, Venesa Alicea, Al D''Elia, Alan Hantman, Roland Charpentier, Jeremy Reed, Domingo Gonzalez; and the City College Engineering Alumni Group.

SPONSORS AND COLLABORATORS

Sciame Construction Co., Inc. (Frank Sciame, Andrew Sciame, FJ Sciame, Larry Best, Darrin McIntyre, Kirsten Leung, Lisa Severin); Empower Solar (David Schieren, Greg Sachs, David Ezzo, Matt Mathosian), New York Power Authority (NYPA); Con Edison; Five Star Electric (Gary Segal, Phil Altheim, John 'Buzz' Buzzell) WDF Inc. (Larry Roman, Denis Limanov); Kohn Pedersen Fox Assoc.; NYSERDA; MZ Woodwork (Martin Zauner); DYAD (Doug Fanning); Nana Wall (Bill Leon); Arnold Glass (Natalie Kopp, Lisa Welch, Alexander von Mezynski); RAICO (Wolfgang Elfner, Philipp Hieber), DRPILLA Assoc. PC (Dominick Pilla, Xiaoli Tong), SolarUS (Steve Elkin), Schneider Electric (Lee Therrien); Solera (Avi Bar), 21st Century Foundation; ACES; Warmboard; Douglaston Development; General Electric (Dave Najewicz, Charles Korman); American Hardwood/Southern Cypress Manufacturer Association (Ian Faight); Tesserae Materials (Erik Larson); Lutron; McNichols; Duravit (Alexandra Bursac); Velux (Dana Haydock, Timothy O'Neill); Turner Construction (Charles D. Avolio, Jorge Franco); STV Inc. (Sam Spata, Sandy Benjamin); WSP Flack + Kurtz; Haefele (Kevin McLee); Zeitraffer (Max Moos); A. Woodner; AmCork (Tom Banner, Linda Hopkins); Armstrong; Island Diversified (Matt Stevens); DOW (James Perling); index-d (Matthew Preston); Epic Management (Dennis Hagerty); Lumicor; Shift Space Design (Mario Gentile); American Standard; Watermark (Jack Abel); Dolan & Traynor (Dave Jackson); Cardella Waste Services; Competition Architectural Metal; Cord Contracting Co. Inc.; New York Roofing Co.; RC Dolner Construction; Hunter Roberts Construction Group; Levine Builders (Jeffrey E. Levine); TDX Construction; IceStone; Alison Downey Interior Design; Arup NY office (Markus Schulte, Adam Courtney); Design Installations (Charlton Girard); Triodetic (Greg Robin, Sue Staniszewski); Multifabricator Inc. (John Cano, David Cano-Bedoya); Permasteelisa Group (John Wolf, John Paradise, Ron Koenig); Forbo; SunPower; Esto Photographics Inc. (Prof. Albert Vecerka, Erica Stoller).

BOOK COLLABORATORS

Preface by Barry Bergdoll (MoMA);
Publishing and Editing by Oscar Riera Ojeda; Graphic Design: Alejandra Román, Leo Malinow; Co-Editing: Kit Maude;
Graphics and Co-Editing Assistance: Rebecca Sterling, Christina Hernandez, Samuel Mikhail, Ross Pechenyy, and John Vlahakis;
Book cover model rendering: Prof. MT Chang;
And: Lynnette Widder and Thilo Widder (Thank you, also for your patience!)

...THANK YOU ALL!

BIBLIOGRAPHY

1 Department of Economic and Social Affairs United Nations, Population Division, "Urban Population, Development and the Environmental 2011", www.un.org/esa/population/publications/2011UrbanPopDevEnv_Chart/urban_wallchart_2011-web-smaller.pdf.
2 Con Edison, "Peak Demand Forecasts," ed. Electric System Long Range Plan (2010).
3 U.S.- Canada Power System Outage Task Force, "Final Report on the August 14, 2003 Blackout in the United States and Canada: Causes and Recommendations," (2004), reports.energy.gov/BlackoutFinal-Web.pdf pg.1.
4 NASA, "System Failure Case Studies: Powerless," 1, no. 10 (2007).
5 Environmental Protection Agency, "Energy Trends in Selected Manufacturing Sectors: Opportunities and Challenges for Environmentally Preferable Energy Outcomes", (2007), www.epa.gov/sectors/pdf/energy/ch2.pdf.
6 U.S. Energy Information Administration, "New Generation by Energy Source: Total" (2012).
7 NYSERDA, "Patterns and Trends: New York State Energy Profiles 1996-2010," (2012), www.nyserda.ny.gov/Programs/Energy-Prices-Supplies-and-Weather-Data/~/media/Files/Publications/Energy-Analysis/EA-2010-pt-r.ashx.
8 Natural Resource and Defense Council, "Nuclear Accident at Indian Point: Consequences and Costs," (2011).
9 Susan Tierney, "Fuel Diversity in the New York Electricity Market," (2008), www.nyiso.com/public/webdocs/newsroom/white_papers/fuel_diversity_11202008.pdf.
10 Carbon Monitoring for Action, carma.org.
11 Sustainable CUNY, "NYC Solar Map," http://www.nycsolarmap.com/.
12 Jonathon Dickinson and Andrea Tenorio, "Inventory of New York City Greenhouse Gas Emissions September 2011", ed. Mayor's Office of Long-Term Planning and Sustainability (2011).
13 New York City Global Partners, "Best Practice" – Planting One Million Trees to Develop the Urban Forest"; www.nyc.gov/html/unccp/gprb/downloads/pdf/NYC_Environment_MillionTrees.pdf.
14 Arbor Environmental Alliance, "Tree Facts," www.arborenvironmentalalliance.com/carbon-tree-facts.asp.
15 U.S. Department of Energy, "Guidelines for Selecting Cool Roofs," Building Technologies Program 1.2 (2010), www1.eere.energy.gov/femp/pdfs/coolroofguide.pdf.
16 National Roofing Contractors Assoc., "Roofing, Energy, and the Environment," White Paper (2003).
17 Gregory McPherson, Paula J. Peper, James R. Simpson, Shelley L. Gardner, Kelaine E. Vargas, Qingfu Xiao, "New York City, New York: Municipal Forest Resource Analysis," (2007), www.nycgovparks.org/sub_your_park/trees_greenstreets/images/NYC_STRATUM_report_2007.pdf.
18 M. Kohler, "Green Roofs in Temperate Climates and in the Hot-Humid Tropics -Far Beyond the Aesthetics," Environment and Health 13 (2002).
19 Stavros E. Carlos, John P. Swaddle, "Increased Avian Diversity is Associated with Lower Incidence of Human West Nile Infection: Observation of the Dilution Effect," PubMed 3, no. 6 (2008).
20 New York City Audubon, "Project Safe Flight," www.nycaudubon.org/project-safe-flight.
21 Agriculture Research Service, "ARS: Honey Bees Colony and Collapse Disorder", www.ars.usda.gov/News/docs.htm?docid=15572.
22 New York City Department of City Planning, "Zoning Background."
23 Ibid.
24 Charles V. Baglie, "$430 a square foot, for air? Only in New York Real Estate.," 2005.
25 Liana; Peter J. Mateyka; Melanie A. Rapino Landivar, "Home-Based Workers in the United States: 2010," (2012).
26 Steve Cohen, "Wasted: New York City's Giant Garbage Problem," The New York Observer (2008), www.observer.com/2008/04/wasted-new-york-citys-giant-garbage-problem/.
27 Ibid.
28 Stu Loeser, "Mayor Bloomberg, Speaker Quinn, Sanitation Commissioner Doherty and Sanitation & Solid Waste Management Chair McMahon Hail Passage of Lndmark Solid Waste Management Plan," News and Press Releases (2006).
29 Keith Kloor, "Wretched Refude," City Limits Magazine 2002.
30 Elizabeth Royte, Garbage Land: On The Secret Trail of Trash (Back Bay Books/ Little, Brown and Company, 2005).
31 Alanna Talty, Samina Iqbal, "Impacts of New York City Waste on the 125th Street BID," Urban Design Lab, The Earth Institute (2007).
32 Reuse and Recycling Bureau of Waste Prevention, "New York City Recycling in Context: A Comprehensive Analysis of Recycling in Major U.S. Cities," www.nyc.gov/html/nycwasteless/downloads/pdf/recycling_context1.pdf.
33 "Edmonton Composting Facility ", www.edmonton.ca/for_residents/CompostingFacility.pdf.
34 City Green Inc./ Gruzen Samton, "Construction and Demolition Waste Manual," NYC Department of Design and Construction (2003).
35 Paul Campbell, "Current Population Reports" Census Bureau, (1997), www.census.gov/prod/2/pop/p25/p25-1131.pdf
36 S. Konopacki, L. Gartland, H. Akbari, and I. Rainer, "Demonstration of Energy Savings of Cool Roofs," Building Technologies Program 1.2 (2010).
37 Vaclav Smil, "Energy at the Crossroads: Global Perspectives and Uncertainties", Cambridge: Massachusetts Institute of Technology, 2003.
38 Green Garage, "Solar Thermal Collectors" www.greengaragedetroit.com/index.php?title=Solar_Thermal_Collectors.
39 Bundesministerium für Umwelt, "Energy Concept: For Environmentally friendly, Reliable, and Afordable Energy Supply" (2010).
40 United States Census Bureau, "Population, Housing Units, Area, and Density," New York - Place and County Subdivision (2010).
41 World Green Roof Congress, "Green Roof Tour and Case Study: Copenhagen", Congress Program(2012), www.worldgreenroofcongress.com.
42 P. Jacobi, A.W. Drescher and J. Amend, "Urban Agriculture, a Response to Crisis? ," UA Magazine 2000.
43 M. Asif, T. Muneer, J Kubie; "Sustainability Analysis of Window Frames", Building Services Engineering Research and Technology, SAGE (2005)
44 "Fire Protection Construction Requirements", ed. Department of Building (2004).
45 "Embodied Energy Coefficients", www.victoria.ac.nz/cbpr/documents/pdfs/ee-coefficients.pdf
46 "Energy Producing Systems: Solar Power," ed. Department of Natural Resources.

47 Garage, "Solar Thermal Collectors".
48 Sarah Goldschmidt. "The Machine Debate" www.cargocollective.com/sarahgoldschmidt/The-Machine-Debate
49 Baruch Spinoza, "A Theological-Political Treatise", tr. by R. H. M. Elwes, Dover Publ. (1951).
50 Gottfried Semper, "Style in the Technical and Tectonic Arts, or Practical Aesthetics", Zürich (1863).
51 Akos Moravanszky, The Pathos of Masonry, ed. Andrea Deplazes (Zurich, 2005).
52 Konrad Wachsmann, "The Turning Point of Building – Structure and Design", Reinhold Publ., pp. 54-70, (1961).
53 Stephen Kieran, James Timberlake. "Loblolly House - Elements of a New Architecture" Princeton Architectural Press, (2008) pg. 83.
54 Sigmund Freud, The Interpretation of Dreams, trans. A. A. Brill (New York: Macmillan 1900).
55 Barry Bergdoll, Peter Christensen, "Home Delivery: Fabricating the Modern Dwelling", MoMA (2008)
56 "The Energy Performace of Buildings – Certificates and Inspections" (England + Wales) Regulations 2007/11 (2007).
57 William McDonough, Michael Braungart, Cradle to Cradle (New York, North Point Press, 2002).
58 S.O.; Joseph C. Lam Cheung, Danny H. Li, C.L. Tsang, "Residential Building Envelope Heat Gain and Cooling Energy Requirements" 30, no. 7 (2005).
59 Building Science Corporation, "National Grid Deep Energy Retrofit Pilot Program, Two-Family Belmont Residence", www.buildingscience.com/documents/case-studies/cold-climate-national-grid-deep-energy-retrofit-belmont/files/111222_MA_NGrid_DER_Brownsberger_Case_Study.pdf
60 Tori Woods, "Aerogels: Thinner, Lighter, Stronger" www.nasa.gov/topics/technology/features/aerogels.html.
61 Lisa W. Foderaro, "A City of Glass Towers, a Hazard for Migratory Birds," N.Y. Region(2011), www.nytimes.com/2011/09/15/nyregion/making-new-yorks-glass-buildings-safer-for-birds.html
62 Roland E. Pittman, "Taxonomy of Learning" PediaPress, 2010.
63 David A. Kolb; Charalampos Mainemelis Boyatzis, "Experiential Learning Theory: Previous Research and New Directions," (1999), www.d.umn.edu/~k gilbert/educ5165-731/Readings/experiential-learning-theory.pdf
64 E.J. Simpson, The Classification of Educational Objectives in the Psychomotor Domain, Washington DC, Gryphon House, (1972)
65 S. Daly, R. Adams, G. Bodner, "What does it mean to Design? A qualitative Investigation of Design Professionals' Experiences", Journal of Engineering Education, Vol. 101, #2, pg. 187 – 219 (2012)
66 Richard Sennet, The Craftsman, Typesetting Inc. (2008)

ILLUSTRATION CREDITS

Every effort has been made to contact the owners of copyright for the images herein. Any omissions will be corrected in subsequent printings. © Farah Ahmad: 2.18; © Aldo Freccero: 3.5; © Architects' Newspaper, Matt Chaban: 1.43; © ASD Westok: 4.12; © Yinery Baez: 2.22, 4.144, 4.146, 4.149; © Benthem Crouwel Architects: 4.8, 4.9; © Bigstock: 1.3, 1.23, 1.34; © John Blackburn for Vanity Fair: 1.7; © Bluesky International: 1.21; © Richard Box: 1.4, © Thorwald Brandwein: 1.26; © Ulrich Buelhoff: T1.1; © 4.216 Adra Bubesi; © Cabot Corp: 4.72; © Michael Catalano + Asher Salzberg: 2.17; © Center for Neighborhood Technology: 1.42; © Mike Cheng: 2.20, 4.117, 4.121, 4.130; © Chicago Infrared Thermal Imaging Inc.: T1.3; © Climate Center at Earth Institute: T1.4; © Amanda Coen: 4.90, 4.148, 4.168, 4.201; © Michael Deng: 3.8; © Mary Doumas: 4.49, 4.97, 4.152, 4.219; © Nigel Dunnett: 1.30; © Earth Pledge Foundation: 2.13; © Energy Research Center of Netherlands: 4.124; © Enrique Browne Architects: 1.27; © Environmental Protection Agency: 1.6, 1.29, 1.46, 1.59; © EPIA Greenpeace, Teske/Hoffmann: 2.5; © Eurostat: 1.53; © George Fahim: 3.14, 3.15, 3.21, 3.22, 3.23, 3.24, 3.25, 4.34, 4.158, 4.208, 5.10, 5.11, 5.14; © Getty Images: 1.35; © Analisa Gerken: 2.25; © Giovanny Giraldo: 4.195; © Arnold Glass: 4.71, 4.86, 4.88, 4.89, 4.95, 4.96; © Yelissa Grullon: 4.204, 4.206, 4.212, 4.221; © Alex Gurevich: 4.243; © Jürgen Heinl: 3.4; © Ralph Hockens: 1.31; © Sven Hoffmann, www.svenhoffmannjournal.com: cover backdrop; © Keya Lea Horiuchi: 4.122, 4.142, 4.211, 4.229, 5.16; © Christopher Hopstock: 4.184, 4.185; © Istockphoto: 3.36; © KAB: 1.50; © Boris Kester, www.traveladventures.org: 1.16a; © Kieran &Timberlake Architects: 3.35; © Jong Soo Kim: 4.224, 4.225; © Moses King: 2.1; © Vincent Laforet: 1.52; © Qi Fu Li: 2.10; © Shwuyi Liu-Ma: 4.73; © Tiffany Liu: Collage, page 10; © Juan Luengas, Maulesh Pathak: 4.127; © René Magritte: 3.2; © Astrit Maksuti: 4.147, 4.154, 4.155, 4.157; © Carlos Matute: 4.13, 4.240, 4.241, 4.252; © Shaun McKinnon: 1.12; © Magnus Manske: 1.40; © Brad McCracken: 1.48; © Samuel Mikhail: 4.2, 4.38, 4.40, 4.44, 4.100, 4.101; © Milstein Division of United States History New York Public Library: 4.3; © Max Moos: 5.27; Cécile de Munck, for J.Wiley + Sons: 1.14; © NASA Landsat: 1.13, 1.25; © NREL: 1.10; © NYC Department of Parks and Recreation: 1.24; © NYC Department of City Planning: 1.36, 1.55; © NYC Department of Design and Construction: 1.57; © NYS Department of Environmental Conservation: 1.58; © Selton Ng: 4.24; © Tim Noble & Sue Webster/ Blain Southern: 1.54; © Elton Nurcellari: 2.6, 2.15, 4.5; © Pablo Patino: 2.11, 3.32, 4.239; © S. Paltera, Department of Energy, Solar Decathlon: 4.137; © Ross Pechenyy: 2.4, 4.68, 4.104, 4.150, 4.151; © Jorge Plazas: 3.38; © Dominick Pilla: 4.27; © Proxy Design Studio, for a+u 2011:10: 1.39; © Ateliers J. Prouvé: 3.34, 4.139, 5.2; © U.S National Park Service: 1.47; © Wen Qu: 4.97; © Rajeevan Ratnanandan: 4.118; © Raico: 4.98, 4.99; © Asher Salzberg: 2.9, 4.69; © Billy Schaefer: 2.8, 2.12, 2.16, 2.28, 4.11, 4.56, 4.58, 4.105, 4.119, 4.120, 4.143, 4.145, 5.17, 5.18, 5.22, 5.41; © Scott B. Smith Photography Inc.: T1.5; © Ricky Shum: 4.10, 4.20; © Shutterstock: 1.28; © SolarUS: 4.182; © Southern City Waste Management: 1.51; © StockExchange: pages 16-17, 100-101; © George Sulley: 4.25, 4.28, 4.29, 4.147; © Peter Sulzer, 3.34; © Sustainable CUNY: 1.9; © Michael Sypkens: 1.20; © Tesserae, Erik Larson: 4.197; © J.Tetro, Department of Energy, Solar Decathlon: 4.153; © Thermasteel: 4.39; © Relbin Thomas: 2.24; © Eddie Toro: 1.33; Eric Touchaleaume: 3.34; © Transport Direct, Scotland: 1.44; © Triodetic: 4.172, 4.176; © Zaman Tushar: 2.26, 4.138; © Tyrone Turner, National Geographic: 1.16b; © Polina Tymoshenko: 2.27; © Albert Vercerka/ESTO: 4.1, 4.74, 4.91, 4.94, 4.102, 4.123, 4.129, 4.140, 4.177, 4.181, 4.253; and all images in prologue (pages 2 and 4), photo sections 1, 2 and 3, and pages 358 and 366. © Teicha Villegas: 4.159, 4.166, 4.198, 4.207; © John Vlahakis: 3.27, 3.28, 3.29, 3.33, 4.4, 4.21, 4.22, 4.48, 4.50, 4.67, 4.76, 4.79, 4.103, 4.228, 4.232; © WarePin: T1.6; © Warner Bothers: 5.1; © Brett Weinstein, www.nrbelex.com: 1.41; © Justin Williams: 2.2; © Wzohaib: 1.8.

GLOSSARY

A

ALBEDO: the proportion of the incident light or radiation reflected by a surface.
AIR FRICTION: the action of friction that slows something moving through air.
ARMATURE: framework that provides strength and stability within construction.

B

BIODIESEL: is a fuel derived from vegetable oil or animal fat, consisting of long chain alkyl esters.
BLOWER DOOR TEST: Special instrument used to measure air leakage of a building's envelope. The equipment is made of a temporary door covering which is installed in an outside doorway and a blower which forces air into or out of the building, identifying how leaky a building is.
BRACING: diagonal members, either temporary or permanent, installed to stabilize a structure against lateral loads.
BUILDING BLOCK: a unit or set of parts that can be used to construct a more complex structure.
BUILDING CODE: a set of legal restrictions intended to assure a minimum standard of health and safety in buildings.

C

CANTILEVER: a beam, truss, or slab that extends beyond its last point of support.
CARBON DIOXIDE (CO^2): a gas formed during combustion of fossil fuels.
CARBON SEQUESTRATION: Process of capture and long-term storage of atmospheric carbon dioxide to mitigate greenhouse gas emission.
CASTELLATED BEAM: a steel wide-flange section whose web has been cut along a zigzag path and reassembled by welding in such a way as to create a deeper section, increasing its structural depth. Nowadays, cellular beams, with round cut-outs dominate the market, due to CNC processing.
CAVITY WALL: Exterior wall system formed from two thicknesses of masonry with a space between them, improving thermal insulation and waterproofing of this element.
CLADDING: a material used as the exterior wall surface of a building.
COLD FORMED STEEL: a steel product made from rolling or pressing thin gauges of sheet steel.
COMPOSITE: materials made from two or more constituent materials that when combined, produce a material with characteristics different from the individual components.
COMPRESSION: a squeezing force.
CONDENSATION: the process of changing from a gaseous to liquid state, especially as applied to water.
CONDUCTION: a heat transfer in a solid.
CONDUIT: a pipe, tube, or tile for protecting electric wires or cables.
CONVECTION: a mode of heat transfer in liquids and gases.
CORNICE: an overhang at the top of building.
CROSS BRACING: a construction technique in which braces are crossed to form an X shape to support a frame.
CURTAIN WALL: exterior wall cladding system suspended from or supported by the structural frame of a building.

D

DAYLIGHTING: Illuminating the interior of a building by natural means.
DEAD LOAD: a load created by the weight of building components.
DECKING: a material used to span across beams or joists to create a floor or roof surface.
DUNNAGE: a structural transition system to transition loads.

E

EMBODIED ENERGY: total amount of energy required to extract raw materials and manufacture a building material.
EGRESS: a path or means to exit.
EXPANSION JOINT: a divider joint that provides space for spreading out.

F

FAÇADE: the exterior face of a building.
FIRE RATING: the ability of a building assembly to endure fire, measured in hours or minutes of time and determined from a standardized full-scale test.
FLANGE: a projecting crosspiece of a wide-flange or channel profile to concentrate material where structurally most effective.
FLOOR AREA RATIO (FAR): a principal bulk regulation controlling the size of a building measuring the ratio of total building floor area to the area of its zoning lot.
FOOTING: the bottom portion of a foundation resting directly on the supporting soil with an area larger than that of the supporting wall or column.
FOSSIL FUEL: a class of materials of biological origin occurring within the Earth's crust that can be used as a source of energy.

G

GASKET: a compressible packing piece, usually of rubber, sandwiched between faces of flanges of a joint to provide a seal.
GAUGE: a measurement of the thickness of sheet material.
GIRDER: a beam of steel, wood, or reinforced concrete, used as a main horizontal support in a building.
GLAZING: an assembly of flat glass and its supporting frame.
GREENHOUSE EFFECT: process by which thermal radiation from a surface is absorbed by atmospheric gases, and is re-radiated in all directions, with the effect of global warming. In architecture, however, the phenomenon describes solar radiation passing through glass and warming an interior space. Since glass is transparent to short- wave radiation, but not to long-wave radiation (heat), heat is captured, overheating the enclosed space.

H

HEAT CAPACITY: the number of heat units needed to raise the temperature of a body by one degree.
HYDROSCOPIC: readily absorbing and retaining moisture.

I

I-BEAM: an American standard section or wide-flange section of hot-rolled steel, composed of top and bottom and web.
IRRIGATION: the artificial supply of waterby means of ditches, pipes, or streams.

L

LAMINATE: a material bond in layers, usually with a better surface.
LATENT HEAT: the heat required to convert a solid into a liquid or vapor, or a liquid into a vapor, without change of temperature.
LIVE LOAD: the weight of snow, people, furnishings, machines, vehicles, and goods in or on a building.
LOUVER: fixed or operable, closely spaced, angled slats attached to a frame to redirect the passage of air or light.

M

METHANE: a colorless, odorless flammable gas, CH4, that is main constituent of natural gas; by-product of decomposition.
MICROCLIMATE: a climate of a smaller specific place within a larger area.
MODULAR: an approach that subdivides a system into smaller parts that can be independently created and then used in different systems to drive multiple functionalities.
MULLION: a vertical or horizontal bar between adjacent window or door units.

N

NATURAL GAS: flammable gas, consisting largely of methane and other hydrocarbons, occurring naturally underground and used as fuel.
NET-ZERO ENERGY: a building with zero net energy consumption and zero carbon emissions annually.
NUCLEAR ENERGY: is the use of sustained nuclear fission to generate heat and electricity.

O

ORIENTED STRAND BOARD (OSB): an engineered wood product formed by layering strands of wood in specific orientations.

P

PARTICULATES: fine solid particles which remain individually dispersed in gases and stack emissions.
PHOTOVOLTAIC PANELS: a method of generating electrical power by converting solar radiation into direct electricity using solar cells containing photovoltaic material.
PLYWOOD: a wood composed of an odd number of layers of wood veneer bonded together under pressure.
POLYSTYRENE: a thermoplastic foam with thermal insulating properties.
PREFABRICATION: manufacture sections of (a building or piece of furniture) to enable quick or easy assembly on site.
PROTOTYPE: a first or preliminary model of something from which other forms are developed or copied.

R

RENEWABLE ENERGY: energy that comes from resources which are continually replenished such as sunlight, wind, rain, tides, waves and geothermal heat.
ROOF MEMBRANE: the entire waterproofing layer on the roof.
R-VALUE: unit of heat resistance, used as a measure of the effectiveness of insulation in stopping heat flow. The opposite of conductance.

S

SHALE GAS: natural gas that is found trapped within shale formations. In order to release it, often a technique called 'fracking' is used, in which typically water is mixed with sand and chemicals, and the mixture is injected at high pressure into a wellbore to create fractures, which form conduits along which fluids such as gas, petroleum, and groundwater may migrate to the well.
SHEATHING: the rough covering applied to the outside of the roof, wall, or floors framing a light frame structure that usually braces the building.
SPACE FRAME: a truss-like, lightweight rigid structure constructed from two directional interlocking struts and ties in a geometric pattern.
SPANDREL PANEL: an approximately triangular surface bounded by the outer curve of an arch and the adjacent wall.
STAINLESS STEEL: a form of steel containing chromium, resistant to tarnishing and rust.
STANCHION: an upright bar, post, or support.
STRUTS: a structural component designed to resist longitudinal compression.
STEEL: a strong and malleable metal alloy made of iron and carbon.
STRUCTURAL INSULATED PANELS (SIPS): panels that consist of an insulating foam core sandwiched between two structural facings, thus a composite.
STUD: one of an array of small, closely spaced, parallel wall framing members.
SUBFLOOR: the surface beneath a finish floor, usually sheathing the assembly.

T

TENSION ROD: an iron rod used as a tension member to strengthen timber or metal framework, roofs, or the like.
THERMAL INERTIA: heat storage capacity defined by the square root of the product of the material's bulk thermal conductivity and volumetric heat capacity.
THERMAL RESISTANCE: measure of the ability of a building material or assembly to resist the flow of heat.
THERMAL TRANSMITTANCE: measure of the inability of a building material or assembly to resist the flow of heat.
TIES: a structural component designed to resist longitudinal tension.
TRELLIS: framed structure, often used to support plants and provide shade.

V

VACUUM INSULATION: is a form of thermal insulation consisting of a nearly gas-tight enclosure surrounding a rigid core, from which the air has been evacuated. Vacuum is without thermal transmission, its performance is suspension.

W

WATERPROOFING: liquid or sheet material applied to the exterior to resist the intrusion of water under hydrostatic pressure; usually bituminous composites.

Z

ZONING ORDINANCE: a law that specifies in detail how land may be used in a municipality.

BOOK CREDITS

Edited by Oscar Riera Ojeda
Creative Direction by Leo Malinow
Graphic design by Alejandra Román
Copy editing by Kit Maude

Original editing and design by AARDVARCHITECTURE LLP.
Copy editing: Rebecca Sterling and Lynnette Widder
Research assistant: Christina Hernandez
Editorial assistant: John Vlahakis

OSCAR RIERA OJEDA
PUBLISHERS

ISBN 978-988-16194-0-2
Published by Oscar Riera Ojeda Publishers Limited
Printed in China by Artron Art Center

Oscar Riera Ojeda Publishers Limited
Unit 6, 22/F.,
Far East Consortium Building,
121 Des Voeux Road Central, Hong Kong
T: +852-5311-1625

Production Offices | China
Suit 19, Shenyun Road
Nanshan District, Shenzhen 518055
T:+86-182-0098-3774

www.oropublishers.com | www.oscarrieraojeda.com
oscar@oscarrieraojeda.com